Advances in Infrared and Raman Spectroscopy

VOLUME 1

Advances in Infrared and Raman Spectroscopy

VOLUME 1

Edited by

R. J. H. CLARK
University College London

R. E. HESTER
University of York

London · New York · Rheine

Heyden & Son Ltd., Spectrum House, Alderton Crescent, London NW4 3XX.
Heyden & Son Inc., 225 Park Avenue, New York 10017, N.Y., U.S.A.
Heyden & Son GmbH, 4440 Rheine/Westf., Münsterstrasse 22, Germany.

ISBN 0 85501 181 5

Set by Eta Services (Typesetters) Ltd., Beccles, Suffolk.
Printed and bound in Great Britain by W. & J. Mackay Ltd., Lordswood,
Chatham, Kent.

CONTENTS

CHAPTER 4: Resonance Raman Spectra of Inorganic Molecules
and Ions—
R. J. H. Clark

CHAPTER 5: The Metal Isotope Effect on Molecular Vibrations—
N. Mohan, A. Müller and K. Nakamoto

CHAPTER 6: Absolute Absorption Intensities as Measured in the
Gas Phase—
D. Steele

LIST OF CONTRIBUTORS

R. J. H. CLARK, Christopher Ingold Laboratories, University College London, 20 Gordon Street, London WC1H 0AJ, United Kingdom (p. 143)

J. R. DOWNEY Jr., Department of Chemistry, Rensselaer Polytechnic Institute, Troy, New York 12181, U.S.A. (p. 1)

B. G. FRUSHOUR, Monsanto Triangle Park Development Center, Inc., Research Triangle Park, North Carolina 27709, U.S.A. (p. 35)

G. J. JANZ, Department of Chemistry, Rensselaer Polytechnic Institute, Troy, New York 12181, U.S.A. (p. 1)

J. L. KOENIG, Department of Macromolecular Science, Case Western Reserve University, Cleveland, Ohio 44106, U.S.A. (p. 35)

T. M. LOEHR, Oregon Graduate Center, Beaverton, Oregon 97005, U.S.A. (p. 98)

N. MOHAN, Institute of Chemistry, University of Dortmund, 46 Dortmund, W. Germany (p. 173)

A. MÜLLER, Institute of Chemistry, University of Dortmund, 46 Dortmund, W. Germany (p. 173)

K. NAKAMOTO, Department of Chemistry, Marquette University, Milwaukee, Wisconsin 53233, U.S.A. (p. 173)

T. G. SPIRO, Department of Chemistry, Princeton University, Princeton, N.J. 08540, U.S.A. (p. 98)

D. STEELE, Department of Chemistry, Royal Holloway College, Egham, Surrey, TW20 0EX, United Kingdom (p. 232)

PREFACE

There are few areas of science which have not already benefited from the application of infrared spectroscopic methods, and progress in this field remains vigorous. Closely related information on chemical and biological materials and systems is obtainable from Raman spectroscopy, though there also are many important differences between the types of information yielded and the types of materials and systems best suited to study by each technique. The close relationship between these two sets of spectroscopic techniques is explicitly recognised in this Series. Advances in Infrared and Raman Spectroscopy contains critical review articles, both fundamental and applied, mainly within the title areas; however, we shall extend the coverage into closely related areas by giving some space to such topics as neutron inelastic scattering or vibronic fluorescence spectroscopy. Thus the Series will be firmly technique orientated. Inasmuch as these techniques have such wide ranging applicability throughout science and engineering, however, the coverage in terms of topics will be wide. Already in the first volume we have articles ranging from the fundamental theory of infrared band intensities through the development of computer-controlled spectrometer systems to applications in biology. This integration of theory and practice, and the bringing together of different areas of academic and industrial science and technology, constitute major objectives of the Series.

The reviews will be in those subjects in which most progress is deemed to have been made in recent years, or is expected to be made in the near future. The Series will appeal to research scientists and technologists as well as to graduate students and teachers of advanced courses. The Series is intended to be of wide general interest both within and beyond the fields of chemistry, physics and biology.

The problem of nomenclature in a truly international Series has to be acknowledged. We have adopted a compromise solution of permitting the use of either English or American spelling (depending on the origin of the review article) and have recommended the use of SI Units. A table on the international system of units is given on p. xi for reference purposes.

August 1975
R. J. H. CLARK
R. E. HESTER

THE INTERNATIONAL SYSTEM OF UNITS (SI)

Physical quantity	Name of unit	Symbol for unit
SI Base Units		
length	metre	m
mass	kilogram	kg
time	second	s
electric current	ampere	A
thermodynamic temperature	kelvin	K
amount of substance	mole	mol
SI Supplementary Units		
plane angle	radian	rad
solid angle	steradian	sr

SI Derived Units having Special Names and Symbols

energy	joule	$J = m^2\,kg\,s^{-2}$
force	newton	$N = m\,kg\,s^{-2} = J\,m^{-1}$
pressure	pascal	$Pa = m^{-1}\,kg\,s^{-2} = N\,m^{-2} = J\,m^{-3}$
power	watt	$W = m^2\,kg\,s^{-3} = J\,s^{-1}$
electric charge	coulomb	$C = s\,A$
electric potential difference	volt	$V = m^2\,kg\,s^{-3}\,A^{-1} = J\,A^{-1}\,s^{-1}$
electric resistance	ohm	$\Omega = m^2\,kg\,s^{-3}\,A^{-2} = V\,A^{-1}$
electric conductance	siemens	$S = m^{-2}\,kg^{-1}\,s^3\,A^2 = \Omega^{-1}$
electric capacitance	farad	$F = m^{-2}\,kg^{-1}\,s^4\,A^2 = C\,V^{-1}$
magnetic flux	weber	$Wb = m^2\,kg\,s^{-2}\,A^{-1} = V\,s$
inductance	henry	$H = m^2\,kg\,s^{-2}\,A^{-2} = V\,s\,A^{-1}$
magnetic flux density	tesla	$T = kg\,s^{-2}\,A^{-1} = V\,s\,m^{-2}$
frequency	hertz	$Hz = s^{-1}$

SOME NON-SI UNITS

Physical quantity	Name of unit	Symbol and definition

Decimal Multiples of SI Units, Some having Special Names and Symbols

length	ångström	$Å = 10^{-10}$ m $= 0.1$ nm $= 100$ pm
length	micron	$\mu m = 10^{-6}$ m
area	are	$a = 100$ m^2
area	barn	$b = 10^{-28}$ m^2
volume	litre	$l = 10^{-3}$ m$^3 = $ dm^3 $= 1000$ cm^3
energy	erg	$erg = 10^{-7}$ J
force	dyne	$dyn = 10^{-5}$ N
force constant	dyne per centimetre	$dyn\,cm^{-1} = 10^{-3}$ N m^{-1}
force constant	millidyne per ångström	$mdyn\,Å^{-1} = 10^{2}$ N m^{-1}
force constant	attojoule per ångström squared	$aJ\,Å^{-2} = 10^{2}$ N m^{-1}
pressure	bar	$bar = 10^{5}$ Pa
concentration	—	$M = 10^{3}$ mol m^{-3} $=$ mol dm^{-3}

Units Defined Exactly in Terms of SI Units

length	inch	$in = 0.0254$ m
mass	pound	$lb = 0.453\ 592\ 27$ kg
force	kilogram-force	$kgf = 9.806\ 65$ N
pressure	standard atmosphere	$atm = 101\ 325$ Pa
pressure	torr	$Torr = 1$ mmHg $= (101\ 325/760)$ Pa
energy	kilowatt hour	$kWh = 3.6 \times 10^{6}$ J
energy	thermochemical calorie	$cal_{th} = 4.184$ J
thermodynamic temperature	degree Celsius[a]	$°C = K$

[a] Celsius or "Centigrade" temperature θ_C is defined in terms of the thermodynamic temperature T by the relation $\theta_C/°C = T/K - 273.15$.

OTHER RELATIONS

1. The physical quantity, the wavenumber (units cm^{-1}), is related to frequency as follows:

$$cm^{-1} \approx (2.998 \times 10^{10})^{-1} \, s^{-1}$$

2. The physical quantity, the molar decadic absorption coefficient (symbol ε) has the SI units $m^2 \, mol^{-1}$. The relation between the usual non-SI and SI units is as follows:

$$M^{-1} \, cm^{-1} = 1 \, mol^{-1} \, cm^{-1} = 10^{-1} \, m^2 \, mol^{-1}$$

The SI Prefixes

Fraction	Prefix	Symbol	Multiple	Prefix	Symbol
10^{-1}	deci	d	10^1	deca	da
10^{-2}	centi	c	10^2	hecto	h
10^{-3}	milli	m	10^3	kilo	k
10^{-6}	micro	μ	10^6	mega	M
10^{-9}	nano	n	10^9	giga	G
10^{-12}	pico	p	10^{12}	tera	T
10^{-15}	femto	f	10^{15}	peta	P
10^{-18}	atto	a	10^{18}	exa	E

Chapter 1

DIGITAL METHODS IN RAMAN SPECTROSCOPY

J. R. Downey Jr. and G. J. Janz

Department of Chemistry, Rensselaer Polytechnic Institute, Troy, New York 12181, U.S.A.

1 INTRODUCTION

The resurgence of Raman spectroscopy in the 1960s has generally been attributed to the development of reliable laser excitation sources. Coupled with the development of these powerful monochromatic sources has been the rapid development of photoelectric recording Raman instruments. During the same time period great strides have been made in the development of digital computers such that they are now readily available (or accessible) to most chemical laboratories. Thus, while as recently as 15 years ago neither was routinely available, many chemical laboratories today have both a photoelectric Raman instrument and a digital computer. The prospect of coupling these two machines has been evident for some time and comprises the subject matter for this chapter.

Our approach generally will be to point out the advantages and problems of coupling a computer to a Raman spectrometer and to discuss the hardware and software necessary to effect the interface. We do not intend this to be a critical review of the field although we will discuss specific Raman–computer interfaces with regard to their advantages and disadvantages.

The discussion will first center on what use may be made of computers in Raman spectroscopy and will then move onto the hardware and software considerations necessary to establish the linkage. Then a brief review will be given of existing Raman–computer interfaces followed by a detailed description of the system in use at Rensselaer Polytechnic Institute.

2 USES OF DIGITAL METHODS IN RAMAN SPECTROSCOPY

2.1 General

Why is it desirable to couple a computer to a Raman spectrometer? What can one do with the resulting system that cannot be done otherwise? Of what

1

use is the computer to a Raman spectroscopist? These are the subjects that will be addressed in this section.

Three general uses of digital computers may be defined with regard to most experiments. They are:

(i) to acquire, average and store data;
(ii) to control the experiment;
(iii) to perform data evaluation or reduction including some means of display.

The first two functions are generally performed under active computer control, i.e. on-line, while the data evaluation step may be performed off-line at a later date. Indeed, the very potent potential of the computer to perform rapid data evaluation seems to be the primary purpose for coupling the computer to a Raman instrument.

2.2   Acquisition, Averaging and Storage of Data

It is this element that usually serves as a deterrent to most people interested in instituting an interface. One method of accomplishing this portion of the system is simply to read the chart recording visually and convert it to a point by point list of intensity versus frequency which may then be manually entered into the computer via teletype, paper punched tape or punched cards. This is, of course, a very trivial and time consuming method and makes no use of the computer for acquiring data. A more advanced method is desirable.

Since many Raman instruments scan at a fixed rate the frequency scale may be entered simply by entering the initial frequency, final frequency (or length of scan), and the increment at which points are taken (or total number of points). These three values may be conveniently entered into the computer via a teletype. The intensity values must generally be entered via some sort of automated process.

One simple way to do this is to feed the analog signal from the chart recorder (or detector output) to an analog to digital converter (ADC) which will digitize the signal at a fixed time interval. The output of the ADC can then be fed automatically to the computer and stored in core. Systems based on this principle are in wide use with a variety of instruments using either commercial ADCs or homemade devices built around digital voltmeters.[1-3]

The system described above is applicable to virtually any instrument whose analog output varies as a function of time. A slightly simpler system is possible with Raman spectrometers that use photon counting as a means of detection. In this case the signal is already digitized and one need merely count pulses for a specified period of time, transfer the number of pulses into core, reset the counter and count again for the next data point. Commercial scalars (ratemeters) are available to perform the counting step and this system has been widely used for Raman instruments.[4-10]

All known Raman–computer interfaces utilize one of the above methods of acquiring the intensity data. Since scan times are generally slow in Raman spectroscopy the rate of data point collection rarely exceeds a few points per second. Thus it is possible to dispense with the computer entirely during collection and simply allow a teletype to punch a paper tape with the corresponding information.[9] The data may then be input to the computer at a later time for evaluation. Some versatility is lost by this procedure but it is adequate in some cases.

Some means of permanent data storage must be provided since the computer core or disk can serve only as a limited temporary storage medium. Common means employed have been paper punched tape, magnetic tape or punched cards. Magnetic tape seems to be preferable but either of the others is satisfactory.

One big advantage of using a computer to acquire data is the possibility of improving the signal to noise ratio (SNR) during collection as shown in Fig. 1. There are several approaches to this problem. One of the most common is simply to collect the same spectrum a number of times and instruct the computer to average these spectra together point by point. This method, usually known as computer averaging of transients (CAT), has been in wide use in NMR spectroscopy for a number of years and is equally applicable to Raman spectroscopy. It is well known that the SNR increases as $\sqrt{N}$ where N is the number of scans that are averaged. Thus the largest improvement takes place in the first few averages.

Another method that can be used with photon counting instruments is to simply count for a longer period of time for each point. Since the SNR increases as $\sqrt{c}$ where c is the total number of counts a significant smoothing of the spectrum results at long count times (high number of counts). This is particularly easy to implement on Raman instruments which use a stepping motor for the monochromator drive rather than a continuous scan motor.

2.3 Experiment Control

In addition to collecting data a computer may also be used to control the course of the Raman experiment. Most systems in use provide, at a minimum, a means for the computer to start the scanning mechanism of the monochromator to ensure that the computer and spectrometer are synchronized during data collection. This can be effected by a simple relay switch activated through a pulse from the computer.

A somewhat more sophisticated control is to allow the computer to reset the instrument to the initial frequency at the end of a scan so that it can average scans without the operator resetting the spectrometer. Since this normally requires scanning past the initial frequency and approaching in a forward direction to remove backlash, it may require scanning through the exciting line. Accordingly it is desirable to have a shutter under the control of the computer as a safeguard for the detector.[7,11]

If the computer is allowed to reset the monochromator drive mechanism it is necessary to ensure that it is precisely reset to the original starting frequency. If not, the peaks in the resulting averaged spectrum will be broadened due to the variation in starting frequency. With monochromators driven by a stepping motor this represents no problem since the computer can easily count steps and reposition to precisely the step desired. With continuous drive monochromators problems arise because merely timing the forward and reverse scans may not be sufficient to ensure adequate reproducibility. In general it is necessary to provide some means of encoding the frequency scale (shaft encoders) so the computer can determine the exact frequency at all times. Frequency/wavelength encoding mechanisms for accomplishing this task have been described.[12,13]

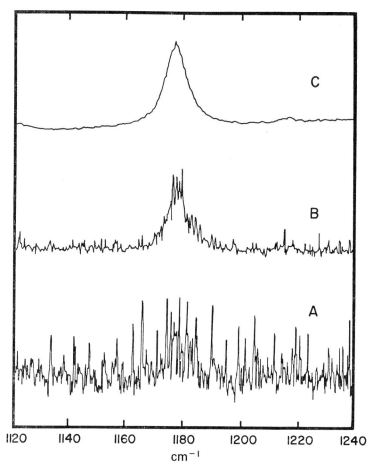

Fig. 1. Computer Averaging of Transients. The increase in signal to noise ratio via computer averaging of transients is illustrated for the 1178 cm^{-1} line of benzene in CCl_4 (3.5×10^{-3} M).[4] A = single scan; B = average of 50 scans; C = average of 900 scans.

Other possible control functions include rotation of half wave plates or analyzers for depolarization ratio measurements as well as on–off functions for other filters, magnetic fields, electric fields, etc.[7] One may also program a sample changer for routine analyses.

2.4 Data Evaluation and Analysis

It is in the area of data evaluation that the computer is probably of most use to the Raman spectroscopist. There are many corrections which should be made to the raw data that have not been made in the past because of lack of a rapid, convenient means to do so. The computer now provides a powerful method of effecting these data manipulations. The topics considered below represent areas of wide interest where a computer may be helpful.

2.4.1 Data Display

How many times has an otherwise perfectly good spectrum been re-run merely to display it in a slightly different manner? With the advent of digital storage of spectra it is easy to re-display the spectra in virtually any configuration with a proper display system. In Fig. 2 the re-display of the spectrum of Na_2S_4, achieved

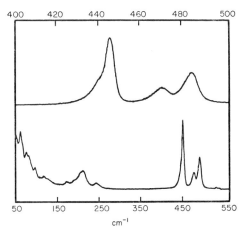

Fig. 2. Display. The spectrum of polycrystalline Na_2S_4 illustrating the versatility of a computer-controlled display system to accomplish various displays. The figure was produced in less than 10 seconds using the RPI CASH-DISPLAY system, see section 5 (laser = 5145 Å, 1.4 watt; slit = 3.5 cm^{-1}; spinning sample cell; digitizing increment = 0.2 cm^{-1}).

in this manner, is illustrated; thus the structural features of a small frequency region can be examined with a re-display of the spectrum, through computer-assisted techniques. The actual means of display, digital $X—Y$ plotter, analog $X—Y$ plotter or oscilloscope, is perhaps not as important as the ability to change scales via the controlling mechanism, the computer.

The display of at least two (preferably more) spectra simultaneously is highly desirable. In Fig. 3, the Raman spectrum for the 1050 cm^{-1} region of the nitrate ion has been displayed by computer control for a series of $AgNO_3$–acetonitrile

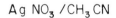

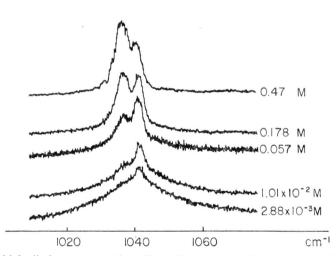

Fig. 3. Multiple display, concentration effects. The spectrum of the nitrate ion in $AgNO_3$/CH_3CN solutions illustrating the increase of the peak due to ion-paired nitrate (1037 cm^{-1}) relative to that for free nitrate (1042 cm^{-1}) as the concentration is increased (laser = 4880 Å, 1.3 watt; slit width = 3.5 cm^{-1}; room temperature, digitizing increment = 0.5 cm^{-1}). Figure produced by the RPI CASH-MULTI system, see section 5.

solutions of increasingly greater dilutions. Comparisons of the spectra for a series of structurally related compounds, or of temperature effects, as illustrated elsewhere in this work are readily achieved with such computer-assisted techniques.

2.4.2 Instrumental corrections

One area of interest in Raman spectroscopy is the comparison of absolute scattering intensities and the calculation of scattering coefficients.[14,15] In order to do this one must know the spectral sensitivity of the particular spectrometer used as a function of frequency and correct the raw data for this response. The instrumental response depends on such features as detection efficiency and grating efficiency. These generally vary as a function of frequency so the overall response of the instrument varies with frequency.

It is a simple matter to measure the instrumental response using a standard

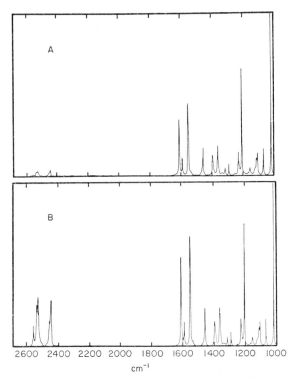

Fig. 4. Instrumental corrections. The Raman spectrum of indene illustrating the effect of instrumental corrections.[2] A = raw digitized spectrum; B = normalized intensity-corrected spectrum.

lamp.[2,4] The computer may then be programmed to correct the raw spectrum for the known instrumental response in order to obtain the true spectrum as shown in Fig. 4.

2.4.3 Isotropic–anisotropic spectra

One of the most powerful applications of Raman spectroscopy is in the area of structural determinations. It is possible to obtain some information about the symmetry of a molecule by studying the polarization characteristics of its Raman spectrum. In many cases a mere comparison of the so-called parallel and perpendicular polarized spectra is sufficient for this purpose. In other cases one would like to obtain the isotropic (which contains only totally symmetric vibrations) and anisotropic spectra. These are generally obtained as a linear combination of the experimentally obtainable parallel and perpendicular polarized spectra.[16] Using polarized exciting light and an analyzer to measure $I_{||}$ and $I_{\perp}$, isotropic ($\bar{\alpha}'^2$) and anisotropic (γ'^2) components are calculated as follows:

$$3\gamma'^2 = I_{\perp}$$
$$45\bar{\alpha}'^2 = I_{||} - 4/3 I_{\perp}$$

Although this calculation would be tedious by hand, a computer can calculate an entire spectrum in a few seconds and display it as shown in Fig. 5.

2.4.4   Difference spectra

Frequently a portion of the spectrum of interest is interfered with by other spectral features such as solvent peaks or Rayleigh line scattering. In order to extract useful information it may be necessary to remove these interfering peaks from the spectrum. One method is curve resolution (see section 2.4.9) but it is usually possible to remove solvent peaks by mathematically subtracting the solvent spectrum from that of the solution. Thus if both spectra are recorded independently the computer may be instructed to do a point by point subtraction so that the interfering peak is removed from the resulting spectrum.[11] A more complex system using a programmed sample changer for alternate positioning

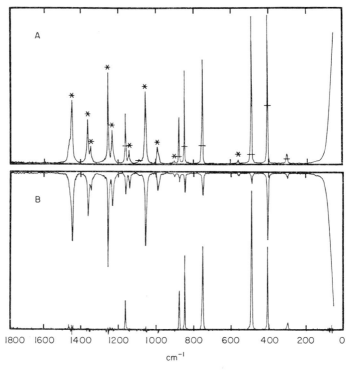

Fig. 5. Isotropic–anisotropic spectra. The Raman spectrum of *trans*-decalin illustrating computer calculation of the isotropic and anisotropic spectra from the observable $I_{||}$ and $I_{\perp}$ spectra. A = the $I_{||}$ spectrum, asterisks indicate depolarized lines where $I_{\perp} = 6/7\ I_{||}$, horizontal lines through the other peaks indicate the intensity of the $I_{\perp}$ component; B = spectra proportional to 45 $\bar{a}'^2$ (bottom) and 7 γ'^2 (upper, reversed scale) as generated from the spectra in A. For the isotropic spectrum note that only polarized peaks appear; also note the increased noise level in the region of depolarized peaks.

of reference and sample cells has been described[6] and the results from this system are illustrated in Fig. 6.

2.4.5 Smoothing of the spectrum

Even though the computer enables one to improve the signal to noise ratio via the CAT procedure or by using longer counting times an alternative procedure to reduce the noise level of the spectrum is desirable. 'CAT-ting' or similar procedures suffer from the disadvantage that they are time consuming from the instrumental standpoint, even if the procedure is automated so as to free the operator. Several mathematical techniques are available to smooth a spectrum via least squares. The procedure described by Savitzky and Golay[17] and adapted by Jones *et al.*[18] to infrared spectra is very efficiently performed using a computer. This smoothing procedure has been adopted by several different laboratories and proves quite satisfactory for Raman spectra.[8,9] The results are indicated in Fig. 7. A brief description of this approach follows.

The Savitzky–Golay procedure is a convoluting procedure whose end result is exactly equivalent to a least squares fit to a polynomial equation. The points in the smoothed spectrum are generated from the original data points according to the formula

$$Y_{j*} = \frac{1}{N} \left[\sum_{i=-m}^{i=+m} C_i Y_{j+i} \right]$$

where Y_{j*} and Y_j represent the intensity data points for the smoothed and unsmoothed data, respectively. The convoluting integers, C_i, are a set of constants and N is a normalizing factor for that particular set of convoluting integers. The integer m specifies the number of original data points $(2m+1)$ used to calculate each new data point. The increase in SNR realized by this procedure is $\sqrt{(2m+1)}$. Sets of convoluting integers are available where m ranges up to 12.[17] Thus an increase in SNR of up to 5 is available via this smoothing procedure.

One advantage of this smoothing procedure over the time constants (RC circuits) used on conventional analog recorders is that there is no bias between present and past data. The smoothing procedure uses points both before and after the point being smoothed in the calculation whereas the RC circuits are based only on data from the past. Thus, the RC circuits bias the data and this can lead to significant errors. These errors are avoided in the smoothing procedure.

2.4.6 Derivatives

The first or second derivative of a spectrum can be most useful in accurately determining peak positions or in detecting the presence of shoulders. These derivative spectra have not been widely used in the past because of the difficulties in obtaining them. The computer allows the capability of rapidly and efficiently calculating these derivative spectra from the normal scattering spectrum. The

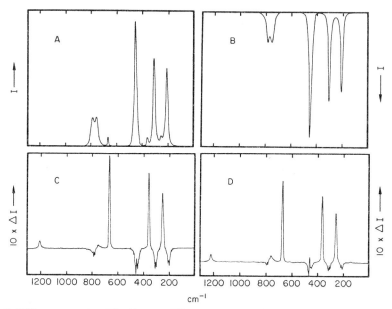

Fig. 6. Difference spectra. The Raman difference spectrum is shown for a 1:10 v/v solution of $CHCl_3$ in CCl_4 versus pure CCl_4.[6] A = sample ($CHCl_3$; CCl_4, 1:10); B = reference (pure CCl_4); C = uncompensated difference spectrum, ordinate expanded tenfold; D = difference spectrum scaled so the ν_1 band maxima are the same for both sample and reference spectra, ordinate expanded tenfold.

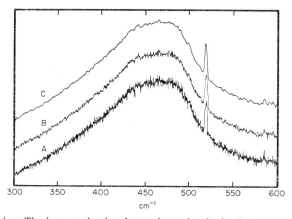

Fig. 7. Smoothing. The increase in signal to noise ratio obtained via smoothing (Savitzky–Golay technique) is illustrated for the Raman spectrum of molten sodium tetrasulfide (365°C). The sharp spike at 520 cm^{-1} is a laser plasma line. Note that the smoothing procedure does not broaden or affect the intensity of this line. A = raw spectrum; B = 9 point quartic smooth; C = 25 point quartic smooth (laser = 5145 Å, 1.4 watt; slit = 3.5 cm^{-1}; digitizing increment = 0.2 cm^{-1}).

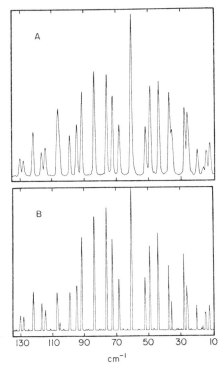

Fig. 8. Derivative spectra. The rotational Raman spectrum of air (A) showing the second derivative spectrum (B) as calculated by the Savitzky–Golay technique (Warren and Ramaley, Ref. 8).

only problem in this calculation appears to be the effect of random noise that is usually present in Raman spectra. For this reason the derivative calculation is usually combined with a smoothing routine.

The Savitzky–Golay convolution method of smoothing is equally applicable to the calculation of derivatives by using a different set of convoluting integers.[17,18] This method has been used in both infrared and Raman[8] spectroscopy. It consists essentially of initial smoothing of the spectrum (optional) with one set of convoluting integers followed by the derivative calculation using another set of convoluting integers. If necessary the resulting derivative spectrum may be smoothed again. The increased resolution obtainable with derivative spectra is illustrated in Fig. 8.

2.4.7 Peak area determinations

In the determination of concentrations of species it is generally the peak area that is used rather than the peak height. Methods using planimeters or weighing of the chart paper have been routinely used but are relatively slow and suffer from inaccuracies. Simple and rapid methods are available, however, once the spectra have been digitized. One of the simplest methods is the trapezoidal method where the area between two adjacent points is given by the frequency interval multiplied by the average intensity. A computer will perform this simple task, requiring less than one second for an entire spectrum.

2.4.8 Fourth power Boltzmann correction

It is well known that the intensity of Raman scattering is a function of the excitation frequency and the Raman shift. The intensity is also a function of the number of molecules in the initial and final states as expressed by the Boltzmann distribution law. The following equation[14] represents this effect where I_0 is the observed intensity, I_t is the true intensity, and K is a constant at constant incident intensity.

$$I_0 = \frac{K(\nu_0 - \Delta\nu)^4}{\Delta\nu\left[1 - \exp\dfrac{h\Delta\nu}{kT}\right]} I_t$$

I_t is, of course, related to the bond polarizability derivatives $\bar{\alpha}'^2$ and γ'^2 where $I_t = 45\bar{\alpha}'^2 + 4\gamma'^2$ if polarized exciting light and an analyzer are used.

The equation above indicates that the observed peaks will be skewed due to the fourth power and Boltzmann distribution effects. The effect will be essentially negligible for narrow Raman lines but can be significant for broad bands such as those observed in hydrogen bonded systems. The effect is illustrated in Fig. 9 for the spectrum of liquid water. For systems such as these the effect is certainly significant and if one wishes to perform curve resolution on this type of spectrum the corrected spectrum should be used as input for the curve resolution technique.

2.4.9 Curve resolution

The problem of the reduction of a complex curve into symmetric component

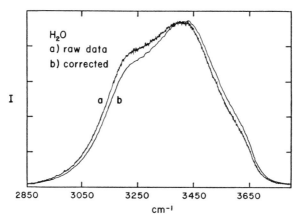

Fig 9. Fourth power Boltzmann correction. The Raman spectrum of water before (a) and after (b) correction for the fourth power Boltzmann factor (laser = 4880 Å, 1.3 watt; slit = 4.0 cm⁻¹; sample at 25°C; digitizing increment = 1.0 cm⁻¹). The corrected spectrum has also been smoothed via a 9 point Savitzky–Golay procedure, accounting for the reduced noise level.

peaks is common to most spectroscopic techniques. A complete discussion of the intricacies of these techniques is beyond the scope of this article but we shall give a brief survey of some of the methods using digital computers.

The first problem that faces one in curve resolution is to obtain the proper spectrum to use as input. We have already discussed the use of the computer to acquire Raman spectra. It is not generally advisable, however, to use the raw spectrum for curve analysis. It should first be corrected for the instrumental response of the spectrometer (section 2.4.2) as well as the fourth power Boltz-mann correction (section 2.4.8). Several of the other sections listed above may also apply before entering the curve resolution step. For example, one may wish to curve resolve the isotropic and anisotropic spectra separately (section 2.4.3) and compare the results.

The curve resolution methods generally fall into two classes, visual comparison of experimental and calculated contours, and least squares analysis. The visual methods which have been described are twofold. The first of these is the use of the DuPont 310 Curve Resolver[19] which is an analog computer. This instru-ment has been widely used in Raman spectroscopy and its main advantage appears to be the rapidity with which one obtains a 'fit' to a spectrum.

The fit is obtained by switching in the desired number of component peaks and varying three controls for each component (position, intensity, halfwidth) until a visually satisfactory match is obtained between the experimental and calculated curves. This instrument has several disadvantages that are generally avoided when using digital computers. Among these are the parallax problem and the requirement that each channel (component peak) be individually formed to the desired shape (leaving room for error in the non-uniformity of component peaks). There is also no provision for viewing the difference curve obtained by subtracting the synthetic matched curve from the observed spectrum.

The second visual system, the CURVER system described by Irish,[20] avoids most of these objections. The CURVER system uses a digital computer to generate the calculated curve, and to display both experimental and calculated curves on an oscilloscope for visual inspection. Otherwise its use is similar to the DuPont instrument. The difference curve may also be displayed, in an amplified form, to assist one in the final stages of curve fitting. In addition a display correspond-ing to the average error is obtained. Thus, most of the disadvantages of the DuPont instrument are avoided. A similar system has been described by Schmid et al.[4] A problem is that such systems are not commercially available.

Several least squares methods of curve resolution have been described. Although details of the programming methods are different all these programs mathematically fit components to a spectrum by minimizing $Q = \sum_i (r_i)^2$ where the r_i's are the point by point differences between the experimental and calculated intensity points. The user must normally supply, in addition to the frequency–intensity data, the number of components desired and initial guesses of the parameters to be determined by least squares analysis, i.e. intensities, positions, halfwidths, percent Lorentzian character, etc. of the component peaks. An objection, frequently voiced, is that the least squares technique may not 'converge' on a solution. One of us (J.R.D.) has had considerable experience with a modified version of the Jones[21] programs with no such problem. It is the authors' opinion that this non-convergence objection is probably outdated and does not apply to most recent least squares fitting programs.

There are several additional considerations that should be noted relative to programs. An extremely large number of computations are required and most of these programs have been written for large high speed computers where large amounts of core and high computational speeds are routine. Therefore they are not ideally suited for operation on the slower and smaller computers found in many chemical laboratories, i.e. the dedicated computers (see section 3.1). Another general observation is that the spectroscopist is left with little 'feel' for the fit since he usually receives only a simple listing of the component parameters (intensities, positions, halfwidths) as well as statistical parameters such as the standard error of estimate or standard deviation. He receives little indication of what effect a slight shift of one of the peak positions, for example, may have on the fit. Such intuitive feelings are usually obtained with the visual methods.

Regardless of the method of curve resolution one is faced with the problem of how many component peaks to use. The minimum number of component peaks can usually be deduced from the number of shoulders in the spectrum. However there may be hidden peaks whose presence is not indicated in the spectrum. The derivative spectra (section 2.4.6) may be useful in determining not only the number of components but also their positions. Butler and Hopkins[22] have given a good discussion of the use of high derivatives (up to the fourth) to resolve optical spectra into their component peaks. Their analysis indicates that some care must be exercised since false components may sometimes be indicated by the derivative spectra.

Some method must also be found to curve resolve spectra where the curve does not return to the same baseline on either side of the peak. Usually one subtracts either a straight sloping line or some more complex line from the spectrum before performing the curve resolution. The computer may be useful here in fitting, via least squares, an exponential curve to points on either side of a peak on the side of a broader curve. The resulting equation may then be used to generate the background curve and subtract it from the experimental spectrum before performing curve resolution.

3 HARDWARE–SOFTWARE CONSIDERATIONS

The hardware–software features which are important in considering a Raman–computer interface may, logically, be divided into three areas: the computer, the computer–spectrometer interface, and spectrometer modifications. While this separation is arbitrary, it should be borne in mind that a systems approach to the problem is absolutely necessary. That is, the scientist must not consider these three areas separately but must, instead, view the overall system and make subtle trade-offs between the spectrometer–interface–computer hardware to obtain the desired result.

A further consideration underlying the preceding is cost. The hardware cost is, of course, the most obvious direct cost; however it can be surpassed by the software cost, i.e. preparation of the desired instructions in programs compatible with the capacity and logic of the hardware system. The latter is a point often not foreseen in the planning stages. Thus a larger computer at a higher initial cost may be a more economical investment if this enables easier programming at a later date.

3.1 The Computer

Several different types of computer are available, each having its own distinct advantages and disadvantages. No one type seems ideally suited for all the problems encountered by the Raman spectroscopist, so one may wish to use more than one type of computer in order to achieve the desired system. For the purposes of this article we shall consider three types of computer: large scale, dedicated use and process control. This classification is not based entirely on the hardware differences, but also on the operational differences of the systems.

The large scale computer is characterized by high computational speeds, a large number of peripherals (magnetic tape units, card readers, line printers, incremental plotters) and large disk and core capacities. The price range is generally such that these machines are only cost effective when utilized for round the clock computational operations. They most frequently see service as the central computing facility in university, governmental, or industrial laboratories, and as such are accessed by a large number of users. Because of this and the cost it is not practical to interface one of these computers to a Raman instrument for data acquisition or control. Such facilities are a great asset for data evaluation, particularly in applications where a large number of computations are required, e.g. as in least squares curve resolution, but not very practical (from the cost and access viewpoints) as data acquisition systems for instruments such as Raman spectrometers.

The dedicated computer is designed for a different application. Dedicated computers are invariably small in size, with limited core capacities (generally not more than 8K), and virtually no peripherals. They are exemplified by the so-called 'minicomputers' which generally have 4K core capacity and (as the only peripheral) a teletype terminal. The cost falls in the price range of $5000 and less. Such computers are ideal for interfacing to instruments for experiment control and data acquisition and storage. They are, thus, entirely dedicated to the instrument being serviced even though the actual control–acquisition functions take only a small amount of the CPU's time. In the latter respect they are wasteful, but this is generally tolerated because of the relatively low cost.

Serious disadvantages are encountered if data analysis as well as data acquisition/storage is a requirement. The relatively low core storage capacity of dedicated computers forces programs to be written in assembly language since high level compilers are generally not available or are expensive (relative to capabilities gained) for these machines. Even with assembly language the typical acquisition–control program will require ~ 1.5–$2K$ of core. Thus only about $2K$ remains available to store the spectral data input from the Raman instrument. This can impose a severe limitation. An input of 3750 data points, for example, completes a scan from 50 cm^{-1} to 3800 cm^{-1} (assuming a 1 cm^{-1} digitizing increment) using computer assisted techniques. Since one may be required to use 'two words' per data point[4] with many minicomputers, the total number of points is immediately limited to ~ 1000, i.e. the capacity is insufficient to achieve such a complete scan in one operation.

Such (small) dedicated computers are thus not as useful for data evaluation as one would desire. The use of published routines, i.e. software packages, is prevented by the assembly language requirements. The limited core capacity generally dictates a piecemeal programming approach,[4] e.g. one program to integrate curves, one to perform smoothing, etc. This is cumbersome and time consuming.

Process control computers are characterized by the capability of simultaneous access by two or more users. In terms of hardware these systems generally have at least 16K of core capacity and enough disk storage capacity to allow facile simultaneous use by at least two users. The primary difference between process control and dedicated use computers is in the monitor system. With the proper hardware and software a process control computer may be used in a dedicated mode simply by loading a different monitor, a task that requires a few minutes at most.

The primary hardware difference between dedicated and process control computers is the necessary presence of a priority interrupt structure in the process control machine. This interrupt structure insures that the highest priority user gains control of the CPU at the required time. For example, most process control machines operate with a time-shared two job background/foreground (B/F) monitor. In this system the foreground job always has priority over the background job. The foreground mode must be used for the data acquisition/ control program while the background mode is used for all other tasks, e.g. program creation, debugging, editing, compilation or execution. Thus if background is being used to execute a program and the foreground operator initiates a scan on the Raman spectrometer, background execution is interrupted so the CPU may process the collection routine. This may require only a few milliseconds every second since Raman data are generally collected at relatively slow speeds, rarely faster than 10 points per second. During the remainder of the time the CPU processes the background job which is interrupted only when the foreground user (via the interface) signals the CPU that its services are needed to process incoming data.

Process control computers have become popular with academic, governmental and industrial laboratories because the foreground job may be used for simultaneous data acquistion from a number of instruments while still allowing background operation. For example, the Mulheim[23] system with 32K of core allows simultaneous operation of 1 fast instrument (mass spectrometer or pulsed n.m.r.) and 20 slow instruments (gas chromatographs, n.m.r.'s, i.r.'s, spectropolarimeters). Other similar systems have been described.[24,25]

Since each of the three types of computers have their own advantages and disadvantages the final choice will depend on what is expected of the Raman–computer system, i.e. what specifications are designed into the system. In fact no one type of computer is ideal for all Raman applications so it may be desirable to design a hierarchal system.[5] In this approach one type of computer may be used to collect data, e.g. a minicomputer, while another type may be used to analyze the Raman data, e.g. process control or large scale computers. In a true hierarchal system the CPU of the larger computer will control the CPU of the smaller computer, and data and control signals will be transferred via a hard link (cable or telephone connection). However the data may be transferred via a soft link (paper punched tape, punched cards, or magnetic tape) and this enables one to use a minicomputer for data acquisition, and an entirely independent and larger computer for data processing and evaluation.

Another consideration, in addition to the type of computer, is the number of bits per word. This may be translated directly into the maximum size integer number that the computer can handle. Thus a 16 bit machine will handle numbers up to $\pm 32,768$ (2^{15} since one bit is normally used for the sign). Such a machine would not be ideally suited for interfacing to a photon counting Raman spectrometer where the number of counts may approach 100,000 thus requiring two core words per data point (double precision). An 18 bit machine, handling numbers to $\pm 131,072$, would be much more suitable since only one core word/point (and thus less core) is required. The questions of peripheral equipment also require attention; i.e. the data storage devices and display systems.

The common data storage devices are magnetic tape, paper punched tape, and punched cards. Magnetic tape would seem to be the preferred storage medium because of its compactness as well as ease and speed of use. The tape drives and tape handlers required for the magnetic tape system, however, make this considerably more expensive than punched paper tape systems.

After the data are collected and possibly modified via smoothing, etc., a graphical display may be the desired output since a simple listing of wavenumber versus intensity is generally not very meaningful to the spectroscopist. Several methods for graphical display are available. The data may be displayed on an analog $X-Y$ or strip chart recorder; this will require a digital to analog converter (DAC) and may be rather expensive. Another method, and more convenient, is the use of an incremental (digital) plotter, directly driven by the computer. Such plotters are relatively inexpensive and are capable of high quality graphical output. Both of the preceding approaches suffer from the disadvantage of slow speed; several minutes may be required per plot. Since some trial and error may be involved before obtaining the proper spectral display this may be very time consuming.

Faster display systems include the use of an oscilloscope and (more recently) cathode ray tube graphics display terminals (GDT). Oscilloscopes are rapid but do not generally provide a hard-copy display except via photography. The GDT terminals combine the features of oscilloscope display with a teletype console for sending commands to the computer. With this arrangement it is possible to obtain the desired display rapidly, since only a few seconds are required per plot. Most of the GDT's are available with an optional 'hard-copy unit'; the latter produces a paper copy of the display on the cathode ray tube upon operator command. The GDT's, while they combine rapid displays with hard-copy capability, are correspondingly more expensive.

3.2 The Interface

The implementation of the interface normally poses the heart of the problem in instituting a Raman–computer system. This arises primarily because many scientists (and electronics technicians) still dwell in the domain of analog signals and remain generally unfamiliar with the techniques and hardware for handling digital signals and performing the analog to digital conversion. For these reasons it is generally wise to use commercially available analog to digital converters or ratemeters unless one either has previous experience in interfacing or is willing to suffer the inevitable lost time and problems in order to learn interfacing techniques. Commercial units are now available from most computer manufacturers and other computer hardware vendors at reasonable prices. Most of the commercial ADCs are versatile enough to handle a variety of instruments, in many instances simultaneously, and this may be a bonus if it is desired to interface other instruments in addition to a Raman spectrometer.

In general the interface should have the capacity of digitizing at least two channels of analog information, the frequency and intensity information. More channels may be desirable in order to handle signals proportional to source intensity, slit width position, etc. In addition, provision for a series of control lines may be desirable to control various shutters or other on/off mechanisms which may be used in the Raman experiment.

One general drawback of most interfaces is that they are often specifically designed for a certain computer and may not be readily adaptable to other models or brands. Thus, if a faster or more versatile computer replaces an older machine it will also probably be necessary to replace or modify the interface. This situation is slowly being rectified as the various computer manufacturers standardize the input/output characteristics of their respective hardware systems.

A generalized interface for 16 bit computers has been described by David and

Schmidlin.[26] The modular nature of this interface makes it easily adaptable to virtually any computer or experiment. Gruber *et al.*[27] describe a versatile interface for the popular PDP-8 series of computers[28] capable of accepting 8 lines of analog information and controlling 6 stepper motors. Perone[29] has reviewed a series of interfaces involving dedicated computers while Shapiro and Schultz[25] have described an interface for process control computers. Dessy and Titus[30-32] have discussed do-it-yourself 'instant interfacing' for mini-computers. Williams and Turner[33] have described an interface featuring a technique of obtaining a d.c. voltage (proportional to intensity) from instruments not already possessing such signals. Many of the references in section 4 contain information relative to specific Raman-computer interfaces (see also the description of the RPI system in section 5).

3.3 The Spectrometer

Few modifications are normally required at the spectrometer end of a system. It is convenient to point out that all commercial Raman spectrometers are somewhat similar in their methods of detection and monochromator drive mechanisms, the two items of highest importance when interfacing the spectrometer.

The monochromator drive mechanisms are generally of two types: continuous scan or stepping motor. The continuous scan mechanisms consist of a synchronous motor and gear arrangement such that the monochromator scans at a fixed time rate. Thus one can measure time rather than some instrumental factor to obtain the frequency information. Though satisfactory for other purposes, this system may not be adequate for repositioning the monochromator when performing a CAT procedure via multiple scans since the monochromator will not stop instantly at the end of each scan. Thus a systematic error (irrespective of any backlash error) will arise if the monochromator is instructed simply to run backwards for an equivalent amount of time to reset for the next scan. This will lead to a broadening of the 'CAT-ted' spectrum. Thus it will be necessary in most cases either to reset the monochromator by hand between scans or to develop some positive means of encoding the frequency information for the computer to prevent this reset error. Several shaft encoding mechanisms have been described in the literature to accomplish the latter task.[12,13,34]

Drive mechanisms which use a stepping motor inherently provide a greater flexibility for controlling the Raman experiment as well as an exact link to the frequency at all times. If the monochromator moves in steps of $1/20 \text{ cm}^{-1}$ and a scan of 100 cm^{-1} is required, the motor will move 2000 steps. Thus, by moving 2000 steps in reverse the precise starting frequency will be reproduced. Backlash errors may be removed by stepping an extra few hundred steps backward and then forward. Since the same pulses that drive the stepping motor may be routed through the computer and counted, the computer constantly keeps track of the position of the monochromator.

The Raman scattering intensity information on commercial instruments is generally obtained via photon counting or some means of d.c. amplification (or means that result in a d.c. signal to a chart recorder). With photon counting instruments the signal is usually picked off after passing through a pulse height amplifier–analyzer system and sent to a digital ratemeter for counting. On instruments using other methods of detection one can normally find some point in the electronics where there is a d.c. signal proportional to the scattering intensity, e.g. the input to the instrumental chart recorder. This may be picked off and sent to an ADC to transfer the information to the computer.

The transference of the frequency and intensity information outlined above represent the two essential items for digital encoding. Other items are sometimes desirable. A second intensity channel, which represents the laser intensity, may be desirable to compensate for source intensity fluctuations. A simple beam-splitter accompanied by a photomultiplier and another channel on the ADC is normally sufficient. In addition, control of the slit mechanism, shutters, or other experimental parameters may occasionally be desirable and require modifications to the instrument and associated equipment. Systems employing some of these modifications are outlined in section 4.

Although an instrument employing any combination of the monochromator drive system–detection system may be interfaced to a computer, the most desirable appears to be the stepping motor drive–photon counting system combination. This combination provides a tremendous amount of flexibility for controlling the Raman experiment by writing the proper software to control the stepping motor–ratemeter combination. Thus one can scan at a constant rate, i.e. count for the same amount of time at each point, simply by supplying fixed rate pulses to the stepping motor; this is a time-based scan, which is the standard for most commercial instruments. One can also count at each point until a fixed number of photons is detected before moving to the next point; this is a count-based scan, which uses the output of the ratemeter to control the stepping motor. Spectra recorded under these conditions have a constant SNR throughout. Combinations of the above are possible, counting to a maximum time limit per point or until a certain number of counts are obtained.

4 · EXISTING RAMAN–COMPUTER SYSTEMS

The range of systems is summarized in Table 1. The examples in this table are illustrative and are not intended as a complete account of the variations advanced in the past decade. Some interesting trends are discernible. Inspection of Table 1 shows that most digital data collection is done with dedicated computers and most of these can be classed in the minicomputer (4K or smaller) category. Several systems using process control computers have been described[2,16] (see also section 5), and two systems use off-line data collection.[3,9] One system with a programmable calculator has been described,[8] and also one which uses a

TABLE 1
Existing Raman–computer systems

	Computer[a]	Spectrometer[b]	Capabilities[c]
Hatzenbuhler[d]	VA (1K, D)	SP (CS, DC)	CAT
Scherer[e]	IBM (32K, PC)	PE or SP (CS, DC)	PF, CL, IC, IA, CAT
Eysel[f]	Off-line	CO (CS, DC)	S, CAT
Schmid[g]	PDP (4K, D)	CO (SM, RC)	CAT, IC, CV, I, SI
Bulkin[h]	PDP (4K, D)[i], Nova (12K, D)	SP (SM, RC)	S, IA, BS
Amy[j]	Off-line	SP (SM, RC)	S, D, PF, AS, I[k]
Ushioda[l]	PDP (4K, D)	SP (SM, RC)	CAT
Warren[m]	HP (4K, D) or Wang 700	SP (SM, RC)	S, CAT, D
Barrett[n]	Off-line	PE (SM, RC)	S[o]
Small[p]	VA (D)	(SM, RC)	CAT
Moore[q]	IN (8K, D)	SP (SM, RC)	CAT, S, I, D, PF, AS
RPI[r]	PDP (24K, PC)	JA (CS, RC)	CAT, S, CL, CV, I

[a] IBM = International Business Machines, PDP = Digital Equipment Corporation, HP = Hewlett-Packard, VA = Varian Associates, IN = Interdata, D = dedicated, PC = process control.

[b] PE = Perkin-Elmer, SP = Spex, CO = Coderg, JA = Jarrell-Ash, SM = stepping motor drive, CS = continuous scan drive, DC = direct current amplification, RC = ratecounter amplification (photon counting).

[c] PF = peak finding routine, CL = least squares curve resolution, CV = visual curve resolution, IC = instrumental corrections, IA = isotropic-anisotropic, CAT = computer averaging of transients (or extended count times/point), S = smoothing, I = integration, SI = standard intensity, BS = background subtraction, D = derivatives, AS = addition-subtraction of spectra.

[d] See Ref. 1.   [l] See Ref. 7.
[e] See Refs. 2 and 16.   [m] See Ref. 8.
[f] See Ref. 3.   [n] See Ref. 9.
[g] See Ref. 4.   [o] Analysis on SDS 9300 computer.
[h] See Ref. 5.   [p] See Ref. 10.
[i] Hierarchal system.   [q] See Ref. 11.
[j] See Ref. 6.   [r] See section 5.
[k] Analysis on CDC 6500 large scale computer.

1024 channel analyzer originally designed to perform 'CAT-ting' of n.m.r. spectra.[1]

Spectrometers from various manufacturers have been used employing virtually every combination of monochromator drive–detection techniques, the only one not represented in Table 1 being the stepping motor drive–d.c. detection combination. By far the most popular, probably for reasons mentioned in section 3.3, are instruments using the stepping motor drive–photon counting combination.

There is a wide variation in the capabilities of the systems in Table 1. More than anything else, this probably reflects the difficulty in developing flexible and comprehensive software, particularly for small core capacity computers. The last column of Table 1 supports this, since a quick glance shows that those systems with the most capabilities also have the larger computers, the only exception being due to Schmid[4] who has developed a comprehensive system using a minicomputer with only 4K of core. Many of the other systems using small computers appear to have been developed merely to increase the SNR in the resulting spectra, either by the 'CAT-ting' or smoothing techniques.

Attention is directed to the work of Scherer and Kint;[2] these authors describe a system using a process control computer capable of accepting data from either of two Raman instruments or one infrared instrument. Amy *et al.*[6] have described a system for obtaining Raman difference spectra utilizing a programmed cell-positioning mechanism. The system described by Ushioda *et al.*[7] contains several signal lines for controlling experimental conditions during the scan. Either of the systems described by Amy or Ushioda seems capable of determining both the parallel and perpendicular polarized spectra during a single scan if the modulation mechanism for rotating the plane of polarization described by Proffitt and Porto[35] is incorporated. Warren and Ramaley[8] and Barrett and Weber[9] have described systems which automatically cancel out the effects of source intensity fluctuations in the course of collecting spectra. Also of interest is a commercially available system, complete with all hardware and software, described by Moore.[11]† The system in use at Rensselaer Polytechnic Institute is described in the next section.

5 THE RPI RAMAN–COMPUTER SYSTEM

The system described in this section was implemented at Rensselaer Polytechnic Institute (RPI) in the summer of 1973. This system was designed with the following criteria as primary guidelines:

    (i) to provide a system capable of collecting (with CAT capability) and storing spectral data in digital form;

    (ii) to provide a versatile display system;

    (iii) to provide software which could readily be expanded as the need arose;

    (iv) to make maximum use of the major hardware items already available: a Jarrell-Ash 25-300 spectrometer and a PDP-15 process control computer.

In its present configuration the system operates in a time-shared mode so that during digital collection of a Raman spectrum the computer remains available to other users or for evaluation/analysis/display of spectra previously collected. The computer is a departmental facility and is currently being interfaced with other instrumentation, including a flash photolysis–kinetics unit, and a PAR 170 electrochemistry apparatus. These can be operated simultaneously with the Raman experiment.

In the following sections a complete description of the hardware and software

† This is not to imply recommendation or approval of this system or manufacturer since similar ones which may be suitable are probably available elsewhere.

as well as the operating characteristics and capabilities of the RPI system is given.

5.1 Hardware

The apparatus shown in pictorial form in Fig. 10 constitutes the RPI system. A complete listing of the major hardware items is given in Table 2. The Jarrell-Ash 25-300 spectrometer uses photon counting detection, so a simple scalar-timer (digital ratemeter) serves as the interface between spectrometer and computer. The output signal to be fed to the digital ratemeter is obtained by placing a simple T-connector between the analyzer and analog ratemeter of the JA 25-300. In this manner a normal chart recording of the spectrum is obtained while the digital intensity information is being transmitted to the computer. The only modification to the spectrometer was the addition of a relay to start and stop the drive mechanism of the monochromator. This relay is controlled by a pulse received from the computer. The spectrometer is operated in the con-tinuous scan mode so that the frequency information is transmitted to the

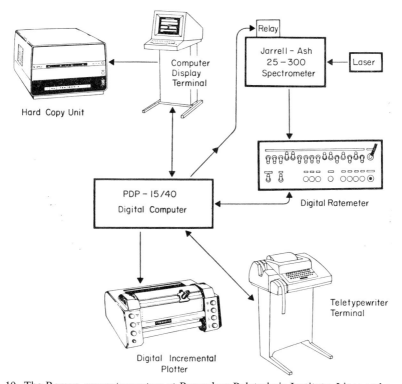

Fig. 10. The Raman–computer system at Rensselaer Polytechnic Institute. Lines and arrow-heads indicate the directional flow of data and control signals. All hardware is commercially available except the relay (a minor modification of the Jarrell-Ash spectrometer) and the digital ratemeter (constructed by an undergraduate student).

computer by supplying the initial frequency, length of scan, and digitizing increment.

The system is laid out with the laser, spectrometer, computer display terminal and hard copy unit in one location. The remaining equipment is in the computer room some 200 feet distant so that several amplifiers, not shown in Fig. 10, are necessary. Their presence does not alter the operational description of the system.

TABLE 2
Major hardware items for the RPI system

Item	Model No.	Source
Raman spectrometer	25-300	Jarrell-Ash
50 MW He–Ne laser	125	Spectra-Physics
4 W argon ion laser	165	Spectra-Physics
Digital ratemeter		RPI[a]
Digital computer	PDP-15/40[b]	Digital Equipment Corporation
Computer display terminal	4010-1	Tektronix
Hard-copy unit	4610	Tektronix
Incremental plotter	565	California Computer Products

[a] See text for description.

[b] The computer is an 18 bit machine and has the following features: core capacity = 24K, two RF-15 DEC disks (total capacity = 500K), four DEC tape transports, high speed paper tape reader and punch, one model 35 and one model 33 teletype, automatic priority interrupt, memory protection, extended arithmetic element.

The ratemeter was designed and constructed at Rensselaer. It is basically a scalar-timer which will count pulses for a specified time period. It consists of a 12 bit binary counter (maximum count of 4096) and a 1 MHz crystal clock gated to an 18 bit flip-flop circuit. The desired count time per device data point is set by the operator using 16 toggle switches on the ratemeter control panel (two flip-flops are hard-wired since it is never necessary to alter their setting). The maximum count time available is 0.262144 s (2^{18} μs). Since longer count times per data point are sometimes desirable, the software has been written so that the intensity data from 12 ratemeter points are added to yield one actual point. Thus in effect a maximum count time of slightly greater than 3 s per point is available.

During data collection the system operates as follows (refer to Fig. 11). After the operator sets the monochromator to the desired starting frequency, he sets the ratemeter toggle switches to settings which reflect the scan speed and digitizing increment that he has selected. At the operator's command a pulse from the CPU simultaneously starts the spectrometer drive system and the ratemeter. The ratemeter counts for the specified time, stores the total in the ratemeter buffer, resets to zero and counts again for the next device point. When 24 device points have been stored in the buffer a pulse signals the CPU and causes it to add these 24 device points into 2 actual points which are placed in the disk buffer. The ratemeter buffer is cleared and the next 24 device points (2 actual points) are collected. Each time the CPU interrupts to write these 2 points on the disk buffer it also checks to see if the disk buffer is full, 250 points. If so, the contents of the buffer are transferred to disk, the disk buffer is cleared and the buffer pointer reset. This process continues until the specified number of points have been collected at which time the CPU again interrupts, transfers all remaining points from the disk buffer to the disk and closes the disk file. The CPU time

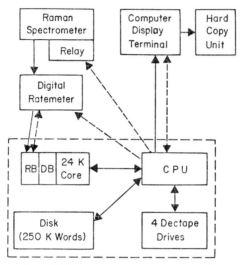

Fig. 11. Block diagram of the RPI Raman–computer system. Solid lines represent data flow while dashed lines indicate the flow of control signals. The area inside the dotted lines is the PDP-15/40 computer (RB = ratemeter buffer, DB = disk buffer; both are segments of the 24K core storage). Items not illustrated are the incremental plotter and a second disk and teletype.

required for each point is approximately 0.5 ms. Since one rarely collects more than 10 points per second in Raman spectroscopy, a maximum of 5 ms per second of CPU time is used in collection. The CPU is available for the background user the remainder of the time.

The disk serves as a temporary storage medium and is capable of storing up to 56 spectral files, depending on the number of points per spectrum. The permanent storage medium is the small set of magnetic tape units designated by the manufacturer as DECtape.[28] These tapes also store up to 56 spectral files.

The third main function of the hardware (collection and storage of data being the first two) is to display the spectral data. The computer display terminal (graphics display terminal) and hard copy unit have proved most satisfactory in this regard. The display terminal requires approximately 2 seconds to plot a spectrum and the display scales can be easily altered to achieve the desired result (see software section). A paper copy of the display may be made in 4 to 5 seconds using the hard copy unit. The quality of these copies is high; with only addition of appropriate scales and labelling these can be used directly for publication.† Although an incremental plotter is available it has not seen much use due to the greater convenience of the display terminal–hard copy system.

5.2 Software

Several software programs are available in the Rensselaer system. They provide three basic capabilities:

(i) control of the ratemeter (to allow data collection);
(ii) control of the computer display terminal and incremental plotter (to allow various modes of display);
(iii) to allow data evaluation and analysis.

Chronologically the first program developed was the Computer-Aided-Spectrum-Handler (CASH) program.[37] This program forms the heart of the Rensselaer software system and fulfilled our initial objectives (section 5.1). It controls not only the ratemeter but also the computer display terminal, thus permitting both data collection and display. It does have certain drawbacks, the main one being that core requirements are large (12.5K) so that when used as the foreground program in the time-shared background/foreground (B/F) operating system little core capacity is left for the background user (the B/F monitor takes up approximately another 10K). Its utility as a data collection program is therefore limited due to other demands on the computer. Thus the BFCOLL program was developed for collecting, averaging, and storing data. Other data evaluation programs were developed as the need arose, i.e. MANDAT and SUMFIT.

The four programs are described below. All, except SUMFIT, are written in a conversational mode. The primary program for collecting spectra is BFCOLL, it being used almost to the exclusion of CASH for this purpose. With BFCOLL in foreground any of the other three programs may be used in background. Thus while one spectrum is being collected (BFCOLL) other spectra may be displayed and evaluated (CASH, MANDAT, SUMFIT).

BFCOLL. This program[38] serves as the data collection, averaging, storage software system. The program is written completely in assembly language to minimize core requirements, which are 1.2K. Using the present B/F monitor system with DECtape handler, approximately 13K of the total 24K core capacity is available for the background user with BFCOLL in foreground. The program serves four functions which may be selected by the user; collecting a new spectrum (C), averaging an additional scan into an old spectrum (A),

† Figures 2, 3, 12 and 13 were produced with the RPI graphics display terminal (labelling added).

storing a spectrum on magnetic tape (S), deleting an unwanted spectrum from the disk (D).

After selecting the desired function, a simple one letter command, the program requests a file name for which the user must supply a 5 character alphanumeric code. If the collect section was selected the program will then ask for 4 vital pieces of information: scan speed (cm^{-1} min^{-1}), increment (cm^{-1}), starting frequency (cm^{-1}), length of scan (cm^{-1}).†

From this information the computer determines how many points are to be collected and for what length of time the ratemeter will collect data. The user then has 9 lines in which to enter descriptive material such as sample description, laser frequency and power, slitwidth, polarization, etc. The notes are terminated by entering a special character (a period at the beginning of a line) and the scan commences upon hitting the *return* key. At the end of the scan the computer sounds a bell to alert the operator and terminates the spectrometer scan. The program stores on disk, under the filename, not only the point by point intensity data but also the 4 pieces of information above and the notes.

If it is desired to store the spectrum permanently the save command (S) is requested, the filename given again, and the complete file will then be transferred to the permanent storage medium (DECtape). If the disk becomes full the delete command (D) may be used to clear it of unwanted spectra, i.e. those already saved on DECtape or non-valid spectra due to poor experimental conditions, laser shutdown, etc.

If averaging ('CAT-ting') is desired, the spectrometer is manually reset to the starting frequency, and the average (A) command is given. The filename which the user gives must be of a spectrum that has already been collected. No further information is needed since the file already contains information on scan speed, increment, etc. The scan starts upon receipt of the *return* command. At the end of the scan the user has the option whether or not to average the current scan into the previously collected scan(s).

CASH. As mentioned above the Computer-Aided-Spectrum-Handler (CASH) program[37] serves as the focal point in the Rensselaer software system. It has the capability of collecting or displaying spectral data as well as allowing for visual curve resolution and outputting spectral data for analysis by other programs (MANDAT or SUMFIT). CASH is designed as a set of seven functionally distinct subsystems and is run in an overlay structure, i.e. only one subsystem at a time is loaded into core. The overlay structure is necessary to reduce the core requirements which are 12.5K with the largest subsystem in core. The overlay structure has not proved particularly burdensome since the maximum time delay when requesting a new subsystem is less than one minute.

Most of the CASH program is written in FORTRAN except for the collection subsystem and portions of the display subsystem which are in assembly language.

---

† Increment, starting frequency and length of scan are entered in terms of tenths of wavenumbers ($cm^{-1} \times 0.1$).

The core requirements could be reduced, of course, by going completely to assembly language, but FORTRAN offers substantial advantages, i.e. the programming can be accomplished without learning a special language and published routines can be used. The FORTRAN routines are particularly useful in the data evaluation sections of the program where frequent changes and/or additions may take place. The subsystem structure of CASH makes it particularly easy to modify existing routines or to add new subsystems as the need arises. The seven present subsystems are Collect; Library; Display; Calculate; Multi; Report; Plot.

The Collect subsystem accomplishes exactly the same purpose as the collect and average sections of BFCOLL with the input being exactly the same. The only difference is that after the scan terminates the spectrum is automatically displayed on the computer display terminal screen. The viewer then sees exactly what the computer has in disk storage and can better judge whether to average in a new scan or change other experimental conditions. This is convenient since often the digitized spectrum may be of higher (or lower) quality than the spectrum that appears on the chart recorder depending on the digitizing increment and scan speed (or its equivalent, the count time per point set on the ratemeter). The operator also sees directly the effect of increasing the SNR as the number of averaged scans is increased. The Collect subsystem is non-operational when CASH is loaded in background since all data must be collected in foreground.

The Library subsystem handles the transfer and storage of spectral files. As mentioned earlier each spectrum is stored under a filename, and consists of

  (i) defining parameters: scan speed, increment, starting frequency, length, number of averages;
 (ii) notes: descriptive information as entered by user at time of collection;
(iii) spectral data: the actual intensity values that comprise the spectrum.

This subsystem enables the user to transfer any spectral file from one storage medium to another. The available media are disk and DECtape. In order for other subsystems to recognize a filename it must be loaded onto disk via Library. Thus old spectra may be recalled from DECtape for display or other operations.

The Library subsystem also allows the following auxiliary operations:

  (i) printout of notes and/or defining parameters;
 (ii) correction and/or addition of notes;
(iii) splitting a file into two or more smaller files;
(iv) deleting spectral files from disk and/or DECtape;
 (v) listing tape and disk directories.

The Display subsystem controls the display of spectra on the computer display terminal and gives the user the capability of manipulating the displayed spectra. The subsystem contains a working area capable of holding two spectra (channels).

The user may selectively view one or the other, or may display both spectra simultaneously for comparison. Four simple commands relative to X scale, Y scale, X range, and height of the baseline, independent for each channel, enable the user to achieve the desired display rapidly. Thus, an entire spectrum may be displayed or a small region may be enlarged to fill the screen. A hard copy may be made as desired. Figure 2 was taken directly from our hard copy unit (labelling added) and shows the same spectral file displayed in both survey and blown-up form.

In addition to the very versatile display capabilities other options are available in this subsystem. The frequency and intensity of any point in the spectrum may be printed using a cursor to locate the point. In addition the area under a peak may be found simply by specifying the frequency interval desired for the integration (trapezoidal method).

The Calculate subsystem enables the user to synthesize Gaussian/Lorentzian contours. The equation used is a sum function

$$Y(\nu) = \sum_i [XL(\nu)_i + (1 - X)G(\nu)_i]$$

where X is the fraction Lorentzian character and $L(\nu)$ and $G(\nu)$ are the Lorentzian and Gaussian functions, respectively, and there are i component peaks. This function has proved satisfactory in fitting a variety of Raman spectra. The user must supply the number of component peaks (i), the fraction Lorentzian character (X), and the height, halfwidth, and position for each component peak. After the spectrum is synthesized the user may switch to the Display system and view both the experimental spectrum and the calculated contour. Thus by switching from Calculate, where selected parameters may be changed, to Display, visual curve resolution may be accomplished as illustrated in Fig. 12.

In practice this system of curve resolution has proven to be very time consuming due to the delay in switching between subsystems as well as the actual computational time. It was mainly for this reason that the least squares analysis curve resolution program (SUMFIT) was developed. The SUMFIT program is now used almost exclusively to provide a curve resolution capability although the Calculate–Display subsystems prove useful in viewing the end result as well as evaluating what effect small changes in the various parameters has on the fit.

The Multi subsystem allows for the simultaneous display of up to 16 spectra to allow easy visualization of temperature and concentration effects as shown in Figs. 3 and 13. Each spectrum may be individually manipulated to achieve the desired X scale, Y scale, and X range. Two modes of display are available: the comparison display where all spectra are displayed relative to the same baseline and the publication display where spectra are separated vertically for clarity.

The Report subsystem outputs the frequency–intensity information of a spectral file to enable further analysis by other programs, i.e. MANDAT and SUMFIT. Output can be routed either to DECtape or disk.

The Plot subsystem outputs the frequency–intensity information necessary

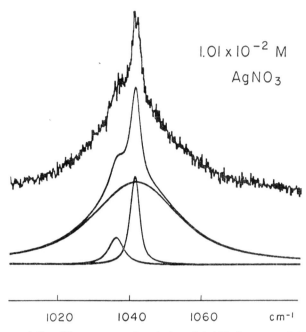

Fig. 12. Curve resolution. The spectrum of a solution of $AgNO_3$ in CH_3CN is illustrated along with a synthetic curve (and its components) obtained via visual means using the RPI system (see section 5). The baseline of the experimental spectrum has been offset for clarity (laser = 4880 Å, 1.3 watt; slit width = 3.5 cm⁻¹; room temperature; digitizing increment = 0.5 cm⁻¹). Figure produced by the RPI system.

to produce spectral plots on the incremental plotter. Rather than directly issuing plotting commands, the Plot system produces a DECtape file which may be plotted at a later time using a standard plotting program.

MANDAT. This program[38] was written to perform data evaluation procedures not currently available in the CASH program. Input for the program comes via the Report subsystem of CASH. Output from the program is available either as a teletype listing of the corrected frequency–intensity data, a disk file of the same data, or a plot on the incremental plotter. The program is written entirely in FORTRAN.

Five options are available to the user; smoothing, fourth power Boltzmann correction, baseline subtraction, scaling and data format changes. Smoothing of data is performed via the Savitzky–Golay technique[17] using the program published by Jones[18] with modifications only in the input–output routines. Three smoothing functions are available: 5 point, 9 point and 25 point. The first two functions were tested by Jones and found satisfactory for infrared spectra. The 25 point function was added to allow a larger increase in the SNR and we find that all three functions are satisfactory for Raman spectra. Our experience has shown that care must be exercised to use a small increment

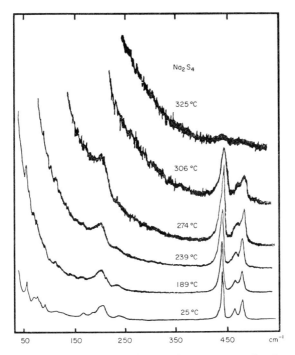

Fig. 13. Multiple display, temperature effects. The spectrum of polycrystalline Na₂S₄ illustrating the effect of temperature. The apparent increase in noise is actually a decrease in signal, the display having been scaled to keep the 445 cm⁻¹ peak at approximately the same intensity throughout (laser = 5145 Å, 1.4 watt; slit = 3.5 cm⁻¹; sample was in a spinning cell; digitizing increment = 0.2 cm⁻¹). Figure produced by the RPI system.

(0.1 cm⁻¹) when using the 25 point smoothing function or some peak distortion may occur.

The fourth power Boltzmann correction and its effects have been described earlier (see section 2.4.8). The user need only supply the exciting frequency and temperature of the example.

The baseline subtraction routine subtracts a straight line from a spectral curve. The line is defined by the user by inputting the frequency–intensity information for any two points. The computer then generates the line and subtracts it from the spectrum. The method is useful for subtracting sloping as well as horizontal baselines.

The scaling routine is required since our least squares curve resolving program, SUMFIT, requires point intensities to be less than 1.0. Since the normal intensities are in terms of thousands of counts a linear scaling is required. This routine accomplishes this task by searching for the most intense point, determining the scaling factor required to convert its intensity to 0.95, and then multiplying each point in the spectrum by that factor.

A digital spectrum may consist of several thousand points covering a wide frequency range. It may be necessary to restrict either the spectral range or number of points (by changing the increment) or both for certain data evaluation techniques such as least squares curve resolution. The data format change section of MANDAT allows the user to stipulate the initial and final frequencies and the increment of data to be transferred to other programs.

SUMFIT. This program[38] is a least squares analysis program designed to fit a Gaussian/Lorentzian sum function to an experimental curve. It is written in FORTRAN and is a modification of Jones's program X.[21] Modifications include changes in I/O routines as well as use of the same Gaussian/Lorentzian sum function mentioned earlier for the Calculate section of CASH.

$$Y = b + \sum_{i=1}^{n} [XL(v)_i + (1 - X)G(v)_i]$$

Parameters determined by least squares analysis include b and X as well as the three defining parameters for each component: intensity, position, and half-width. The program is limited to a maximum of 6 component peaks ($n = 6$) and 200 data points. Input data points are taken from the output of MANDAT. Other necessary input information is the number of peaks as well as initial guesses for the parameters.

The program output consists of the least squares values of the parameters along with statistical information such as the sum of squares of errors, standard error of estimate, largest intensity deviation and its frequency, and area of component curves. The above information is virtually the same as in the original Jones program. We have added to this a teletype plotting routine which plots one symbol for the experimental point, another symbol for the calculated point, and a third symbol if the points are coincident within the same printer location. This visual display has proven quite helpful in determining the 'goodness of fit'.

Our success in fitting Raman spectra with this program has generally been good. We have encountered no convergence problems, only an occasional instance of the computer 'throwing away' a peak by making the halfwidth extremely narrow or shifting the frequency well away from the spectral range of data points. The CPU time required to obtain a fit approaches 30 minutes for the maximum number of points and components. This places a heavy demand on computer access, especially with a time-shared departmental instrument, and from this viewpoint, additional improvements in the computational efficiency of the program are desirable.

5.3 Operating Procedure

As mentioned above the standard data collection program is BFCOLL, normally used in a time-shared mode of operation. The BFCOLL program may be operated either from the computer display terminal or a teletype. If it is desired to use

CASH simultaneously (i.e. as the background program) BFCOLL must be controlled from a teletype so as to free the computer display terminal (see Fig. 10) for the CASH operated display system. When used in this configuration virtually all of the 24K core capacity of the computer is required. Our experience has shown that the most time consuming operation is data collection with data display and analysis (via CASH, MANDAT, and SUMFIT) requiring much less time. Thus while we are collecting spectra the background mode of the computer (approximately 13K core capacity) is normally available for other departmental users.

The system may also be operated with the computer in a dedicated mode (using the advanced operating system). The only advantage to this method of usage is reduced hardware requirements. The maximum hardware required in this configuration is: 16K core capacity, one RF-15 DEC disk, two DEC tape transports. Automatic priority interrupt and memory protection capabilities are not required (see Table 2). Thus any of the four programs may be operated on reduced hardware requirements if the time-shared capability is not required. The dedicated mode is particularly inefficient when used with BFCOLL since a very small program completely ties up the computer for long periods of time. For this reason the time-shared monitor is always employed when collecting data with BFCOLL although the other programs are sometimes employed in a dedicated mode if there is no other departmental demand for the computer.

5.4 Recommendations

The data collection and storage functions perform satisfactorily and no changes are anticipated in these areas in the near future. However, certain software improvements and additions appear desirable in the analysis/evaluation area to make the system more useful. The primary change, which is currently in progress, is to merge the five MANDAT functions into the CASH program as a new subsystem. This is desirable for two reasons. It will cut down on the number of data file transfers and facilitate data evaluation by making it more rapid and convenient. It will enable the use of the convenient CASH operated display system to compare and contrast the various spectral manipulations inherent in MANDAT. At the present time this is possible only via the incremental plotter.

Simultaneous with the MANDAT–CASH merge a routine is being inserted to add and subtract spectral files. This will enable not only the calculation of difference spectra (section 2.4.4) but also the synthesis of isotropic-anisotropic components (section 2.4.3) as well as the implementation of a routine for instrumental corrections (section 2.4.2). Thus when these modifications are complete CASH will be a versatile program for spectral data collection, analysis, and display.

It is envisaged that BFCOLL will remain as the primary data collection/storage program while CASH will be the primary analysis/display program and may be used for collecting data if desired. SUMFIT will most likely remain as a separate program (although it could also be merged into CASH) because of the amount of CPU time required for its operation.

REFERENCES

(1) D. A. Hatzenbuhler, R. R. Smardzewski and L. Andrews, *Appl. Spec.* **26**, 479 (1972).
(2) J. R. Scherer and S. Kint, *Appl. Optics* **9**, 1615 (1970).
(3) H. H. Eysel and K. Lucas, *Z. Naturforsch.* **25A**, 316 (1970).
(4) E. D. Schmid, G. Berthold, H. Berthold and B. Brosa, *Ber. Bunsenges Phys. Chem.* **75**, 149 (1971).
(5) B. J. Bulkin, E. H. Cole and A. Noguerola, *J. Chem. Ed.* **51**, A273 (1974).
(6) J. W. Amy, R. W. Chrisman, J. W. Lundeen, T. Y. Ridley, J. C. Sprowles and R. S. Tobias, *Appl. Spec.* **28**, 262 (1974).
(7) S. Ushioda, J. B. Valdez, W. H. Ward and A. R. Evans, *Rev. Sci. Instrum.* **45**, 479 (1974).
(8) C. H. Warren and L. Ramaley, *Appl. Optics* **12**, 1967 (1973).
(9) J. J. Barrett and A. Weber, *J. Opt. Soc. Amer.* **60**, 70 (1970).
(10) E. W. Small, B. Fanconi and W. L. Peticolas, *J. Chem. Phys.* **52**, 4369 (1970).
(11) J. F. Moore, *The Spex Speaker*, **XIX**, September 1974.
(12) F. J. Wunderlich, W. J. Hones and D. E. Shaw, Villanova University, personal communication, 1973.
(13) H. Longerich and L. Ramaley, *Anal. Chem.* **46**, 2067 (1974).
(14) R. E. Hester in *Raman Spectroscopy; Theory and Practice*, Vol. 1 (H. A. Szymanski, Ed.), Plenum Press, New York, 1967, Chapter 4.
(15) H. J. Bernstein and G. Allen, *J. Opt. Soc. Amer.* **45**, 237 (1955).
(16) J. R. Scherer, S. Kint and G. F. Bailey, *J. Mol. Spectrosc.* **39**, 146 (1971).
(17) A. Savitzky and M. J. E. Golay, *Anal. Chem.* **36**, 1627 (1964).
(18) R. N. Jones, T. E. Bach, H. Fuhrer, V. B. Kartha, J. Pitha, K. S. Seshadri, P. Venkataraghavan and R. P. Young, *National Research Council of Canada Bulletin No. 11*, Ottawa, 1968.
(19) DuPont 310 Curve Resolver, *Chem. Eng. News* **43** (Nov. 15), 50 (1965).
(20) A. R. Davis, D. E. Irish, R. B. Roden and A. J. Weerheim, *Appl. Spectrosc.* **26**, 384 (1972).
(21) J. Pitha and R. N. Jones, *National Research Council of Canada Bulletin No. 12*, Ottawa, 1968.
(22) W. L. Butler and D. W. Hopkins, *Photochem. and Photobiol.* **12**, 439 (1970); **12**, 451 (1970).
(23) E. Ziegler, D. Henneberg and G. Schomburg, *Anal. Chem.* **42**, 51A (1970).
(24) J. W. Frazer, *Chem. Instrum.* **2**, 271 (1970).
(25) M. Shapiro and A. Schultz, *Anal. Chem.* **43**, 398 (1971).
(26) J. E. David and E. D. Schmidlin, *Chem. Instrum.* **4**, 169 (1973).
(27) K. Gruber, J. Farrer, E. Zopfi and H. H. Günthard, *J. Phys. E. Sci. Instrum.* **6**, 666 (1973).
(28) Digital Equipment Corporation, Maynard, Mass. 01754, USA.
(29) S. P. Perone, *Anal. Chem.* **43**, 1288 (1971).
(30) R. E. Dessy and J. Titus, *Anal. Chem.* **45**, 124A (1973).
(31) R. E. Dessy and J. Titus, *Chemtech.* **436** (July 1973).
(32) R. E. Dessy and J. Titus, *Anal. Chem.* **46**, 294A (1974).
(33) R. C. Williams and T. J. Turner, *Rev. Sci. Instrum.* **43**, 1207 (1972).
(34) H. Drommert, E. Kleimon and W. Pilz. *Exp. Tech. Physik* **17**, 493 (1969).
(35) W. Proffitt and S. P. S. Porto, *J. Opt. Soc. Amer.* **63**, 77 (1973).
(36) G. Jakobsche, Rensselaer Polytechnic Institute, personal communication, 1972.
(37) L. J. Berman and P. V. Shaffer, Software Consultants, personal communication, 1973.
(38) J. R. Downey Jr, Rensselaer Polytechnic Institute, unpublished work, 1974.

Chapter 2

RAMAN SPECTROSCOPY OF PROTEINS

B. G. Frushour

Monsanto Triangle Park Development Center, Inc., Research Triangle Park, North Carolina 27709, U.S.A.

J. L. Koenig

Department of Macromolecular Science, Case Western Reserve University, Cleveland, Ohio 44106, U.S.A.

1 GENERAL INTRODUCTION

The determination of protein structure and elucidation of enzyme mechanisms have long been a major goal of biochemists. These efforts have required a battery of physico-chemical techniques that are sensitive to various levels of protein structure; from the size and shape of the protein molecule down to the conformation of individual peptide units. This review will focus on the conformation of the polypeptide chain in proteins, which is often termed the secondary structure. Several well established techniques are sensitive to the polypeptide chain conformation, including X-ray diffraction, infrared spectroscopy, optical rotatory dispersion (ORD) and circular dichroism (CD). Establishing yet another technique to measure secondary structure, laser excited Raman spectroscopy, may seem redundant. The Raman technique, however, complements the other methods in several important considerations and for the first time a biochemist can characterize the secondary structure of a protein in practically any environment or phase.

Let us consider the current techniques for protein analysis. The most complete analysis of a protein structure can be achieved by X-ray analysis. This technique, however, is laborious and must be applied to protein single crystals for which suitable isomorphous derivations in the native state can be prepared. Infrared spectroscopy has been very valuable in determining the chain conformation of oriented fibrous proteins but has limited use for aqueous systems of enzymes and other globular proteins. Water is a poor solvent for infrared measurements by virtue of its intense absorbance. Finally ORD and CD are most applicable to protein solutions of low concentrations.

Some of the advantages offered by Raman spectroscopy will be considered. Raman spectroscopy is a light scattering technique and the scattering intensity is proportional to the change in molecular polarizability induced by a normal vibration. Since the conformationally sensitive vibrations of the polypeptide backbone involve the stretching and bond angle deformation of polarizable C—N and C—C bonds, the Raman lines corresponding to these vibrations are often intense. These same vibrations may appear only weakly in the infrared spectrum because they induce only small changes in the permanent dipole moment of the molecule. Being a scattering technique, Raman spectroscopy may be applied to practically any sample that has sufficient surface to scatter a laser beam. Fibers, films, powders, gel and crystalline solids of biopolymers have been examined with success. Water is an excellent solvent for Raman spectroscopy because water is a weak scatterer in the Raman effect. Very often the effect of hydration on polymer conformation cannot be determined because few conformationally sensitive techniques are equally applicable to the solution and solid states. Raman spectroscopy, however, can usually be applied to both states and many important solution processes can be studied. One section in this review deals with the differences in protein conformation in the crystalline, lyophilized and solution states.

This review consists of three parts: an introductory section on protein structure, a discussion of Raman theory and experimental techniques and the review of the current literature involving globular and fibrous proteins. A great amount of work with the homopolypeptides has provided many of the structure-frequency correlations used in the analysis of protein Raman spectra. The results of these investigations are summarized and tabulated in the third section. The reader should consult review articles by Koenig[1,2] and Frushour and Koenig[3] for more information concerning the polypeptide studies.

2 PRINCIPLES OF PROTEIN STRUCTURE

Proteins consist of linear polypeptide chains made by condensation polymerization of amino acids. The amino acids common to most proteins number twenty-four and differ by the chemical functionality of the side chain. The variation in the chemical structure of the side chain encompasses most of the common organic functional groups. A brief description of the relationship between the amino acid sequence and the resultant protein structure will now be given.

Many types of intermolecular forces stabilize the structure of a protein.[4] These forces exist between amino acids on the same polypeptide chain and on adjacent chains. Primary or covalent bonds form the polypeptide chain and are also found in the cystine disulfide crosslinks, and these bonds have energies on the order of 50 to 100 kcal mol^{-1}. However protein denaturation, a conformational transition leading to a loss of biological activity, often proceeds without the rupture of any primary bonds.[5] The forces that stabilize the native conformation of a protein fall in the realm of secondary bonds, i.e., those bonds with energies less than 10 kcal mol^{-1}. Examples of these forces are hydrogen bonds, van der Waals interactions and hydrophobic bonds.[6] Even though these bonds are weak, their accumulated effect in a large molecule, where thousands of these interactions exist, is quite significant.

In discussing the relationship between the amino acid sequence and the native conformation of a protein, it is convenient to consider various organization levels of a protein and the relations between them. Linderstrom-Lang[7] defined three organizational levels of protein structure: the primary, secondary and tertiary structures. The primary structure is the amino acid sequence. The secondary structure refers to the stable conformations of the polypeptide chain, such as the alpha helix and antiparallel beta sheet. The tertiary structure refers to the folding of the polypeptide chains into the final three dimensional shape of the protein. Enzymes have ellipsoidal or globular tertiary structures and are classified as globular proteins. Structural proteins like collagen and keratin are examples of proteins with a fibrous tertiary structure.

In any polymer, synthetic or biological, the stable chain conformations correspond to potential energy minima in the configurational space of the polymer.[8] The conformation of minimum potential energy for a polypeptide chain is achieved by simultaneously maximizing the van der Waals, hydrogen bonding and other attractive forces while minimizing the steric repulsion forces. The polypeptide chain can assume many isomeric structures through free rotation about the C_α—N and C—C_α bonds; designated as the phi and psi rotation angles, respectively, in Fig. 1.[9] However, the partial double bond character of the amide C—N bond will allow only small degrees of rotation and this greatly restricts the number of possible chain conformers.[10] Once potential energy functions are constructed that incorporate the various intermolecular forces the free energy of the polymer chain can be calculated as a function of the ψ and ϕ bond rotation angles.[8] Of the many possible structures available to the polypeptide chain, only a restricted number of structures have been found to exist in nature and these structures satisfy the criteria for minimum potential energy discussed above.

Consider polyglycine. Glycine is the simplest amino acid since the side chain consists of a single hydrogen atom. Conformational energy minima are found at ψ, ϕ, angles that correspond to three ordered secondary structures: the alpha helix, the beta sheet, and the 3_1 helix.[11] These structures appear in Fig. 2. The alpha helix contains 18 amino acid residues in 5 turns of the helix axis and is stabilized by NH$\cdots$O=C intramolecular hydrogen bonds aligned approximately parallel to the chain axis.[9] In the antiparallel beta sheet structure the polypeptide chain is highly extended and contains two amino acid residues per turn. Consequently the N—H and C=O bonds extend perpendicular to the chain axis. Adjacent chains are joined by intermolecular hydrogen bonds between amide groups and sheet like aggregates are formed.[9] The 3_1 helix contains 3 amino acid residues per turn and is also stabilized by intermolecular

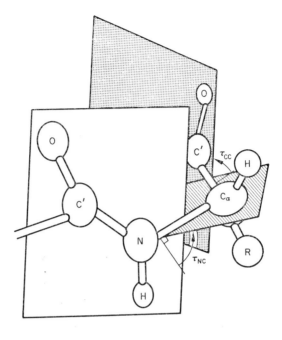

Fig. 1. Geometry of the peptide bond.

hydrogen bonds perpendicular to the chain axis.[9] X-ray diffraction and infrared dichroism show that polyglycine can be prepared in either the antiparallel beta sheet or 3_1 helical forms which are often termed the PG I and PG II conformations, respectively.[11]

The size, shape and polarity of the side chain is crucial in determining the secondary conformation of the polypeptide chain. Homopolypeptides containing a primary or secondary carbon atom at the beta position of the side chain form the alpha helix. Examples are poly-L-alanine, poly-L-leucine and poly-L-methionine.[12] Branching at the beta carbon atom destabilizes the alpha helical conformation. However the bulky side chain can be accommodated easily in the beta sheet conformation.[11] Poly-L-valine has a *gem* dimethyl group at the beta carbon atom and forms the beta sheet rather than the alpha helix.[11] The

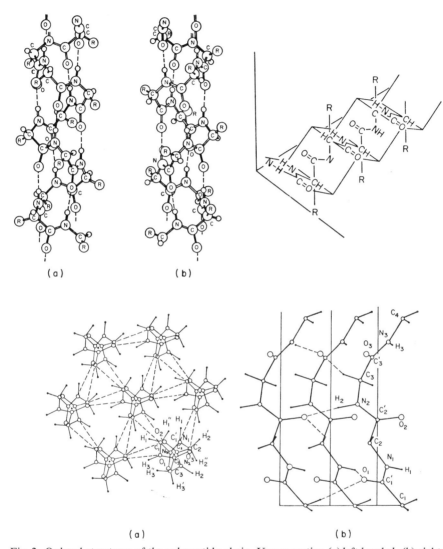

Fig. 2. Ordered structures of the polypeptide chain. Upper portion (a) left-handed, (b) right-handed α-helix, three dimensional β-sheet. Lower portion (a) 3_1 helix—viewed along axis (b) 3_1 helix—side view.

stability of the antiparallel beta conformation is predicted by potential energy calculations.[11,13] For isolated molecules the calculations predict that the alpha helix and beta sheet conformations are equally stable but in the solid state where the effect of intermolecular hydrogen bonds becomes important the beta sheet is the only observed conformation.[11] In conclusion, the most stable secondary conformations correspond to minima in the potential energy functions and a maximum in the strength of the hydrogen bonding.

The relationship between protein structure and function is demonstrated in the tertiary structure. Consider a globular protein, such as an enzyme, which exhibits chemical activity in aqueous solution. The tertiary structure must satisfy two basic requirements. First of all, the protein must be stable under appropriate physical conditions and resist denaturation reactions such as unfolding, aggregation and precipitation. Secondly, on the protein surface and accessible to the aqueous environment must be an active site.[9] Usually the active site is a crevice in the protein surface containing several polar amino acid side chains arranged to catalyze a very specific reaction on a substrate molecule. Surprisingly, only a few amino acids form the active site and the remainder of the protein serves as a stable bulk that keeps the active site in solution.

The polypeptide chains are folded into a globular shape that varies for different proteins from 50 Å to several hundred ångstroms.[9] An uninterrupted sequence of a basic conformation such as the alpha helix cannot be extensive because the polypeptide chain must eventually bend or fold in order to assume the globular shape. Instead, short segments of several basic secondary conformations are found in most globular proteins. The folding of a chain allows amino acid side chains to interact and stabilize the structure. For example, salt bridges between carboxylic anions and amino cations may be formed upon chain folding.[6] Cross beta sheet structures are formed in ribonuclease as the polypeptide chain folds back on itself 180 degrees thus achieving an antiparallel arrangement of adjacent chains.[9] Many other secondary interactions of this type are outlined by Scheraga.[6] Identification of these specific interactions and the evaluation of their contribution toward stabilization of the native protein conformation remains a challenging research area in protein chemistry.

One area of intense activity has been the study of enzyme activity and treatments leading to the loss of this activity called denaturation. Conformation transitions between the native and denatured states of proteins interests the biochemist for several reasons. Some biological processes may involve protein denaturation. Control of denaturation is vital in processing of food and other natural products. By learning the mechanism of denaturation achieved through chemical and thermal methods one can evaluate the role of specific interactions in stabilizing the native protein structure. Denaturation encompasses the entire spectrum of conformational transitions and its study requires many different techniques. Changes in the overall size and shape of the protein can be measured by techniques such as light scattering and hydrodynamic properties.[14] Most denaturation mechanisms, however, proceed through a change in the secondary structure of the protein and techniques sensitive to the chain conformation must be employed.

Denaturation is one of many phenomena involving possible changes in the conformation of proteins that is of interest. Several other examples will be given here. The most detailed models of protein structure are obtained by X-ray diffraction of single crystals[9] and the tacit assumption is often made that the crystalline structure remains invariant upon dissolution in water. This assumption must be further tested. Interactions between proteins and other macromolecules are thought to involve changes in the secondary and tertiary structures of the protein. The DNA-histone protein interactions and the antibody reactions are examples where the mechanism of these interactions is not well understood. Detecting these changes in structure requires very sensitive techniques and in many cases Raman will complement the established techniques (X-ray diffraction, infrared spectroscopy, ORD and CD) that have provided the bulk of relevant data to date.

3 THEORY AND EXPERIMENTAL TECHNIQUES

The vibrational selection rules for a polymer chain can be applied rigorously only to systems having at least one dimensional crystallinity. An isolated chain in an ordered conformation such as the alpha helix satisfies this requirement. A chain containing N unit cells with M atoms per unit cell has $3 \times N \times M$ degrees of freedom and after subtracting 3 translational degrees of freedom and the single rotation about the chain axis there remains $[(3 \times N \times M) - 4]$ vibrational motions.[15] Obviously this represents a large number of vibrations, but in order to have infrared or Raman activity the phase angle between identical vibrations in neighboring unit cells must be zero. This selection rule was derived by Born and von Karmen[16] and is discussed in detail by Krimm.[17] Therefore we need only consider the vibrations within a single unit cell. One should not identify the chemical repeat with the unit cell; only in special cases are they identical and generally the number of chemical repeats per unit cell is that number required to complete an integral number of turns of the helix axis. The number of chemical repeats per one dimensional unit cell of polypeptide chain in a planar zig-zag, 3_1 helix, and alpha helix (18_5) is 2, 3, and 18, respectively.

Molecular symmetry governs the infrared and Raman activity of normal vibrations. Zbinden[18] has thoroughly discussed the symmetry aspects of polymer chain vibrations. Infrared activity requires that the vibration produce a change in the dipole moment of the molecule. The dipole moment is a vector quantity and has components in the x, y, z direction, where z is conventionally taken as the chain axis. Any vibration that transforms under the symmetry operations of the polymer in the manner of a unit vector in the x, y, or z direction will be infrared active. Vibrations that transform like the z unit vector are said to have parallel polarization since they absorb radiation most strongly when the electric vector of the radiation lies parallel to the z or chain axis. Similarly, vibrations that transform according to the x and y unit vectors have perpendicular polarization. By measuring infrared band intensities of oriented polymer samples with perpendicular and parallel polarized infrared radiation the symmetry species to which the band belongs may be assigned, or in the case of fibrous proteins,[19] the conformation of the polypeptide chain may be determined.

Raman activity requires that the vibration induce a change in the molecular polarizability. The Raman scattered light from a random array of molecules will be polarized for vibrations that are totally symmetric. All other types of Raman active vibrations will depolarize the scattered light. Classification of Raman lines according to their polarization assists the spectroscopist in assigning Raman lines to their normal vibrations. Raman polarization measurements on oriented solids are potentially quite useful but suffer from some experimental limitations. Snyder[20] has derived the required expressions for the technique and Fanconi et al.[2] have measured the polarized Raman scattering from oriented fibers of poly-L-alanine.

Figure 3 shows a schematic of a typical Raman scattering experiment. The

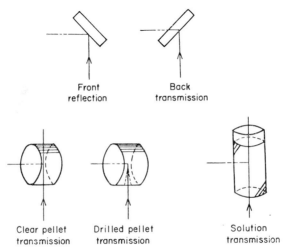

Front reflection   Back transmission

Clear pellet transmission   Drilled pellet transmission   Solution transmission

Fig. 3. Typical geometries for Raman scattering.

beam of light from a continuous wave laser is scattered off the sample at right angles to the monochromator. When examining solid samples a number of arrangements may be used depending upon the size, shape and transparency of the solid. Liquids are enclosed in small sample vials and mounted one of two ways. Micro quantities of sample can be examined in a glass capillary (melting point tube) mounted so that the laser beam enters the capillary at 90° to the axis. Perhaps an easier method, suitable for larger quantities, is to use a flat bottomed cylinder and pass the laser beam up through the bottom of the tube. The latter arrangement often makes sample alignment much easier.

The laser power required to achieve suitable signal to noise ratio depends upon the laser light frequency, sample purity and concentration, quality of the monochromator and sensitivity of the detection system. Assuming a good double monochromator, photon counting detection system and a very pure sample, a protein or polypeptide spectrum can be obtained at $4\,\text{cm}^{-1}$ resolution with

50–100 milliwatts of 4880 or 5145 Å laser light from an argon ion laser. More power will be required with longer wavelength excitation because the Raman scattering intensity and photomultiplier tube sensitivity decreases with increasing wavelength. The most severe requirement in sample preparation is purity. Impurities that absorb the incident radiation can cause sample heating and also produce a strong background that can easily overwhelm the Raman signal. This background has been attributed to fluorescence but other mechanisms may be operative. Often the background decays with time until a suitable Raman signal to noise level is achieved. Obviously this procedure of simply allowing the laser beam to 'burn out' the impurities must be applied with caution to enzymes and other globular proteins that denature on heating. Fortunately, routine bio-chemical purification techniques including chromatography, dialysis and recrystallization will usually render proteins sufficiently pure for Raman measure-ments. In the authors' laboratory the spectra of lysozyme and beta lactoglobulin single crystals have been obtained immediately after placing the sample in the beam with little or no decaying background observed. Assuming that fluorescence of absorbing impurities causes the background, a number of electronic techniques have been devised to eliminate this problem. Yaney[23] has utilized the difference in relaxation times between Raman scattering and fluorescence emission. Since Raman scattering is several orders of magnitude faster than fluorescence emis-sion, a detection system can in principle be designed to reject the fluorescence signal after passing the Raman signal. Current approaches use a pulsed laser and gated detector to pass the Raman signal and reject the fluorescence. Applications of this technique to biological systems have not yet been published. Another technique, devised by Bulkin,[24] employs the difference in band shape of the Raman and fluorescence signals. The fluorescence background is much broader than the Raman lines and therefore can be separated from the Raman scattering in the Fourier transform of the total spectrum. Retransforming yields the Raman spectrum with reduced background. At the present writing neither technique has emerged as the general solution to the fluorescence problem encountered with polymers. Until these techniques are perfected and offered commercially the experimenter must devise suitable methods to remove the impurities or quench their emission.

4   RAMAN SPECTRA OF PROTEINS

4.1   Polypeptides

The Raman spectra of proteins are very complicated due to the many amino acid side chain lines and the coexistence of several chain conformations. Since the homopolypeptides contain a single side chain and can be prepared in a single conformation their spectra are much easier to interpret and can be used to establish structure-frequency correlations. Also the homopolypeptides undergo conformational transitions that are analogous to protein denaturation. The Raman spectra of the homopolypeptides have been studied extensively and reviewed by Koenig[1,2] and Frushour and Koenig;[3] therefore only their salient features will be discussed here.

The amide group has several characteristic vibrations that appear in the infrared and Raman spectra and are designated amide X. These vibrations for N-methylacetamide are shown in Fig. 4. The amide I and II vibrations were

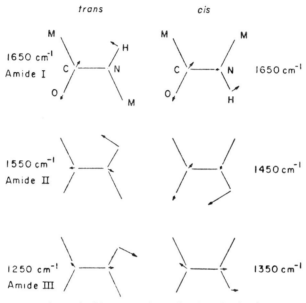

Fig. 4. Amide I, II and III vibrations (Ref. 41).

utilized in pioneering days of protein structure analysis to determine the chain conformation of fibrous proteins from the infrared dichroic spectra.[19] The remaining amide III—VI infrared bands are only marginally useful for conformational analysis. Raman investigations have centered on the amide I and III modes. The amide I mode consists of the carbonyl stretching vibration with small contributions from the C—N—H in plane bending and the C—N stretching vibrations. The potential energy distribution over these three components has been calculated by a number of authors.[25,26] Work by Krimm and coworkers on monomeric amides and nylons[27,28] and polyglycine I[29] and II[30] represent the most detailed analysis.

The Raman active amide I mode for the alpha helical conformation appears at the same frequency as in the infrared spectra, i.e. from 1650–1657 cm^{-1}. Strong splitting of the amide I mode is observed for the antiparallel beta sheet conformation. A strong amide I line appears in the Raman spectrum near 1670 cm^{-1} while two lines appear near 1630 and 1685 cm^{-1} in the infrared spectrum for this conformation. Miyazawa[31] developed classical perturbation equations (that were based on the weakly coupled oscillator model) to explain the splitting in the infrared spectrum. The equations were modified by Krimm and Abe[32] to account for the recent Raman data and the results of their normal coordinate analysis of polyglycine I. There are four peptide units in the two dimensional unit cell of polyglycine I (and other beta sheet polypeptides if the side chain is ignored); therefore one expects four amide I vibrations. The allowable intrachain phase angle, δ, and interchain phase angle, δ', are 0 and π and the four combinations are listed below, where the first number in the parenthesis refers to δ and the second to δ':

$\nu(0, 0)$ Raman active, infrared inactive, polarized in Raman;

$\nu(\pi, 0)_\perp$ Raman active, infrared active, perpendicular i.r. dichroism;

$\nu(0, \pi)_{||}$ Raman active, infrared active, parallel i.r. dichroism;

$\nu(\pi, \pi)_\perp$ Raman active, infrared active, perpendicular i.r. dichroism.

These four vibrations are drawn in Fig. 5. In practice the $\nu(0, 0)$ mode appears in the Raman spectrum near 1670 cm^{-1}. The $\nu(\pi, 0)$ mode appears in the infrared spectrum near 1630 cm^{-1} with strong intensity and perpendicular polarization and the $\nu(0, \pi)$ mode also appears in the infrared spectrum near 1680 cm^{-1} with weak intensity and parallel polarization. The amide I mode for the disordered conformation appears near 1665 cm^{-1} in the Raman spectrum and near 1655 cm^{-1} in the infrared spectrum. By using the amide I mode in the Raman spectrum one can easily differentiate between the alpha helical and beta sheet or disordered conformations but cannot easily differentiate between the latter two conformations.

The amide III mode consists of the C—N—H in plane bending and C—N stretching modes but may be coupled to other motions that occur in this frequency region. This vibration is not localized in the amide group to the extent of the amide I mode and therefore may be more directly sensitive to the conformation of the polypeptide chain. The amide I mode derives its sensitivity to conformation primarily through dipole-dipole interactions among the carbonyl groups that are functions of the chain geometry.[32] The antiparallel beta sheet can be immediately distinguished by an intense amide III line at 1235 ± 5 cm^{-1} in the Raman spectrum. High resolution spectra may occasionally reveal another amide III component near 1270 cm^{-1}. The disordered conformation appears near 1245 cm^{-1} in the amide III region of the Raman spectrum. Only weak scattering in this region can be observed for the alpha helical conformation and lines appearing from 1260 to 1295 cm^{-1} have been assigned to this mode. This remarkable decrease in intensity in the amide III region upon going from the disordered or beta sheet conformation to the alpha helix has been attributed to a hypochromic effect by Koenig and coworkers[33,34] and is analogous to the Raman hypochromism observed by Peticolas and coworkers[35] and Thomas[36] during the coil to helix transitions of the polynucleic acids. Lack of any strong Raman lines from 1200–1300 cm^{-1} in the spectrum of a polypeptide is strong evidence for the alpha helical conformation.

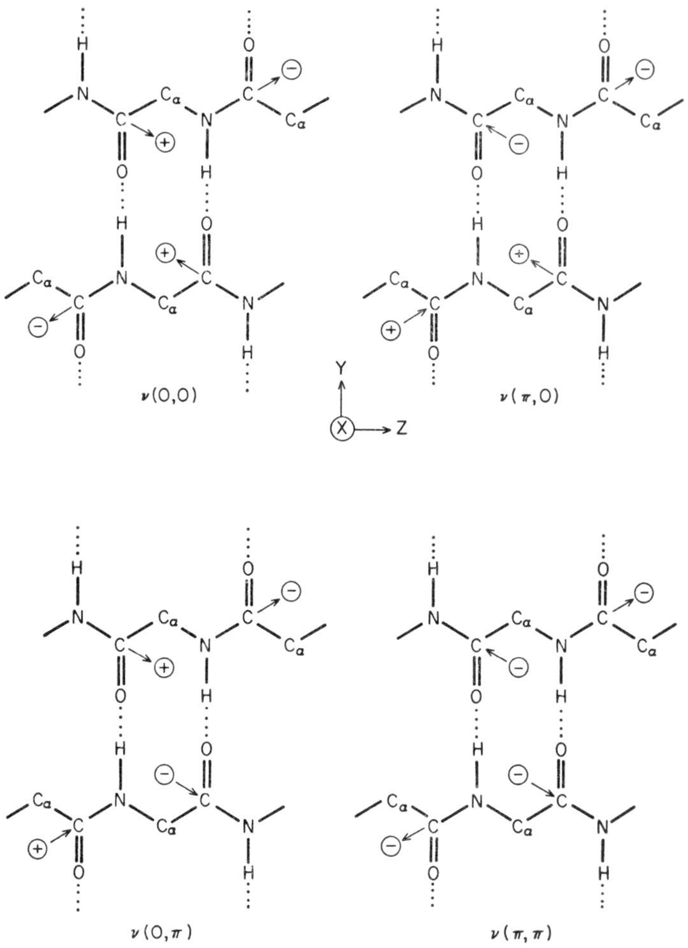

Fig. 5. A schematic representation of the vibrational modes of the antiparallel chain pleated sheet (Ref. 31).

The alpha helical conformation is also characterized by a strong line appearing near 900 cm^{-1} in the Raman spectrum that has been assigned to a skeletal stretching vibration.[33,37] This line is either absent or weak in the infrared spectrum. No strong Raman lines appear in this region of the spectrum for the beta sheet polypeptides, and in cases where an alpha helix to beta sheet transition may be induced this line becomes very weak. The intensity of this line for the disordered conformation is not well behaved. During the pH induced helix to coil transition of ionizable polypeptides[33,37] the frequency shifts from near 930 cm^{-1} to 950 cm^{-1} but does not decrease in intensity. However a large decrease in intensity during the solvent induced helix to coil transition of poly-β-benzyl-L-aspartate has been observed.[38]

The frequencies for the amide I, III and 900 cm^{-1} skeletal modes of the Raman spectra for all of the polypeptides studied to date are summarized in Table 1. In order for the reader to assimilate these data, a histogram of the frequencies appears in Fig. 6. The relative line intensities in the histogram are

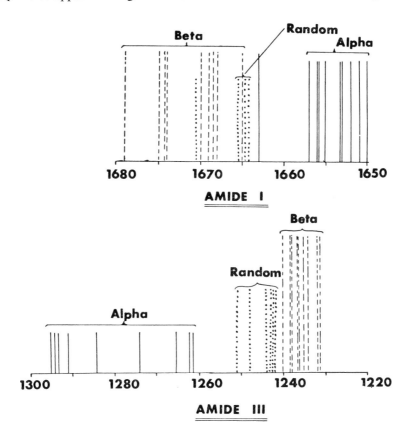

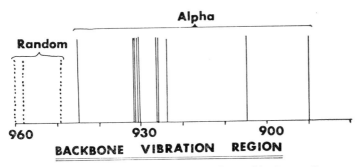

Fig. 6. Histogram of conformationally sensitive Raman lines.

TABLE 1
Conformationally sensitive Raman lines of polypeptides

Polypeptide	Amide I	Amide III	Skeletal vibrations (850–950 cm^{-1})	References in footnotes
Alpha helix				
Poly-L-alanine	1655–1659 s[a]	Triplet 1284 vw 1274 vw 1262 vw	905 s	[g-k]
Poly-γ-benzyl-L-glutamate (PBLG)	1650–1652 s	1294 w	931 s	[k,l]
Poly-γ-methyl-L-glutamate	1656 s	1250–1350 w	924 s	[j]
Poly γ benzyl-L-aspartate[b]	1663 s	?	911 m, 890 s	[m]
Poly-γ-methyl-L-aspartate	1661 s	1294 w		[n]
Poly-L-leucine	1653 s	1261 w ⎱c 1294 w ⎰	931 s	[i,j,l]
Poly-L-methionine	1652 s	1264 w	907 s	[o]
Poly-L-glutamic acid, solid	1656 s	1246 wbr	926 s	[p]
Poly-L-glutamic acid, pH 4.8		1238 m	926 s	[p]
Poly-L-lysine, HCl, solid at 50% R.H.[d]	1655 m	1295 w	945 m	[k]
Poly-L-lysine, pH 110	1645 m (H$_2$O) 1632 m (D$_2$O)	1200–1300 w (H$_2$O)	945 s (H$_2$O) 950 s (D$_2$O)	[q,o]
Random copolymer of L-leucine and L-glutamic acid	1653 s	1200–1300 w	931 s	[i]
Random copolymer of L- and D-leucine	1658 s	1258 w ⎱c 1294 w ⎰	931 m	[r]
Racemic blend of PBLG and PBDG	1650 s	1291 w	931 s	[s]
Anti-parallel pleated sheet				
Polyglycine I	1674 s	1234 s, 1221 w	884 w	[t,u]
Poly-L-alanine	1669 s	1243 s, 1231 m	909 s	[i,v]
Poly-L-valine	1666–1671 s	1231 s, 1277 w[e]	959 w, 942 w	[i-l]
Poly-L-lysine gel	1670 s	1240 s	1002 m	[r]
Poly-L-serine	1674 s	1235 s	894 m	[l,o]
Poly-β-benzyl-L-aspartate	1679 s	1236 m	911 m	[m]
Poly-β-methyl-L-aspartate	1668 s	1230 m		[n]
Poly-S-methyl-L-cysteine	1675 s	1238 s	940 m	[o]
Poly-(Ala-Gly)	1665 s	1238 s, 1271 w	925 m, 890 m	[m]
Poly-(Ser-Gly)[f]	1668 s	1236 s	983 m	[m]
Random coil				
Poly-L-glutamic acid, pH 7.0	1665 s	1248 m	949 s	[p,w]
Poly-L-lysine, pH 7.0	1665 s (H$_2$O) 1660 s (D$_2$O)	1243 s	958 m	[q]
Poly-L-ornithine, pH 7.0	1665 s (H$_2$O)	1242 s	960 m	[p]
Copolymers of L-glutamic acid and L-tyrosine, pH 10.0		1251 m	930 s	[x]
Copolymers of L- and D-lysine at pH 7.0	1671 s (H$_2$O)	1243 s	959 m	[r]

TABLE 1—*continued*

Polypeptide	Amide I	Amide III	Skeletal vibrations (850–950 cm⁻¹)	References in footnotes
		3_1 *Helix*		
Polyglycine II	1654 m	1283 w, 1261 m, 1244 m	884 s	*t,y*
		Poly imino acids		
		Carbonyl stretching		
Poly-L-proline I	1650 s (solid)			*z,aa*
Poly-L-proline II	1650 m (solid), 1630 m (in H_2O)			*z,bb,cc*
Poly-L-hydroxyproline	1628 m (solid), 1630 s (in H_2O)			*dd,ee*

[a] Abbreviations used here and in following tables : s (strong), m (medium), w (weak), v (very), br (broad), sp (sharp), sh (shoulder), db (doublet), tr (triplet); all frequencies are in cm⁻¹ and refer to the energy shift from the exciting radiation.

[b] This polypeptide is unusual because the chain of L-residues forms a left-handed alpha helix rather than a right-handed helix.

[c] The amide III mode is most likely one of these two lines.

[d] R.H. (relative humidity).

[e] Amide III assignments based on deuteration studies of Ref. *i*; Ref. *k* is in slight disagreement.

[f] The sample of poly(Ser-Gly) used in Ref. *m* contained a disordered chain fraction.

[g] J. L. Koenig and P. L. Sutton, *Biopolymers* **8**, 167 (1969).

[h] B. Fanconi, E. Small and W. L. Peticolas, *Biopolymers* **10**, 1277 (1971).

[i] B. G. Frushour and J. L. Koenig, *Biopolymers* **13**, 455 (1974).

[j] L. Simons, G. Bergstrom, G. Blomfelt, S. Fores, H. Stenback and G. Wansen, *Commentat. Phys. Math., Soc. Sci. Fenn.* **42**, 125 (1972).

[k] M. C. Chen and R. C. Lord, *J. Amer. Chem. Soc.* **96**, 4750 (1974).

[l] J. L. Koenig and P. L. Sutton, *Biopolymers* **10**, 89 (1971).

[m] B. G. Frushour and J. L. Koenig, *Biopolymers*, in press (1975).

[n] V. Lin and J. L. Koenig, *Biopolymers*, submitted for publication (1975).

[o] B. G. Frushour, unpublished results.

[p] J. L. Koenig and B. G. Frushour, *Biopolymers* **11**, 1871 (1972).

[q] T. J. Yu, J. L. Lippert and W. L. Peticolas, *Biopolymers* **12**, 2161 (1973).

[r] B. G. Frushour and J. L. Koenig, *Biopolymers*, in press (1975).

[s] W. T. Wilser and D. B. Fitchen, *Biopolymers* **13**, 1435 (1974).

[t] E. W. Small, B. Fanconi and W. L. Peticolas, *J. Chem. Phys.* **52**, 4369 (1970).

[u] Y. Abe and S. Krimm, *Biopolymers* **11**, 1817 (1972).

[v] B. Fanconi, B. Tomlinson, L. A. Nafie, W. Small and W. L. Peticolas, *J. Chem. Phys.* **51**, 3993 (1969).

[w] R. C. Lord and N. T. Yu, *J. Mol. Biol.* **50**, 509 (1970).

[x] B. G. Frushour and J. L. Koenig, *Biopolymers*, in press (1975).

[y] Y. Abe and S. Krimm, *Biopolymers* **11**, 1841 (1972).

[z] A. G. Walton, W. B. Rippon and J. L. Koenig, *J. Amer. Chem. Soc.* **92**, 7455 (1970).

[aa] A. M. Dwivedi and V. D. Gupta, *Chem. Phys. Lett.* **16**, 109 (1972).

[bb] M. Smith, A. G. Walton and J. L. Koenig, *Biopolymers* **8**, 173 (1969).

[cc] V. D. Gupta, R. D. Singh and A. M. Dwivedi, *Biopolymers* **12**, 1377 (1973).

[dd] M. J. Deveney, A. G. Walton and J. L. Koenig, *Biopolymers* **10**, 615 (1971).

[ee] R. B. Srivastava and V. D. Gupta, *Biopolymers* **13**, 1965 (1974).

only approximate since the data originated from different laboratories using various types of instrumentation.

In the previous paragraphs we summarized the conformationally sensitive Raman lines of the homopolypeptides. The globular and fibrous proteins differ from the simple polypeptides in at least two basic regards. First, proteins are complicated copolymers containing approximately 24 different amino residues that have aromatic, aliphatic and reactive functional groups as side chains. Several of these side chain vibrations scatter strongly in the Raman spectrum and to a much lesser degree in the infrared spectrum. Fortunately, these lines do not appear with strong intensity in the conformationally sensitive regions of the Raman spectrum, i.e. the amide I and III regions. Without this fortuitous 'rule of exclusion', the prospects of protein studies with Raman spectroscopy would be greatly diminished. The Raman lines due to side chains are themselves of interest because of their sensitivity to the local environment of the side chain. Some of the amino acid residues are part of enzyme active sites such as tryptophan and tyrosine and have strong conformationally sensitive Raman lines. One can foresee studying enzyme activation and denaturation via the Raman scattering of these amino acid residues. A second major difference between the polypeptides and proteins is the distribution of secondary structures found in the latter. A typical globular protein like lysozyme will contain a mixture of alpha helical, beta sheet and disordered chains. Usually the ordered chain conformations in proteins are not very extensive but are short and often distorted relative to the homopolypeptides (and fibrous proteins). Defects in the ordered structures are the rule rather than the exception. Here the term defect refers to nonlinear hydrogen bonds, distorted helices, breaks in helical segments to accommodate the cornering or folding of the polypeptide chain and other similar features. The selection rules for the normal vibrations of the ordered polypeptide structures were derived assuming zero defects and infinitely long structures. Relaxation of either assumption implies a loss of symmetry; therefore frequency shifts, line broadening and the appearance of new lines are possible consequences. Some of these effects are observed for synthetic polymer chains containing defects and have been treated by Zerbi.[39,40]

The scope of this review is biased towards the Raman spectra of globular proteins for the following reasons. The literature published on the Raman spectra of fibrous proteins has been limited by experimental difficulties, though wool keratin has been studied in detail. Globular proteins are usually studied in the aqueous or hydrated crystalline phases and infrared spectroscopy has been of limited use in these investigations due to the strong absorption spectrum of H_2O. The reader should consult the review by Suzuki[41] on this subject. Recent advances in Fourier transform infrared spectroscopy, a technique offering improved signal to noise ratio over dispersion instruments, promises to enhance infrared spectroscopy of aqueous proteins since the water spectrum can be removed by computer subtraction.[42,43]

4.2   Raman Spectra of Globular Proteins

4.2.1   Lysozyme

Hen eggwhite lysozyme is a globular protein of molecular weight 14,600 g mol^{-1}, with 129 amino acids in a single polypeptide chain and four disulfide bridges. Lysozyme attacks many bacteria by lysing, or dissolving, the mucopolysaccharide structure of the cell wall. Lysozyme was the second protein and first enzyme to have its detailed molecular structure worked out by X-ray analysis, by Phillips, North, Blake, and coworkers.[44] The most striking feature of the molecule is a crevice running across the waist of the egg-shaped molecule. It divides the molecule into two parts, a predominantly helical core containing residues 1–40 and 101–129 and a more irregular wing containing a three-stranded antiparallel beta sheet structure.

In 1958 Garfinkel and Edsall[45] published the Raman spectrum of lysozyme, the first Raman spectrum of a naturally occurring biopolymer. Their spectrum was obtained using mercury excitation and revealed 14 faint lines of which several could be assigned to common group vibrations. The CO_2^- symmetrical stretching, CH deformation, CH stretching and a skeletal vibration were among the identifiable vibrational modes. Tobin[46] published the first laser excited Raman spectrum of lysozyme, along with several other proteins in 1968, utilizing helium–neon 6328 Å and argon 5145 Å wavelength laser radiation with photon counting detection. A total of 21 lines were reported of which about half had been found by Garfinkel and Edsall.[45] The work of Lord and Yu[47] with concentrated aqueous solutions of lysozyme represents the first attempt to interpret the entire protein spectrum. The use of solutions greatly reduced the background often encountered with solid samples. The authors recognized that their first task was the separation of Raman lines from the polypeptide backbone and side chains. A superposition technique was used whereby the spectrum of a mixture of amino acids corresponding to the protein amino acid composition was compared to the actual protein spectrum. In Fig. 7 these spectra are compared. Many side chain lines, particularly those emanating from the aromatic amino residues tyrosine, tryptophan and phenylalanine can be readily identified in the two spectra. One should note the weak scattering in the amide III region of the superposition spectrum. This small level of side chain scattering will not severely interfere with the amide III vibrations in the protein spectrum. Lines present in the protein spectrum but absent in the superposition spectrum may be confidently assigned to the backbone. Vibrations of the amide group involve motions of the N—H bond. The rather labile amide hydrogen rapidly exchanges with deuterium when the protein is dissolved in deuterium oxide, thus decreasing the frequency of the amide vibrations. This simple technique was applied to lysozyme by Lord and Yu[47] in order to assign the amide lines. Table 2 includes the Raman line assignment for lysozyme dissolved in water and deuterium oxide.

X-ray diffraction analysis of lysozyme single crystals reveals alpha helical,

TABLE 2
Raman spectra of native lysozyme in water and deuterium oxide
(250 to 1800 cm⁻¹)

Frequency (cm⁻¹) H₂O	D₂O	Tentative assignment	Frequency (cm⁻¹) H₂O	D₂O	Tentative assignment
311(0)			1035(2sh)	1035(2)	Phe
354(0)			1055(0)	1060(0)	
429(2br)			1078(1)	1080(0) ⎫	
	450(3br)		1109(4)	1110(1) ⎬ ν(C—N)	
460(0)			1128(3)	1128(1) ⎭	
491(0)	489(0)		1160(1)	1161(0)	
509(5sp)	509(5sp)	ν(S—S)	1179(1)	1183(1)	Tyr; O-deuterated Tyr and N—D bending of the deuterated indole
320(1)	524(0)				
544(1)	544(1)	Trp; N-deuterated Trp	1198(3db)	1200(1sh) ⎫ Tyr and Phe	
577(1)	574(1)	Trp; N-deuterated Trp	1210(2db)	1211(2) ⎭	
603(0)?			1240(4)		amide III
624(0)	625(0)	Phe		1253(1)	
646(0)	646(0)	Tyr; O-deuterated Tyr	1262(5)		amide III
661(0)	662(0)	ν(C—S) (disulfide)	1274(2)		amide III
700(1)	698(1)	ν(C—S) (Met)		1284(1)	
724(1)	723(1)		1290(0)?		
761(10sp)	760(10sp)	Trp; N-deuterated Trp	1304(0)		
800(0)?	804(1)		1338(8)	1337(6)	Trp; N-deuterated Trp
820(0)			1363(5sh)	1362(2)	Trp
836(1)	835(2)	Tyr; O-deuterated Tyr		1385(2sh)	N-deuterated Trp
	857(2)	N-deuterated Trp		1420(2sh)	symmetrical CO₂⁻ str.?
858(0)		Tyr	1432(4sh)		N—H bending vibration of the indole ring
	873(2)		1448(9)	1447(8sh) ⎫ C—H deformation	
879(5)		Trp	1459(8sh)	1459(10) ⎬ vibration	
900(2)	907(2)	⎱ ν(C—C)		1482(1sh) ⎭	
936(5)	934(7)	⎰ ν(C—C)	1494(0)?		His?
	950(6br)	amide III′	1553(8sp)	1553(8sp)	Trp; N-deuterated Trp
964(0)	965(1)		1582(2)	1580(2)	Trp; N-deuterated Try
984(1)			1622(4B)	1620(2sh)	Trp; Tyr and Phe
1006(7Sdb)	1006(7Sdb)	Phe	1660(10)	1658(10)	amide I; amide I′
1014(8Sdb)	1014(8Sdb)	Trp; N-deuterated Trp			

Abbreviations used here and in the following tables: Trp (tryptophan), Phe (phenylalanine), Tsr (tyrosine), Cys (cystine), Met (methionine), Pro (proline), Hypo (hydroxyproline). ν(X—Y) is a frequency assigned to an X—Y bond-stretching vibration. Figures in parentheses are relative intensities based on a value of 10 for the strongest line in each spectrum. These intensities have been corrected for the frequency dependence of instrument response (determined with a standard lamp) by multiplying the observed intensity with the following factors: 250 to 500 cm⁻¹, 1.0; 500 to 750 cm⁻¹, 1.1; 750 to 1000 cm⁻¹, 1.2; 1000 to 1250 cm⁻¹, 1.3; 1250 to 1500 cm⁻¹, 1.5; 1500 to 1800 cm⁻¹, 1.8 (Ref. 47).

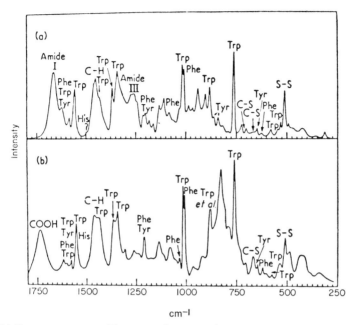

Fig. 7. (a) Raman spectrum of lysozyme. Lysozyme in water, pH 5.2. (b) Superposition of Raman spectra of the constituent amino acids, pH 1.0 (Ref. 47).

antiparallel beta sheet and disordered regions of the polypeptide chain[44] and one might expect these features to be reflected in the Raman spectrum. Only a single amide I line is observed and the frequency is 1660 cm^{-1} in H$_2$O (and 1658 cm^{-1} in D$_2$O). However, the amide III line centered near 1260 cm^{-1} has separate components at 1240, 1262 and 1274 cm^{-1} and the authors suggested the splitting may reflect the different chain structures. Eventually specific assignments of the separate amide III components were made through denaturation studies and comparison with model polypeptides. The chemical and thermal denaturation of lysozyme was thoroughly investigated by Lord and colleagues.[48–50] The first successful denaturant studied by Raman spectroscopy was LiBr.[48] Addition of large amounts of LiBr to a solution of lysozyme induces a conformational transition to a random coil. No change in the Raman spectrum was observed as the concentration of LiBr was increased from 1 to 4 M LiBr but at 5 M the shoulder at 1273 cm^{-1} weakened as seen in Fig. 8 and at 6 M the peak center shifted to 1245 cm^{-1}. The shoulder at 1273 cm^{-1} almost completely disappeared at this concentration and the 1238 cm^{-1} line was no longer visible. The Raman spectra of ionizable polypeptides in the randomly coiled conformation exhibit a single amide III line near 1245 cm^{-1}; therefore the shift in the amide III line in lysozyme to this frequency can be attributed to disordering by the LiBr. The denaturation was reflected in other regions of the spectrum. The disulfide S—S stretching vibration at 509 cm^{-1} broadened and decreased in intensity. Lines assigned to the C—C and C—N stretching motions also broadened. When the disulfide S—S bonds were cleaved by a mercaptoethanol reagent and capped with acrylonitrile, drastic changes were observed in the spectrum which reflect the conformational changes that occur during denaturation.[49] These changes are apparent in Fig. 9 and Table 3. The most pronounced feature in the spectrum is the large intensity increase in the amide I and III regions. The amide III mode has shifted from near 1260 to 1243 cm^{-1}, which is characteristic of the random coil and is also observed after denaturation with LiBr[48]. The amide I mode could not be observed during the LiBr denaturation due to the interference of the deformation mode of water appearing at 1640 cm^{-1} but in Fig. 9 the amide I shifts from 1660 to 1672 cm^{-1} after the disulfide cleavage. Reference to the previous discussion of polypeptides and to Table 1 reveals that the amide I mode for the alpha helix occurs from 1650 to 1657 cm^{-1}. Both the antiparallel beta sheet and random coil conformations have amide I modes appearing from 1665 to 1675 cm^{-1} in the Raman spectrum. Therefore the shift in the amide I line upon denaturation of lysozyme indicates the destruction of the alpha helical regions. The authors point out that a transition from an ordered to disordered polypeptide chain may lead to a hydrogen bond weakening which would increase the frequency of the amide I line. However the equally high amide I frequency for the beta sheet conformation can be ascribed to direct coupling between non-bonded carbonyl groups.[32] One sees that the beta sheet and random coil conformations cannot be easily distinguished using only the amide I mode but in the amide III region these two conformations appear near 1235 and 1245 cm^{-1}, respectively.

Changes in the intensities of the aromatic side chains occur during LiBr and chemical denaturation. Upon increasing the LiBr concentration from 5 to 6 M the tryptophan lines at 1338 and 1363 cm^{-1} decrease in intensity (Fig. 8) which indicates a disturbance in the environments of these residues. Similar changes are noted after cleavage of the disulfide bonds. Less pronounced changes are observed in a number of the tyrosine and phenylalanine lines. A specific mechanism for the environmental change of tryptophan 62 and 63 is proposed by the authors.

The reversible and irreversible thermal denaturation of lysozyme has been investigated by Brunner and Sussner[51] and Chen, Lord and Mendelsohn.[50] Brunner and Sussner followed the intensity of the S—S stretching vibration of the disulfide bond at 504 cm^{-1} from room temperature through the reversible denaturation temperature of 76°C and claimed that the continuous intensity decrease with no change in half-width resulted from the thermal cleavage of the disulfide bonds. According to this interpretation all disulfide bonds are cleaved above 76°C, which is contrary to n.m.r. evidence.[52] The above changes in the line intensity at 504 cm^{-1} were not detected by Chen, Lord and Mendelsohn,[50] who concluded that the disulfide bonds remain intact. Brunner and Sussner measured the line intensity on a steeply sloping background that could make quantitative measurements of line intensity and half-width difficult. Possibly the

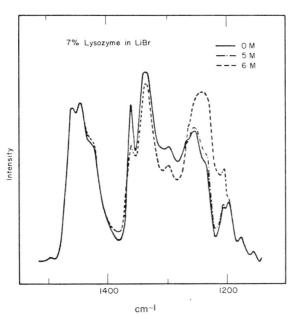

Fig. 8. Redrawn Raman spectra (1150–1500 cm⁻¹) of 7% lysozyme in 0, 5 and 6 M LiBr, pH 4, normalized to the intensity of the methylene deformation mode at 1448 cm⁻¹, after correction for the water background (Ref. 49).

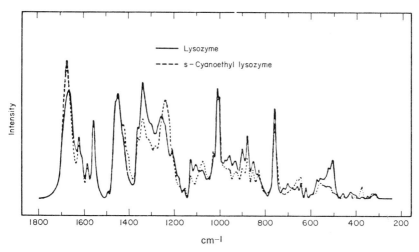

Fig. 9. Redrawn Raman spectra of lysozyme powder and S-cyanoethyl lysozyme normalized to the intensity of the 1448 cm⁻¹ band, after correction for the background (Ref. 49).

TABLE 3
Raman frequencies and intensities of lysozyme powder and solid S-cyanoethyl lysozyme

⌐——Frequency, cm^{-1}——⌐		
Lysozyme	S-cyanoethyl lysozyme	Tentative assignment
315(1)		
333(0sh)	333(1)	
353(0)		
375(0)		
	377(1)	
408(0sh)	408(0)	
429(1)	429(1)	
462(1)	462(0)	
	492(1)	
507(4)		⎫
	515(1sh)	⎬ ν(S—S)
525(1sh)		
543(0)	543(0)	Trp
562(1)	562(1)	
574(1)		Trp
598(0)	598(0)	
622(1)	622(1)	Phe
645(1)	645(2)	Tyr
660(1)	660(2)	ν(C—S)Cys
700(1)	700(1)	ν(C—S)Met
720(1)		
760(9)	760(7)	Trp
	832(2)	⎫
834(1sh)		⎬ Tyr
854(2)	854(2)	⎭
878(6)	878(4)	Trp
900(5)	900(3)	⎫
933(3)	933(2)	⎬ ν(C—C)
960(3)	960(2)	
978(4)	978(3)	
1005(5sh)	1005(5sh)	Phe
1012(11)	1012(11)	Trp
1030(1sh)	1030(1sh)	Phe
	1070(4)	
1076(3)		⎫
	1079(4)	⎬ ν(C—N)
1107(3)	1107(0)	⎬
1129(2)	1129(4)	⎭
1154(1)	1154(1)	
1176(1sh)	1176(0)	Tyr
1200(0sh)	1200(0sh)	⎫ Tyr and Phe
1208(1)		⎭
	1223(0)	

TABLE 3—*continued*

┌──────Frequency, cm⁻¹──────┐		
Lysozyme	S-cyanoethyl lysozyme	Tentative assignment
1238(2sh)		
	1243(8)	
1254(5)		Amide III
	1263(0)	
1271(1sh)		
	1281(0)	
1300(0)		
	1308(1sh)	
1327(0sh)		
1338(11)	1338(8)	Trp and δ(C—H)
1362(2)	1362(2)	
	1405(0)	
1427(1sh)	1427(1sh)	δ(N—H) indole rings
1448(10)	1448(10)	δ(C—H₂)
1459(5sh)	1459(5sh)	
1490(0)	1490(0)	
1553(8)	1553(8)	Trp
1582(3)	1582(3)	
1607(4sh)	1607(4sh)	
1622(5)	1622(5)	Trp, Tyr, and Phe
1660(10)		Amide I
	1672(13)	

^{a} ν(A—B) means A—B stretching vibration. δ(C—D) = C—D bending vibration. Numerical figures in parentheses are relative peak intensities with the 1448 cm⁻¹ line taken as 10 (Ref. 49).

observed decrease in intensity was actually an increase in the background slope. Both groups report that reversible denaturation does not significantly change the shape or intensity of the amide III line, indicating that the conformation of the polypeptide backbone is not appreciably altered.

Heating a lysozyme solution above 80°C for an extended period of time induces irreversible changes in the Raman spectrum resembling those reported from studies of denaturation by chemical means.[51] The center of the amide III line shifts from near 1260 to 1247 cm^{-1} and increases in intensity, while the amide I line shifts from 1660 to 1670 cm^{-1}, increases in intensity and narrows. The frequency increase and narrowing of the amide I line suggests that the hydrogen bonding may be weaker but more regular in the denatured protein than in the native state. As before, these spectral changes reflect the destruction of the ordered chain conformations. A number of changes in the intensity of the aromatic side chains occur. Three of the tryptophans, numbers 62, 63 and 104, are located in the active site. Of these three, numbers 63 and 62 are, respectively, adjacent to and once removed from cystine 64—80, whose S—S bond is the one most closely associated with the active site. The authors speculate that irreversible denaturation may rupture the S—S bond with the attendant decrease in the S—S stretching intensity at 504 cm^{-1} and increase in the C—S stretching intensity at 650 cm^{-1}. At the same time, alteration in the environment of tryptophanes 62 and 63 induces the intensity changes in the Raman lines at 1338 and 1363 cm^{-1}.

In summary, the three basic chain conformations present in lysozyme are the alpha helix, beta sheet and random coil, and these are possibly reflected in the splitting of the amide III line. Yu and Jo[53] have recently reviewed the lysozyme denaturation studies and the separate components observed at 1240, 1262 and 1274 cm^{-1} were tentatively assigned to the beta sheet, random coil and alpha helical conformations.[53] The amide I and III modes can be used to study denaturation induced by heat, salt, and chemical cleavage of the disulfide bonds. All three denaturing agents drastically reduce the content of ordered chain conformations and this is indicated by the single amide III line near 1245 cm^{-1} and amide I line near 1670 cm^{-1} which are characteristic features of the random coil. The intensities of several tyrosine and tryptophan lines vary considerably after denaturation and reflect local environmental changes of these amino acids. The location of tryptophanes in the active site of lysozyme suggests the exciting prospect that Raman spectroscopy may be useful in studying enzyme mechanisms. First, however, detailed studies with model compounds will be needed to interpret these intensity changes.

4.2.2 Ribonuclease

Raman studies of the polypeptide chain conformation in ribonuclease are assisted by the high resolution models of the tertiary structure available from X-ray analysis[54,55] and the absence of tryptophan which yields many interfering Raman lines. The Raman spectrum of this enzyme has been published by several investigators but to date no studies of denaturation have appeared. Yu and Lord[56] analyzed the spectrum of native ribonuclease in aqueous solution in a manner very similar to their interpretation of lysozyme spectrum discussed previously, i.e. the superposition spectra of the amino acids and deuteration of the amide N—H were used to separate and identify the side chain and backbone vibrations. From Fig. 10 and Table 4 one can discern only a single amide I line at 1667 cm^{-1} while two well separated amide III lines appear at 1240 and 1262 cm^{-1}. Before considering the assignments proposed for these two lines, we will briefly discuss the recent model of ribonuclease obtained with X-ray diffraction analysis. Carlisle et al.[55] have published the structure of ribonuclease down to 2.5 Å resolution using the method of isomorphous replacement. The single polypeptide chain of 124 amino acid residues twists backwards and forwards over a length of about 90 Å forming a U-shaped structure enclosing a cleft, within which lie some of the residues responsible for the enzymatic activity. A stereoscopic view of the molecule showing only the α-carbon atoms appears in Fig. 11.

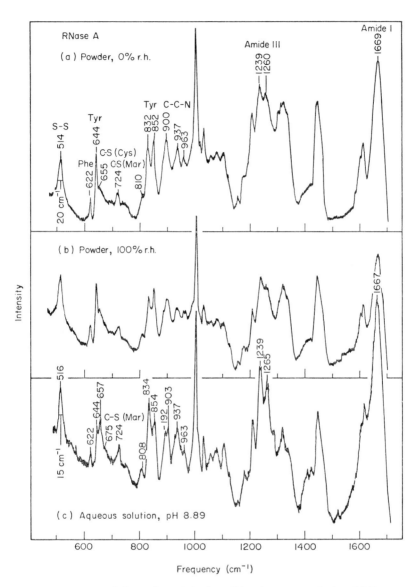

Fig. 10. Raman spectra of ribonuclease A in the solid and aqueous solution. (a) Spectrum of the lyophilized powder of RNase A in 0% relative humidity (r.h.). The powder was dried *in vacuo* at 25°C over phosphorus pentoxide for 2 h before the experiment. The flask containing the sample and phosphorus pentoxide was then filled with dry nitrogen at 1 atm and used for laser Raman scattering work. (b) Spectrum of the lyophilized powder of RNase A in 100% rh. The powder was equilibrated at room temperature with saturated H_2O vapor pressure (about 24 mm) for 5 h. (c) Spectrum of RNase A in aqueous solution at pH 8.89; $\Delta\sigma$, 4 cm^{-1} (200 μ). The solution sample at 200 mg/ml concentration and pH 8.89 was obtained by dissolving the required amount of the powder used in (a) and (b) without adding HCl or NaOH solution (Ref. 57).

TABLE 4
Raman spectra of native ribonuclease in water and in deuterium oxide (250 to 1800 cm⁻¹)

Frequency (cm⁻¹)		Tentative assignment	Frequency (cm⁻¹)		Tentative assignment
H_2O	D_2O		H_2O	D_2O	
380(1)	373(0)		1060(2)	1056(2) ⎫	
445(2br)	440(2br)		1083(4)	1078(2) ⎬ ν(C—N)	
487(1)	487(1)		1106(4)	1106(3) ⎭	
516(3)	516(3)	ν(S—S)	1123(2sh)		NH₃⁺ rock.?
561(1)			1160(0)	1158(0)?	
624(1)	624(1)	Phe	1181(3)	1181(3)	Tyr and Phe;
647(3)	647(3)	Tyr; O-deuterated Tyr			O-deuterated Tyr and Phe
659(3br)	659(3br)	ν(C—S) (disulfide)	1211(4)	1211(4)	Tyr and Phe;
725(2)	725(2)	ν(C—S) (Met)			O-deuterated Tyr and Phe
	754(0)				
	804(3)		1240(5)		amide III
811(2)				1242(2)	
36(5)	833(5)	Tyr; O-deuterated Tyr		1256(2)	
			1262(5)		amide III
855(4)	854(4)	Tyr; O-deuterated Tyr		1305(1)	
			1317(4)		
899(4br)	889(3) ⎫		1337(4)	1332(4br)	
939(2)	937(2) ⎬ ν(C—C)			1409(5sp)	N-deuterated imidazo- lium of His
	950(6br)	amide III′			
968(0)			1416(3)		sym. CO₂⁻ str.
983(8sp)	983(8sp)	SO₄²⁻	1451(8br)	1451(8br)	C—H deformation
1004(9sp)	1004(9sp)	Phe		1588(1)	Tyr and Phe
1031(2)	1031(2)	Phe	1609(1sh)	1609(2)	Tyr and Phe
			1617(3)		Tyr
			1667(10br)	1663(10br)	amide I; amide I′

The two arms of the U are held in position by an extensive network of hydrogen bonds which are involved in the alpha helix formation, near-helical formation, beta sheet formation and various secondary interactions. The secondary structure in Fig. 11 can be adequately traced, and total alpha helical content is only about 19 percent.[55] The rest of the polypeptide chain consists of unordered regions that can be broadly interpreted as helical but to a large extent are beta sheet.

The interpretation of the splitting in the amide III region has not been resolved. Yu et al.[57,58] suggest that the 1240 cm⁻¹ line may be assigned to the beta sheet component. This is consistent with the homopolypeptide studies. Recall that in the spectra of beta sheet homopolypeptides the amide III line appeared with strong intensity at 1235 ± 5 cm⁻¹. These authors assign the 1262 cm⁻¹ component to the alpha helical conformation that comprises 19 percent of the residues and this assignment is based in part on the Raman analysis of crystalline

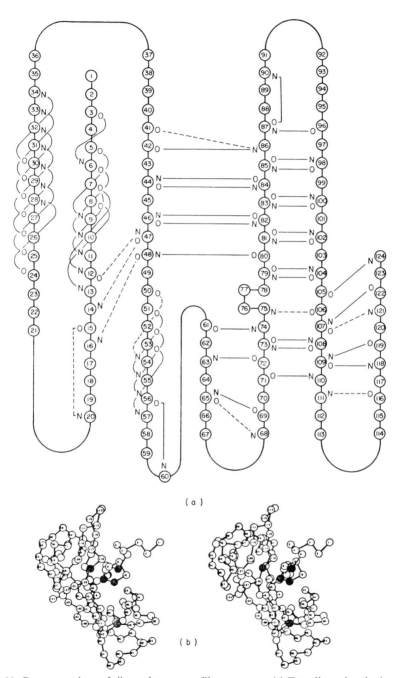

Fig. 11. Representations of ribonuclease crystalline structure. (a) Two dimensional schematic showing hydrogen bonds in RNase A. Broken lines represent those bonds present in RNase S but not RNase A. (b) Three dimensional view (Ref. 55).

glucagon,[59] a hormone that contains a large alpha helical fraction. Frushour and Koenig[60] have commented on the latter assignment. Their studies with alpha helical polypeptides[33,61] and tropomyosin,[60] a muscle protein having a helical content of 90 percent indicate that the amide III mode for the alpha helical conformation is very weak and therefore inconsistent with the medium intensity 1262 cm^{-1} component in the ribonuclease spectrum. As seen above, X-ray diffraction analysis reveals a significant fraction of unordered residues in ribonuclease. Greenfield and Fasman[62] have calculated from circular dichroism the following percentages of secondary structures in ribonuclease: alpha helix, 9–12 percent; beta sheet, 33–43 percent; and disordered, 38–44 percent. Therefore Frushour and Koenig suggest that the 1262 cm^{-1} component was more likely due to the disordered chain fraction.

However, recently the spectrum of ribonuclease in a saturated D_2O atmosphere has been obtained.[114] By heating D_2O solutions of this protein, freeze drying the product and then allowing exchange with D_2O vapor, it was possible to exchange all the amide N—H groups. The line at 1260 cm^{-1} was still in evidence as shown in Fig. 12. Clearly this is not an amide III mode but can probably be

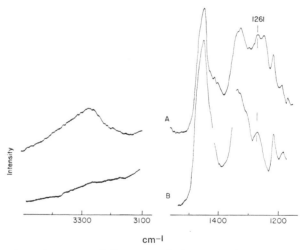

Fig. 12. Comparison of spectrum of ribonuclease in fundamental N—H stretching region and amide III region of the spectrum (A) with a fully deuterated sample of the protein (B).

assigned to tyrosine. The amide I line for the beta sheet and random coil appears from 1665–1674 cm^{-1} in the spectra of the homopolypeptides but appears near 1655 cm^{-1} for the alpha helical conformation in the homopolypeptides and the proteins: bovine serum albumin,[63,64] tropomyosin[60] and keratin.[65] The single line appearing at 1667 cm^{-1} in the ribonuclease spectrum is probably the superposition of unresolved components from the beta sheet and disordered conformations.

The S—S and C—S stretching frequencies of the four disulfide bonds in native ribonuclease appear respectively at 516 and 659 cm^{-1} in Fig. 10 which is somewhat different from 509 and 667 cm^{-1} in the superposition spectrum.[56] However, the intensity ratio I(C—S)/I(S—S) is about one in both cases. In the spectrum of native lysozyme by Lord and Yu[47] the corresponding lines appear at 509 and 661 cm^{-1} and the intensity ratio is considerably smaller, i.e. ~0.1. This suggests that the conformations of the C—S—S—C cross-links in ribonuclease are different from those in lysozyme. On the basis of model compound studies (discussed earlier in connection with lysozyme) Lord and Yu[56] conclude that the C—S—S angles are smaller in ribonuclease than in lysozyme.

4.2.3 Chymotrypsinogen and α-chymotrypsin

α-Chymotrypsin is one of several proteolytic enzymes produced by the pancreas. Actually the pancreas produces chymotrypsinogen which consists of a single polypeptide chain of molecular weight 25,000 g mol^{-1} and is biologically inactive. Trypsin and other activating enzymes cleave the chain in two places leading to α-chymotrypsin which has three separate polypeptide chains and is biologically active. The tertiary and secondary structures of chymotrypsinogen and chymotrypsin have been examined by X-ray analysis and appear to be similar.[66,67] There are no extensive ordered regions of the polypeptide chains, but mostly disordered regions and short beta sheet–like aggregates. The alpha helical content is less than ten percent.[66]

The Raman spectra of native chymotrypsinogen A[68] and α-chymotrypsin[56] have been analyzed and are quite similar. A single amide I line is observed at 1669 cm^{-1} though two amide III lines appear at 1245 and 1260 cm^{-1}. The reversible thermal denaturation of chymotrypsinogen was studied by Koenig and Frushour.[68] A 3 cm^{-1} decrease in the frequency of the amide I mode was observed when the protein was dissolved in D$_2$O and heated through the reversible denaturation transition temperature. This shift was attributed to a perturbation of the native structure of sufficient degree to allow permeation of D$_2$O into the protein but absence of change in the amide III region (when dissolved in H$_2$O) suggests that the denaturation mechanism involves little or no change in the secondary structure.

4.2.4 Insulin

The Raman investigation of native and thermally denatured insulin has been completed successfully by Yu and his colleagues.[69-71] This relatively small enzyme is an excellent model protein for determining the sensitivity of Raman spectroscopy to conformational analysis for three reasons. First, absence of tryptophan, an aromatic amino acid having many strong Raman lines, reduces the complexity of the spectrum. Second, the well-studied globular to fibrous transformation grossly rearranges the tertiary structure and is accompanied by definite spectral changes. Finally, the X-ray analysis by Adams *et al.*[72] provides a good structural base for interpreting the spectra. The primary structure of insulin consists of two chains, A and B, containing 21 and 30 residues, respectively, and connected by two disulfide bonds. Chain A also contains an intramolecular disulfide bond. The A chain is a compact unit around which chain B is wrapped. Chain A appears highly folded and contains short alpha helical segments. The center of chain B consists of 3 turns of the alpha helix and the two chain ends are rather extended.

The spectrum of native insulin as a crystalline powder was obtained by Yu, Liu and O'Shea[71] and appears at the top of Fig. 13. The amide I mode appears

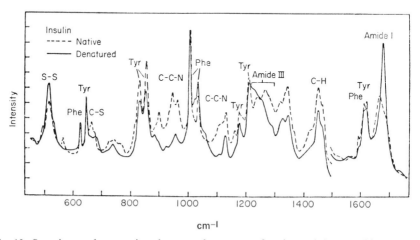

Fig. 13. Superimposed comparison between the spectra of native and denatured insulin. The line at 624 cm⁻¹ due to the in plane ring vibration of phenylalanine residues is known to be conformation-independent and is used as an internal reference This line has nearly equal intensity in both spectra (Ref. 71).

to be split with the major peak located at $1662 \ cm^{-1}$ and a shoulder at $1680 \ cm^{-1}$. The $1662 \ cm^{-1}$ component is assigned to the alpha helical conformation and the $1685 \ cm^{-1}$ component to the random coil. It should be interjected that alpha helical polypeptides[33,61] and predominantly alpha helical proteins[60,63−65] have amide I lines near $1655 \ cm^{-1}$. In the conformationally sensitive amide III region of the insulin spectrum the scattering becomes indistinct, which is also characteristic of proteins containing a large alpha helical content.[60,63−65] The major peak at $1270 \ cm^{-1}$ and the very weak shoulder at $1288 \ cm^{-1}$ are assigned to the alpha helical conformation while the $1239 \ cm^{-1}$ shoulder may arise from the random coil conformation with a small contribution from the beta sheet. In order to define a good baseline for the amide III region Yu, Jo, Chang and Huber[73] obtained the Raman spectrum of completely deuterated insulin crystals. The line at $1260 \ cm^{-1}$ remaining after deuteration comes from tyrosine. A corrected amide III region for native insulin was constructed by subtracting the spectrum of the deuterated crystal from the non-deuterated one. Several frequency shifts occur after subtraction and the authors contend that only the frequencies in the difference spectrum should be regarded as 'true'. The newly discovered line at 1303 cm^{-1} in the difference spectrum along with the lines at 1269 and 1284 cm^{-1} are assigned to the alpha helical conformation. The splitting of the alpha helical amide III vibration into several components is attributed to different strengths of hydrogen bonds and/or irregularities in the alpha helical structure.

Denaturing insulin by boiling in water at pH 2.42 until the protein precipitates has a drastic effect on the Raman spectrum as shown in Fig. 13. The spectra of the native and denatured protein are normalized to the phenylalanine vibration at 1032 cm^{-1} to accentuate the differences. Also, the frequencies and intensities of the Raman lines in the native and denatured spectra are listed in Table 5. Upon denaturation the amide I line has shifted to 1672 cm^{-1} and sharpened considerably. Recall that Lord and coworkers[47-50] observed a similar shift of the amide I frequency after thermal and chemical denaturation of lysozyme and attributed this shift to the weakening of amide hydrogen bonds by formation of a random coil. However, we know from work with homopolypeptides summarized in Table 1 that the $v(0, 0)$ totally symmetric component of the amide I mode for the antiparallel beta sheet conformation also appears near this frequency. One also expects the $v(\pi, 0)$ mode for this conformation to appear near 1630 cm^{-1} in the infrared spectrum. In fact, an infrared band is seen at 1635 cm^{-1} in the spectrum of denatured insulin[74] confirming the presence of antiparallel beta sheet after denaturation. Changes in the amide III region of the Raman spectrum also reflect the formation of antiparallel beta sheet conformation. The line appearing at 1230 cm^{-1} for the denatured sample agrees with the amide III frequency for the model compounds known to have antiparallel beta sheet structure.

Further evidence for conformational change can be seen in regions of the spectrum where precise structure–frequency correlations are now beginning to be established. Two lines of medium intensity appear at 946 and 934 cm^{-1} in the spectrum of the native protein that are assigned to C—C stretching motions in the polypeptide backbone. Studies by Koenig and coworkers[33,60,61,64] show that strong lines are observed in this region for alpha helical polypeptides and proteins containing a large alpha helical content, and disruption of the helices leads to a decrease in intensity of these lines. From Fig. 13 we see an analogous decrease in intensity of the 946 and 934 cm^{-1} lines after denaturation as the beta sheet conformation appears at the expense of the alpha helical regions.

Lord and Yu[47,56] have previously shown that the intensity ratio of the S—S and C—S stretching frequencies near 500 and 650 cm^{-1}, respectively, may depend upon the C—S—S angle of the disulfide bonds. In native insulin the S—S frequencies of the three disulfide linkages appear as an unresolved broad line centered at 515 cm^{-1} while the C—S frequencies appear at 668 and 678 cm^{-1}. The intrachain disulfide bond in native insulin may possibly have a different local conformation than the two interchain bonds, resulting in two separate C—S stretching frequencies. When insulin is denatured the intensity of the S—S

TABLE 5
Raman spectra of insulin (bovine) (200 to 1800 cm⁻¹)

Frequencies in cm⁻¹		Tentative assignments[a]
Native (crystals)	Denatured (solid)	
	265(0.9)	Skeletal bending
333(0.9)	325(0.8)	
410(0.8)	420(0.5)	
467(0.8)	460(0.4)	
495(1.2)	480(0.3)	
515(3.2)	513(4.4)	ν(S—S)
	532(1.5sh)	Skeletal bending
563(1.0)		
624(2.0sp)	624(2.0sp)	Phe
644(3.6sp)	644(3.6sp)	Tyr
668(2.0)	657(1.1)	ν(C—S) of the C—S—S—C group
678(1.0sh)	680(1.3)	
725(0.8tp)		Skeletal bending
747(0.8tp)	737(0.9br)	
770(0.8tp)		
814(1.4sh)		
832(4.4db)	830(3.9db)	Tyr
854(5.5db)	853(4.5db)	
900(2.0)	882(1.5)	ν(C—C)
934(2.0sh)	922(0.8)	
946(3.2db)		
963(2.9db)	956(1.6br)	
1004(10.0sp)	1004(10.0sp)	Phe
	1020(2.5)	ν(C—N)
1032(3.3sp)	1032(3.3sp)	Phe
	1057(1.3sh)	ν(C—N)
1112(1.5sh)		
1128(1.8)	1127(1.7)	
1162(0.9)	1161(0.5sh)	
1177(2.4)	1175(2.4)	Tyr
1212(4.6sp)	1214(4.9)	Tyr and Phe
	1227(4.3)	Amide III (β-structure)
	1252(4.0sh)	
1239(5.0sh)		Amide III (random-coil)
1270(5.3)		Amide III (α-helical)
1288(4.7sh)		Amide III (α-helical)
1322(2.0sh)	1327(2.0db)	CH deformation
1344(4.0)	1343(3.1db)	
1367(1.6sh)		
	1407(0.4sh)	
1425(2.5sh)	1422(1.1sh)	Symmetrical CO_2^- stretching
1450(5.0)	1450(3.8)	CH_2 deformation
1462(4.6sh)	1462(3.1sh)	
1587(1.3)	1587(1.0sh)	Phe
1607(3.6db)	1607(3.5db)	Phe and Tyr
1615(3.6db)	1615(3.5db)	Tyr
1662(4.6)		Amide I (α-helical structure)
	1673(8.6sp)	Amide I (β-structure)
1685(4.0sh)		Amide I (random-coil)
	1735(0.4br)	—COOH

[a] Ref. 71.

stretching motion at 515 cm^{-1} has increased while the C—S stretching motion at 668 cm^{-1} has shifted to 657 cm^{-1} and decreased in intensity. The authors contend that the geometry of the two interchain disulfide linkages have changed while the single intrachain linkage remains unchanged.

Ambrose and Elliott[74] originally proposed that orientation of the denatured insulin leads to the cross-beta sheet conformation. In this conformation the antiparallel chains are arranged in a regular fashion so that the chain axes are perpendicular to the direction of orientation. Recently Burke and Rouguie[75] confirmed these findings utilizing electron diffraction. It should be noted, however, that reversibility of the insulin denaturation leads to the assumption that the cross-beta conformation in insulin consisted of a linear aggregate of only slightly distorted molecules. Contrary to this view, the Raman studies clearly show that denaturation involves a gross rearrangement of the tertiary and secondary structure.

Since insulin consists of two chains connected by disulfide bonds, investigators, including Givol et al.,[76] speculated that insulin may result from the in vivo activation of a single chain precursor. In this way, insulin production would resemble the activation of the zymogens, i.e. the conversion of the inactive, single chain chymotrypsinogen to active chymotrypsin discussed earlier. A single chain precursor was found by Steiner and Oyer[77] in humans and labeled proinsulin. In proinsulin the carboxyl terminal end of the A chain is connected to the amino-terminal end of the B chain by a polypeptide chain called the C-peptide. The length of the C-peptide varies among species and is 30 residues long in bovine proinsulin.

Since Raman spectroscopy is sensitive to small changes in protein structure, Yu et al.[71] examined the spectra of porcine proinsulin to see if the insulin moiety exists in a conformation nearly the same as that of free insulin. The Raman spectra of insulin and proinsulin are normalized to the 647 cm^{-1} phenylalanine intensity in Fig. 14. The agreement in frequency and intensity of conformationally sensitive lines numbered one through eleven led the authors to conclude that the conformation of insulin remains essentially unchanged upon activation and that the differences in the amide III and I regions of the two spectra can be attributed to the C-peptide. A graphical subtraction in the amide I region shown in the insert of Fig. 14 reveals a large peak centered at 1663 cm^{-1} and a shoulder at 1685 cm^{-1}, both attributed to the C-peptide. X-ray crystallographic studies of Fullerton, Potter and Low[78] reveal two short segments of alpha helix in the C-peptide unit of proinsulin while Frank and Veros[79] determined from the circular dichroic spectra that the C-peptide took on a random-coilcon formation in solution. Yu et al.[71] associate the 1663 cm^{-1} line and 1685 cm^{-1} shoulder in the difference spectrum to the alpha helical and random coil conformations, respectively, and conclude that the C-peptide contains a considerable alpha helical fraction. A note of caution must be interjected here. As mentioned above, the amide I mode for polypeptides and proteins containing large alpha helical fractions occurs from 1652–1658 cm^{-1}. On the other hand ribonuclease[56] with a helical content of less than 20 percent and α-casein[60] with no ordered structure have their amide I lines at 1667 and 1668 cm^{-1}, respectively. Therefore association of the 1663 cm^{-1} line in the difference spectrum with the alpha helix may be questionable. Perhaps both the 1663 and 1685 cm^{-1} lines may be associated with the random coil, making the Raman results consistent with the circular dichroic studies of Frank and Veros.[79]

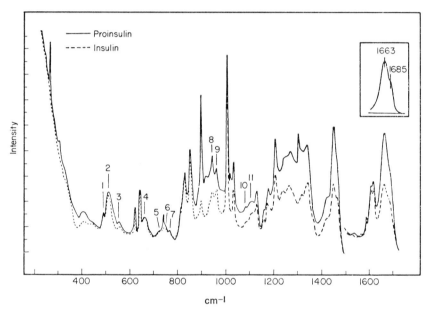

Fig. 14. Comparison of the Raman spectra of proinsulin and insulin in the solid state. Both compounds are of porcine origin. The insert shows the graphical subtraction between the two spectra in the amide I region (Ref. 71).

4.2.5 Tropomyosin

In the investigations discussed thus far proteins with well resolved X-ray structures allowing good analysis of the secondary structure were chosen so that the Raman spectra could be more easily interpreted. As we have seen, these proteins tend to have very complicated Raman spectra due to the coexistence of several basic chain conformations (which has little to do with the ability to form single crystals). Perhaps more desirable for establishing structure–frequency correlations are proteins known to have a preponderance of one conformation but can be denatured to another conformation. Tropomyosin fits this description because in the solid state and at neutral pH 90 percent of the amino acid residues are in the alpha helical conformation[80] but an increase in pH will bring about a rapid helix to random coil transition. Then the Raman spectrum of the randomly coiled state can be compared to the spectra of randomly coiled ionizable poly-peptides or largely disordered proteins such as α-casein.

The helix to coil transition of rabbit muscle tropomyosin was investigated by Frushour and Koenig.[60] This protein has a molecular weight of 54,000 g mol^{-1} and forms a double stranded alpha helical coiled-coil about 340 Å long and 14 Å in diameter.[81] Approximately 24 percent of the amino acid residues are acidic and 16 percent are basic, therefore, the molecule is very soluble at neutral pH. Lowey[80] has determined the helical content as a function of pH from optical rotatory dispersion (ORD) studies. The Moffit-Yang[82] treatment of ORD data yields a parameter, b_0, that has a value of -630 for a completely helical polypeptide chain. At the top of Fig. 15 the b_0 value for tropomyosin solutions, expressed as percent alpha helix, appears as a function of pH. Raising the pH above 8.0 ionizes the acidic side chains and increases the electrostatic repulsions

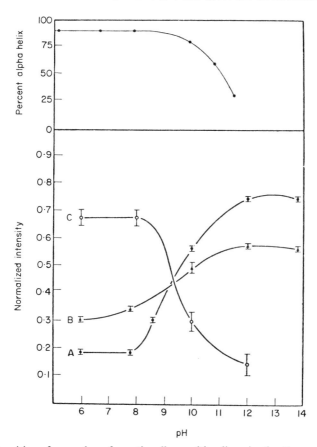

Fig. 15. Intensities of several conformationally sensitive lines in the Raman spectrum of tropomyosin plotted as a function of pH. The intensities were measured as peak heights and normalized to the 1451 cm^{-1} line. (A) Amide III mode, 1254 cm^{-1}, (B) ν_a(COO$^-$), 1402 cm^{-1}, (C) skeletal vibration, 940 cm^{-1} (Ref. 60).

between side chains which causes the helix to unfold. Therefore, the magnitude of b_0 drops above pH 8.0. Analogous pH–induced helix to coil transitions are observed for poly-L-glutamic acid and poly-L-lysine and have been investigated using Raman spectroscopy.[33,37] The Raman spectra of native tropomyosin in the freeze-dried solid and aqueous solution states are shown in Fig. 16 and the

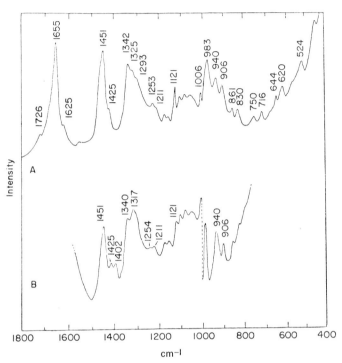

Fig. 16. Raman spectra of native tropomyosin. (A) Freeze-dried solid: slit width 4 cm⁻¹, scan rate 10 cm⁻¹ min⁻¹, laser power at sample 200 mW at 5145.3 Å, time constant 4 s. (B) 10% solution at pH 7.5, 0.1 M NaCl: conditions same as above except laser power is 400 mW (Ref. 60).

assignments are listed in Table 6. Notice that the amide I line appears at 1665 cm⁻¹ in the spectrum of the solid (the water Raman line at 1640 cm⁻¹ interferes in the spectrum of the solution). This frequency agrees with the value observed for the alpha helical polypeptides as summarized in Table 1. Very weak scattering is observed in the amide III region from 1200–1300 cm⁻¹. The weak line at 1253 cm⁻¹ may arise from the ten percent of residues not in the alpha helical conformation and the 1211 cm⁻¹ line comes from tyrosine. Strong scattering is observed from 900–940 cm⁻¹ (the intense 983 cm⁻¹ line is residual sulfate from purification). The two lines appearing at 940 and 906 cm⁻¹ may correspond to strong Raman lines found in this region of the homopolypeptide spectra and assigned to the C—N and C—C stretching motions of the polypeptide backbone.

TABLE 6
Raman lines of tropomyosin and α-casein[a]

Tropomyosin (freeze-dried)	α-Casein (freeze-dried)	Assignment
1726 vw		ν(C=O)
1655 s	1668 s	Amide I
1625 w		
	1617 m	Tyr
	1608 vw	Tyr, Phe
	1590 vw	Tyr, Phe
	1557 vw	Trp
1460 sh	1470 sh	
1451 s	1451 s	δ(CH$_2$), δ(CH$_3$)
1425 w	1422 w	
	1408 vw	ν_s(COO$^-$)
1342 s	1343 s	$\left\{ \begin{array}{l} \delta\text{(CH)}, \gamma_w\text{(CH}_2\text{)}, \end{array} \right.$
1325 s	1324 s	$\left. \begin{array}{l} \gamma_t\text{(CH}_2\text{)}, \text{Trp} \end{array} \right\}$
1293 m	—	
	1270 s sh	Amide III
1253 w	1254 s	Amide III
1211 w	1211 w	Tyr
1172 w	1178 vw	
1156 w	1161 vw	Tyr
1121 w	1126 w	
1106 m	1101 vw	ν(C—N)
1087 m	1081 vw	
1054 m		
	1034 w	Phe
1006 w	1006 s	Phe
	996 vw	
983 s		SO$_4^{2-}$
	962 w	Skeletal
940 m	938 w	Stretching
906 w	928 vw	
	883 w	Tyr
861 w	856 w	Tyr
830 w	832 w	
750 w	762 w	Trp
716 w	716 w	ν(C—S)
644 w	646 w	Tyr
620 w	625 w	Phe
557 w		
524 w		
448 w		

[a] Abbreviations: ν (stretching coordinate), δ (deformation coordinate, γ_w (wagging coordinate), γ_t (twisting coordinate).

Significant spectral changes occur during the pH–induced helix to coil transition. The intensity of the amide III line at 1254 cm^{-1} increases as seen in Figs. 15 and 17. Very similar behavior in the amide III intensity was observed during the helix to coil transitions of poly-L-glutamic acid[33] and poly-L-lysine.[37] There is no change in the shape of the band envelope around 1300 cm^{-1}–1400 cm^{-1}; a change might be expected if the alpha helical conformation had an amide III line of appreciable intensity in this region of the spectrum. One must conclude that the amide III line for the alpha helical conformation in polypeptides and proteins has weak intensity. The skeletal stretching vibrations at 940 and 906 cm^{-1} have decreased in intensity after denaturation. Similar observations have been reported for the denaturation of bovine serum albumin[64] and alpha keratin[65] which have large helical contents.

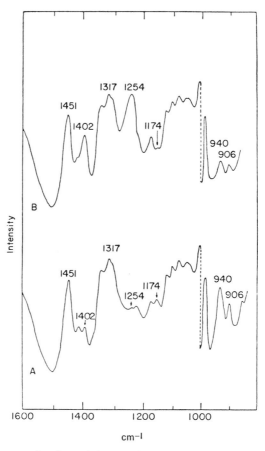

Fig. 17. Raman spectra of native and denatured tropomyosin. (A) Native, 10% concentration at pH 7.5, 0.1 M NaCl: instrument settings same as Fig. 16B. (B) Denatured, instrument settings same as above but pH 12.0 (Ref. 60).

The weak scattering in the amide III region of the native protein further establishes that side chain Raman lines do not necessarily interfere with the true contour of the amide III line. Therefore, careful measurement of the amide III intensities may be used to determine the secondary structure of proteins in the solid state and aqueous solution. This can be done with the tropomyosin spectrum. In Fig. 15 the normalized intensities of three lines: the amide III line at 1253 cm^{-1}, the stretching motion of the ionized carboxyl at 1402 cm^{-1} and the skeletal mode at 940 cm^{-1} were plotted against pH and for reference the b_0 value from the Moffit-Yang determination of helical content by Lowey[80] was included. Clearly the amide III line and 940 cm^{-1} skeletal modes change in intensity according to the alpha helical content and could be used to estimate the percent alpha helix at any pH.

4.2.6 α-Casein

Milk contains many proteins[83] and one of them, α-casein, has no measurable ordered secondary structure.[84] This feature may aid in the digestion of the protein by infant mammals. Frushour and Koenig[60] used the protein to characterize the Raman scattering of the disordered state in proteins. The spectra in both the solution and aqueous states are compared in Fig. 18. The amide I and III modes appear at 1668 and 1254 cm^{-1}, respectively. The latter frequency agrees well with the amide III frequency in denatured tropomyosin.

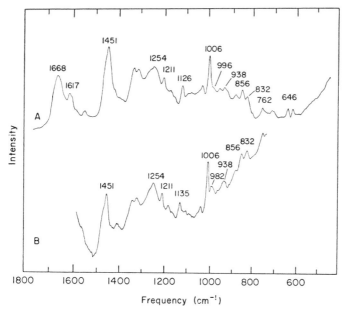

Fig. 18. Raman spectra of alpha-casein. (A) Freeze-dried solid, slit width 5 cm^{-1}, scan rate 20 cm^{-1} min^{-1}, laser power at sample 400 mW, time constant 4 s. (B) 5 % aqueous solution at pH 8.0, same as above except laser power is 600 mW (Ref. 60).

However, in the ionizable polypeptides in the random coil conformation the amide III frequency ranges from 1243 to 1249 cm^{-1}. This difference between the proteins and polypeptide amide III frequencies could mean that the average local conformation of the polypeptide chain, i.e. the distribution of rotation angles about the C—C$_\alpha$ and C$_\alpha$—N bonds, may differ. A weak line can be detected at 938 cm^{-1}. Recall that the spectrum of native tropomyosin exhibited an intense line at 940 cm^{-1} that decreased in intensity during the helix to coil transition. Perhaps the spectra of disordered proteins will be generally characterized by weak lines in this region of the spectrum.

4.2.7 Serum albumin

Bovine serum albumin (BSA) is a globular protein of molecular weight 66,000 g mol^{-1} and crosslinked with 18 cystine residues. The molecule's secondary structure consists of approximately 55 percent alpha helix and 45 percent disordered structure as determined by optical rotatory dispersion.[85] Below pH 4.3 and above 10.5 the prolate ellipsoid shaped molecule undergoes reversible expansion, largely electrostatic in origin.[93] Conformational studies of human serum albumin (HSA) and BSA upon denaturation at high pH and high temperature have been controversial. While most of the participants agree that a partial helix to coil transition is involved, the appearance of the beta sheet conformation in the denatured protein has been questioned.

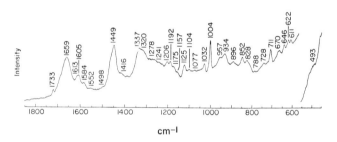

Fig. 19. Raman spectrum of native BSA in solid state (Ref. 64).

The Raman spectra of native BSA in aqueous solution was first interpreted by Bellocq, Lord and Mendelsohn.[63] They noticed the weak intensity of the amide III line and strong intensity of a skeletal mode at 938 cm^{-1}. As we have discussed previously for tropomyosin,[60] these features are characteristic of proteins having a large fraction of alpha helix. Lin and Koenig[64] obtained both the solution and solid state spectrum of the native protein and also investigated the pH and thermal denaturation. The spectrum of the native protein in the solid state appears in Fig. 19. Upon dissolution of BSA in water the conformationally sensitive 938 cm^{-1} line increased in intensity approximately 40 percent while the amide III line sharpened slightly. The authors contend that

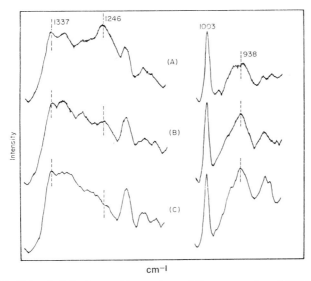

Fig. 20. Amide III and skeletal stretching regions of Raman spectra of 5% BSA solutions heated at (A) 70°, (B) 62°, and (C) 38°C for 3 h but spectra recorded at room temperature (Ref. 64).

the helical content may increase slightly or the helical perfection may increase upon dissolution. The solid state protein used in this study was not crystalline but a freeze-dried powder and Raman studies by Yu *et al.*[57] have shown that freeze-drying may partially dehydrate the protein and induce conformational changes. It would be interesting and more conclusive to ascertain if similar spectral changes were observed upon dissolving crystalline BSA.

The spectral changes that occur after heat denaturation are summarized in Fig. 20. Upon heating the aqueous protein solution the amide III line near 1250 cm^{-1} increases in intensity while the 938 cm^{-1} decreases, and these changes reflect the helix to coil transition. In Fig. 21 the intensities of these two lines are

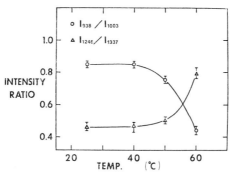

Fig. 21. Dependence of intensity ratios I_{938}/I_{1003} and I_{1247}/I_{1337} on temperature. 5% BSA solutions heated at various temperatures for 3 h and then cooled to room temperature prior to recording of Raman spectra (Ref. 64).

plotted against temperature. These spectral changes are partially reversible. When the BSA solution is heated above 60°C and cooled to room temperature a clear gel forms and the spectrum now reveals a line at 1240 cm⁻¹, which may be associated with the beta sheet conformation. This line appears more strongly when the gel is formed by cooling a 90°C solution. The appearance of the 1240 cm⁻¹ line along with gel formation strongly suggest that thermal denaturation of BSA involves the formation of both intermolecular bonds and beta sheet polypeptide chains. The complete mechanism of thermal denaturation proposed by the authors appears in Fig. 22. This mechanism features a reversible unfolding

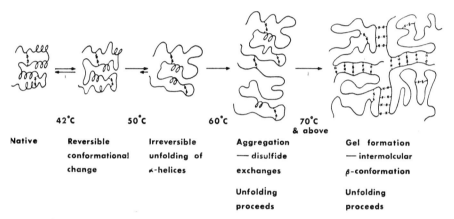

42°C	50°C	60°C	70°C & above
Native	**Reversible**	**Irreversible**	**Aggregation**	**Gel formation**

Native

Reversible conformational change

Irreversible unfolding of ∝-helices

Aggregation — disulfide exchanges — Unfolding proceeds

Gel formation — intermolcular β-conformation — Unfolding proceeds

Fig. 22. Mechanism for heat denaturation of bovine serum albumin (Ref. 64).

of the alpha helices from 42° to 50°C, irreversible unfolding from 50° to 60°C as well as the formation of the beta sheet and intermolecular bonds above 70°C.

The authors also followed the acidic and alkaline denaturation of BSA. Both processes involve a partial helix to coil transition but not to the extent observed during thermal denaturation.

4.2.8 Beta lactoglobulin

This is a milk protein of 35,000 g mol⁻¹. Below pH 3.5 the protein dissociates into two dimers having half the original molecular weight. Aschaffenburg and Drewry[86] and Bell[87] identified three genetic variants designated A, B and C that differ only slightly in amino acid content. Optical rotatory dispersion (ORD)[88,89] and circular dichroism (CD)[90] studies reveal a secondary structure similar to that for ribonuclease discussed earlier. The alpha helical fraction is approximately 10–20 percent, the disordered fraction near 50 percent and the rest is beta sheet.[90] All beta lactoglobulin variants undergo an increase in levorotation when exposed to alkaline solution. ORD and CD spectra indicate that the alpha helical content remains nearly constant as the pH goes above 6.0, but near pH 9.0 the beta sheet region is converted into the random coil.

The Raman spectrum of beta lactoglobulin was first published by Bellocq, Mendelsohn and Lord.[63] A spectrum of the native protein appears in Fig. 23

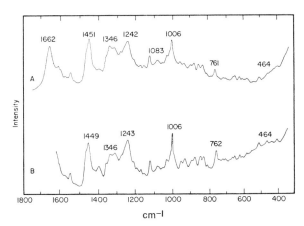

Fig. 23. Raman spectra of native beta lactoglobulin. (A) Single crystals of beta lactoglobulin AB: slit width 6 cm⁻¹; scan rate 5 cm⁻¹ min⁻¹, laser power at sample 250 mW at 5145.3 Å, time constant 4 s. (B) 10% solution of beta lactoglobulin AB in 0.1 M NaCl, pH 6.0: slit width 4 cm⁻¹, scan rate 10 cm⁻¹ min⁻¹, laser power 400 mW, time constant 4 s (Ref. 91).

for the solution and freeze-dried solid states. The amide I line appears at $1662 \ cm^{-1}$ in the spectrum of the solid protein but cannot be observed in the solution spectrum due to water interference. This value agrees with the $1667 \ cm^{-1}$ amide I line observed for ribonuclease and reflects the large percentage of residues in the beta sheet and disordered conformations. Only a single amide III is observed, at $1242 \ cm^{-1}$, whereas two distinct lines could be resolved in the ribonuclease spectrum at 1239 and $1265 \ cm^{-1}$. This difference in the amide III regions of the two spectra may reflect the sensitivity of this mode to the degree of uniformity of each separate chain conformation. The amide III lines for the beta sheet and disordered conformations in proteins occur from $1220-1240 \ cm^{-1}$ and $1245-1265 \ cm^{-1}$, respectively. Perhaps if we relax the order of a beta sheet segment and allow it to approach progressively the random coil state its amide III frequency will continuously shift upwards in frequency until it cannot be spectroscopically differentiated from the random coil. One could conceivably prepare random copolymers of beta sheet forming residues and randomizing residues that would introduce defects into the beta sheet structure and in this way follow the effect of defects on the normal vibrations. The inability to resolve two lines in the beta lactoglobulin spectrum may be a consequence of an abnormally high beta sheet amide III line that overlaps with the amide III line for the disordered regions of the proteins. If the beta sheet chains were only several residues long, had weak or distorted hydrogen bonds, or perhaps other defects, an increase in frequency of this mode over the value normally observed for highly regular beta sheet structures might occur.

Frushour and Koenig[91] followed the pH denaturation of beta lactoglobulin. As mentioned previously raising the pH above 9.0 destroys the beta sheet regions of the protein while not affecting the small alpha helical content. The spectra of the protein in the native state at pH 6.0 and the denatured state at pH 11.0 are compared in Fig. 24. Denaturation broadens the amide III line, shifts the

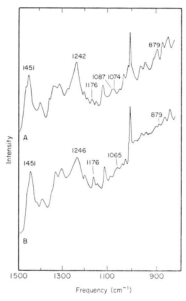

Fig. 24. Raman spectra of native and denatured beta lactoglobulin B. (A) Native protein at pH 6.0 in 0.1 M NaCl, 5% concentration. (B) Denatured protein, same as above except pH is 11.0.

maximum to 1246 cm^{-1} and decreases the intensity. Since the amide III line for the disordered conformation is generally observed from 1245 to 1265 cm^{-1} in globular proteins and is less intense than for the beta sheet, the spectral changes upon denaturation are consistent with the disordering of the beta sheet regions. The amide III intensity and half-width are plotted in Fig. 25 as a function of pH and the onset of denaturation near pH 9.0 can be readily discerned. When the pH is raised further to 13.5 the onset of hydrolysis can be detected by the yellowing of the solution as the remaining ordered structure collapses.[90] The amide III line now shifts upwards to 1257 cm^{-1} and resembles in this region the spectra of alpha casein and denatured tropomyosin.

The effects of denaturation extend to the local environments of the tyrosine and tryptophan residues. In the spectrum of the native protein three lines of medium intensity appear at 883, 862 and 832 cm^{-1}. The line at 883 cm^{-1} arises from tryptophan and the other two are from tyrosine. Yu *et al.*[92] have shown that the relative intensities of the two tyrosine lines depend upon the local environment of the residue. If a tyrosine has a hydrophobic environment, i.e. is buried within the protein core away from solvent, then the 832 cm^{-1} line appears slightly more intense than the 862 cm^{-1} line. After a denaturation that perturbs the residue and possibly makes the environment more hydrophilic, the intensity ratio reverses. From Fig. 24 one sees that the denaturation of beta lactoglobulin causes the 832 cm^{-1} line to increase in intensity relative to

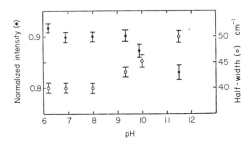

Fig. 25. The effect of pH on the intensity and half width of the amide III line in the Raman spectrum of beta lactoglobulin AB, 0.1 M NaCl. The intensities were measured as peak heights and normalized to the 1451 cm^{-1} line (Ref. 91).

the 862 cm^{-1} line. This suggests that the tyrosine environments become more hydrophobic after denaturation, and may be the result of aggregation known to occur at high pH. The intensity of the tryptophan line at 883 cm^{-1} decreases in intensity drastically after denaturation and this may also reflect a change in the local residue environment. Studies with specific model peptides will be required to understand the mechanism of the intensity changes for the tyrosine and tryptophan lines.

4.3 Fibrous Proteins

Fibrous proteins occur in a large number of animals, from insects to man, when load-bearing elements are required. Examples are cocoon silk, spider web, insect cuticle, horn, hair and internal structures like tendon, bone and blood vessels. These proteins are highly ordered and anisotropic so that large tensile strengths can be achieved along the fiber axis. Usually these proteins consist of a major secondary structure rather than a distribution of several structures as was the case with the globular proteins. They are potentially excellent systems for interpreting the Raman spectra of globular proteins and establishing structure-frequency correlations. Unfortunately, good spectra are very elusive because of the high background emission that can be attributed to various impurities.

4.3.1 Keratins

Lin and Koenig[65] analyzed the Raman spectra of several α-keratins including wool, porcupine quill, human hair, silk and feather. The alpha helix to beta sheet transition induced by the stretching or LiBr contraction were followed by Raman spectroscopy. The α-keratins contain alpha helices in the form of microfibrils which are imbedded in an amorphous matrix.[94] The X-ray diffraction pattern of α-keratins shows the 1.5 Å residue repeat characteristic of the alpha helix.[95] The infrared spectra of these proteins reveal an amide I mode near 1650 cm^{-1} with parallel dichroism and an amide II mode near 1550 cm^{-1} with perpendicular dichroism.[96] Four types of beta sheet or β-keratins can be differentiated, based upon the wide angle X-ray residue repeat distance,[97-100]

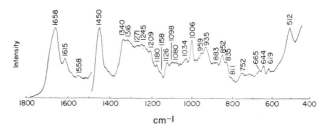

Fig. 26. Raman spectrum of wool in native state.

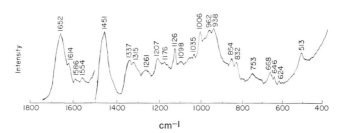

Fig. 27. Raman spectrum of wool in deuterium oxide (Ref. 65).

which ranges from 3.1 Å in feather to 3.5 Å in silk. The infrared amide bands characteristic of the beta sheet are found in the β-keratins, i.e. a perpendicular dichroic amide I band at 1630 cm^{-1} and a parallel dichroic amide II band at 1535 cm^{-1}.[96]

The Raman spectra of wool in the native state and immersed in D$_2$O are shown in Figs. 26 and 27, respectively, and the line assignments are listed in Table 7. The amide I mode appearing at 1658 cm^{-1} reflects the high alpha helical content of native wool, though the asymmetry of the line shape probably arises from the disordered chains in the amorphous matrix. As expected, for predominantly alpha helical proteins, the amide III region is broad and not well defined. After deuteration the amide III region becomes weaker with a broad peak centered at 1261 cm^{-1}. The authors attribute this line to alpha helices that have not been completely deuterated because they reside in hydrophobic regions of the microfibril.[101] The new intense line appearing at 962 cm^{-1} in the spectrum of the deuterated wool is assigned to the amide III mode of the

TABLE 7
Frequencies and tentative assignments of Raman spectra of wool in native state and in deuterium oxide[a]

Wool	Wool in D$_2$O	Tentative assignments
1658 vs		Amide I
	1652 vs	Amide I
1615 m	1614 m	Tyr and Trp
	1586 vw	Phe
1558 vw	1554 vw	Trp
1450 vs	1451 vs	CH$_2$ and CH$_3$ bending mode
1340 sdb	1337 sdb	CH bend, Trp
1316 sdb	1315 sdb	C$_\alpha$H bend
1271 mvbr		
	1261 wvbr	Amide III (α-helical)
1245 mvbr		Amide III (disordered)
1209 m		Tyr and Phe
	1207 m	D$_2$O
1180 w	1176 w	Tyr
1158 vwsh		Skeletal stretch (β)
1226 m	1126 m	
1098 wbr	1098 wbr	C—N stretch
1080 wbr		
1034 w	1035 w	Phe
1006 s	1006 s	Phe and Trp
	962 sdb	Amide III'
959 w		CH$_2$ rock
935 m	938 sdb	Skeletal stretch (a), residue C—C stretch
883 wbr		Skeletal stretch (β)
852 mdb	854 mdb	Tyr
835 mdb	832 mdb	Tyr
811 vw		
752 wbr	753 wbr	Trp
665 vw	668 w	Cys C—S stretch
644 w	646 w	Tyr
619 vw	624 vw	Phe
512 m	513 m	Cys S—S stretch

[a] Ref. 65.

deuterated, disordered chains in the amorphous regions. Spectral differences between the two spectra also show evidence of swelling by D_2O. The intensities of the lines at 852 cm^{-1} and 644 cm^{-1} relative to the 835 cm^{-1} line decrease after deuteration. These three lines are tyrosine ring vibrations. Yu and co-workers,[57,92] as mentioned previously, attribute similar tyrosine line intensity changes to a perturbation of the residues in hydrophobic regions of the polymer. The environment of the tyrosine residues would be expected to change upon swelling in D_2O.

In Fig. 28 the spectra of porcupine quill, wool and human hair are compared

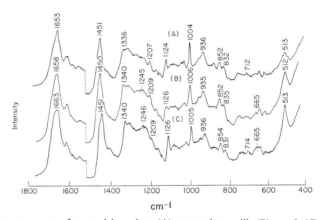

Fig. 28. Raman spectra of several keratins: (A) porcupine quill, (B) wool, (C) human hair (Ref. 65).

and the amide I frequencies appear at 1653, 1658 and 1663 cm^{-1}, respectively. For this same series the sulfur content (in the disulfide bonds) is 2.7, 3.7 and 3.9 percent, respectively.[100] Since addition of disulfide bonds increases the amorphous content of the keratin, the progressive increase in amide I frequency can be interpreted as a decrease in the alpha helical content. The 1653 cm^{-1} Raman frequency for the amide I line of quill agrees with the frequency in the infrared spectrum reported by Ambrose and Elliott.[102] Hair has the least amount of alpha helical structure and an infrared amide I band appearing at 1663 cm^{-1} and assigned to the disordered structure[96] agrees well with the Raman frequency. A similar correlation between the amide III line and the helical content may be derived. The amide III line for human hair is the most intense of the three keratins and the distinct line at 1246 cm^{-1} may be assigned to disordered structure.[48,57,60,103]

Wool undergoes an alpha helix to beta sheet conformational transition upon stretching.[104] The Raman spectra of wool stretched 30 percent and 60 percent are compared to the unstretched state in Fig. 29. The amide I line shifts from 1658 to 1672 cm^{-1} and the latter frequency is associated with both the beta sheet and disordered conformations. The half width of the amide I line

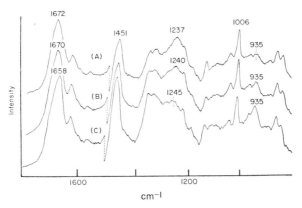

Fig. 29. Raman spectra of progressively elongated wool. (A) 60%, (B) 30%, (C) 0% elongation (Ref. 65).

increases approximately 25 percent upon stretching. Contrary to this observation, recall that the half width of this mode decreased considerably after the alpha helix to beta sheet transition of insulin.[71] The authors suggest that an intermediate conformational state may be introduced by stretching which leads to the observed broadening. During the course of the transition the amide III line increases in intensity and decreases in frequency from 1245 to 1237 cm^{-1}. The latter frequency can be assigned to the beta sheet conformation. Finally, the intensity of the 935 cm^{-1} skeletal mode decreases upon stretching. This line behaved in a similar manner during the alpha helix to coil transitions in tropomyosin[60] and BSA[64] previously discussed. Apparently the line has strong intensity for proteins containing large fractions of alpha helix, and conversion to the disordered conformation or beta sheet reduces the intensity of the line. In keratins, the authors contend that the 935 cm^{-1} band contains a component assigned to a side chain C—C stretching vibration that is not conformationally sensitive.

Supercontraction of wool with steam leads to the cross-beta conformation containing folded chains that results in polypeptide chains placed perpendicular to the fiber axis. Except for the folds and reversal of orientation, this conformation is essentially the typical antiparallel beta sheet arrangement. The infrared and Raman spectra of these two conformations are very similar, and need not be discussed. However, supercontraction by LiBr leads to a different Raman spectrum. The amide I line appears at 1666 cm^{-1}, instead of 1671 cm^{-1} for the steam contracted sample, and is much broader. The amide III line appears at 1249 cm^{-1} instead of 1237 cm^{-1}. These features indicate that LiBr introduces much disorder in the polypeptide chain conformation, and is therefore similar to the effect on lysozyme discussed previously.

Bombyx mori silk fibroin in the native state contains a large beta sheet fraction. High uniformity of the beta sheet regions would be expected from the chemical analysis[105] and X-ray pattern.[99] Suzuki[106] has identified the antiparallel

beta sheet conformation from the infrared spectrum. The amide I mode appears
at 1667 cm^{-1} in the Raman spectrum (Fig. 30) with a half width of 27 cm^{-1}, as
compared to 76 cm^{-1} for β-keratin, and reflects the uniformity or narrow
distribution of chain conformations in silk. In the amide III region an intense
line appears at 1229 cm^{-1} and is assigned to the $\nu(0, 0)$ amide III mode on the

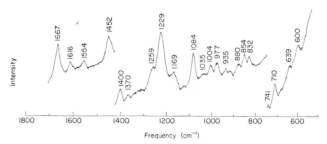

Fig. 30. Raman spectrum of *Bombyx mori* silk fibroin (Ref. 65).

basis of the intensity.[32] Frushour and Koenig[38] obtained the Raman spectra
of two sequential polypeptides, i.e. poly(Ala-Gly) and poly(Ser-Gly), which are
models for beta sheet regions of silk. The amide I lines of these two polypeptides appeared at 1665 and 1668 cm^{-1}, respectively, in good agreement with
the frequency for silk.

In summary, the Raman spectra of the keratins and globular proteins are
similar in their response to the conformational transitions of the protein and the
frequency and intensity of the side chain lines. The conformationally sensitive
Raman lines of the keratins are summarized in Table 8.

TABLE 8
Conformationally sensitive Raman modes of fibrous proteins

Predominant conformation	Amide I	(cm^{-1}) Amide III	Skeletal modes
α-helical			
Porcupine quill	1653 s (sp)	~1260 w	935 s
Wool	1658 s (bd)	1261 w (a')	935 s
		1245 m (disordered)	
β-form			
β-Keratin	1672 s	1237 s	
Cross-β	1671 s	1239 s	
(steam supercontracted wool)			
Silk	1667 s	1229 s (β)	880 w (β)
		1259 w (disordered)	1169 m (β)
Feather	1669 s	1240 s (β)	879 w (β)
		1274 m (disordered)[a]	1162 m (β)
Disordered			
LiBr supercontracted wool	1666 s	1249 s	

[a] See text for explanation of discrepancy (Ref. 65).

4.3.2 Collagen

This fibrous protein is the major constituent of the load-bearing organs including tendon, bone and blood vessels. Approximately 30 percent of the amino acid residues are glycine and another 20 percent are either proline or hydroxyproline.[11] These latter two residues restrict the conformational freedom of the chain because the prolidine ring prevents rotation about the C_α—C bond (ψ rotation angle) and thereby restricts the collagen molecule to certain conformations. The basic structural unit of a collagen fibril is the tropocollagen molecule that has dimensions of 15×3000 Å and is formed by twisting together three separate polypeptide chains into a three stranded coiled coil with a right handed screw sense.[11] The three (alpha) chains appear to be very similar. They form a 3_1 left-handed helix that is stabilized by the proline and hydroxyproline residues. The amino acid sequence and electron microscope staining pattern of the alpha chains reveals a biphasic distribution of polar and nonpolar amino acids.[107] After a short teleopeptide peptide every third residue is glycine but in the non-polar regions of the chain one of the other two residues is likely to be proline or hydroxyproline. In the polar regions ionizable residues such as lysine or glutamic acid may be found with the glycine. Only the non-polar regions are thought to form the 3_1 helix while the polar regions are possibly unordered.[107]

The Raman spectra of the collagen in ox tibia bone[108] and tooth dentyne[109] was published by Rippon, Walton and coworkers. Using a relatively weak laser only low resolution spectra could be obtained. Attempts to obtain the spectrum of purified collagen were unsuccessful. Frushour and Koenig[103] were able to obtain good quality spectra of collagen and gelatin from several sources using much improved equipment. The spectra of collagen and gelatin appear in Fig. 31 and the line assignments are listed in Table 9. Recourse to the superposition

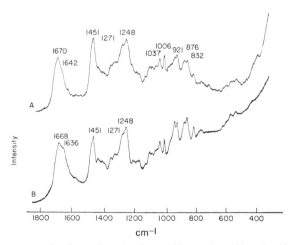

Fig. 31. Raman spectra of collagen from bovine achilles tendon (A) and calf skin gelatin (B): slit width 8 cm^{-1}, scan rate 10 cm^{-1} min^{-1}, laser power at sample 300 mW in (A) and 600 mW in (B), time constant 10 s (Ref 103).

TABLE 9
Raman lines in collagen and related spectra

Collagen (B.A.T.)[a]	Gelatin (10% aqueous solution)	Mixture of all amino acids pH 13	Mixture of all amino acids pH 2	Mixture of aromatic amino acids pH 13	Mixture of non-aromatic amino acids pH 13	Gelatin (10% in D_2O)	Assignment
			1746 s				ν(C=O)
1670 s	1668 s					1664 s	Amide I
1642 s sh	1686 s sh					1645 s sh?	Amide I
	1608 w	1611 m	1611 m	1604 m	1601 m	1611 w	Phe, Tyr
	1566 w	1589 m	1589 m	1585 m	1576 m		Pro, Hypro
				1483 w			
1464 s sh	1464 s sh		1460 s		1457 s	1464 s	δ(CH$_3$, CH$_2$)
1451 s	1451 s	1450 s	1438 s	1451 s			δ(CH$_3$, CH$_2$)
1422 m	1422 m						in pl. bend of carboxyl OH
		1415 vs		1396 s	1408 s	1415 m	ν_s(COO$^-$)
1392 m	1399 m						
	1389 m						
1343 m	1347 m	1353 s	1353 m	1330 m	1350 m	1347 m	γ_w(CH$_3$, CH$_2$)
1314 m	1320 m	1323 m	1324 m	1320 m	1320 m	1330 m	γ_t(CH$_3$, CH$_2$), δ(C$_\alpha$—H)
1271 s	1271 s		1271 w	1274 w		1274 w	Amide III
1248 s	1248 s	1245 w	1235 m	1238 w		1247 w	Amide III
1211 w	1211 w	1211 w	1211 m	1211 m		1211 s (D$_2$O)	Hypro, Tyr
	1198 w						
1178 w	1182 w	1188 w	1188 w	1188 m	1178 w	1185 w	Tyr
1161 w	1165 w						
1128 w	1128 w				1145 w	1135 w	NH$_3^+$

							Assignment
1067 w	1064 w					1057 w	o.pl. bend of carboxyl OH
	1051 w			1051 w			Pro
			1044 s				
1037 m	1037 m	1037 w		1034 m			
					1024 w		
1006 m	1006 m	1006 m	1006 m	1006 s		1006 s	Phe
						993 m	Amide III'
		986 w		982 m			
966 w	969 sh		969 m			966 m	Amide III'
938 m	942 s		942 s		942 s	942 s	ν(C—C) of residues
921 m	925 s					925 s	ν(C—C) of protein backbone
918 w		907 vs	918 m	918 s	908 s		ν(C—C) of Pro ring
890 w	890 w						
876 m	880 s		870 vs	873 s		884 s	ν(C—C) of Hypro ring
856 m	863 s			859 s	866 s		ν(C—C) of residues
		856 vs				856 s	ν(C—C) of Pro ring
821 w	818 m		825 vs				ν(C—C) of residues
		783 m		786 w	783 w	814 m	ν(C—C) of backbone
769 w	769 w	769 w	741 s		759 wBr	759 wBr	Phe
			650 m	646 w	668 w		
622 w	625 w	629 w	629 m	629 w			
		593 w	593 w		593		
568 w	572 w		575 w			575 w	Pro
533 w	536 w	543 s	522 m		540 w	540 w	Hypro
		518 m	504 s		522 w		
472 w		447 w		479 m			
				447 m			
	425 w		414 m			407 w	Pro
396 w				345 w			Pro

a B.A.T. (Bovine Achilles tendon). Ref. 103.

technique (discussed previously for lysozyme) was necessary to assign the Raman lines since the amino acid composition of collagen differs significantly from those proteins previously studied.

Two lines appear in the amide I and amide III regions. Superposition and deuteration spectra show that both the 1271 and 1248 cm^{-1} lines can be assigned the amide III mode. The similarity of the amide III regions for collagen and gelatin is surprising since the gelatin should have no significant amounts of order. However the collagen sample, which was fibrous, could have been denatured by the laser beam. In an attempt to avoid this possibility the authors examined the spectra of water soluble calf skin collagen in aqueous solution before and after conversion to gelatin by thermal treatment. Small but definite changes in the spectra were detected. In the amide III region of the collagen solutions shown in Fig. 32 the line at 1248 cm^{-1} has shifted to 1251 cm^{-1} and decreased in intensity slightly upon conversion to gelatin. The amide III line for polyglycine in the 3_1 helical conformation[110] appears at 1245 cm^{-1} and from 1243 to 1250 cm^{-1} for the random coil forms of ionizable polypeptides.[33,37,47] Therefore the authors tentatively concluded that the 1248 cm^{-1} line in the collagen spectrum can be assigned to the amide III mode of the 3_1 helix. However, Raman spectroscopy apparently cannot easily distinguish between the 3_1 and random conformations in collagen. The amide III line at 1270 cm^{-1} did not change upon denaturation, and the presence of two lines in the amide III region

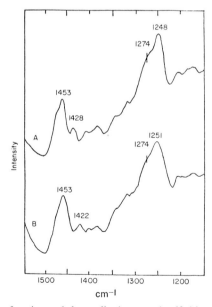

Fig. 32. Raman spectra of native and thermally denatured calf skin collagen. (A) Native, 2% concentration at pH 4.0, 25°C. (B) Denatured, same as above but heated at 70°C for 2 h and cooled to 25°C (Ref. 103).

of the denatured product (gelatin) remains puzzling. The circular dichroism spectrum revealed only the random coil. Gelatin will still have a biphasic distribution of amino acids, however, and presence of proline and hydroxyproline in the non-polar regions will restrict the number of available conformations over those available to the polar regions. Perhaps such a difference in local conformation, or more precisely, local distribution of ϕ and ψ angles, leads to the splitting of the amide III region in gelatin with the 1251 cm^{-1} line assigned to the disordered, proline-rich non-polar regions and the 1270 cm^{-1} line to the proline-poor, polar regions.

4.3.3 Elastin

This fibrous protein occurs in conjunction with collagen when elastic properties of a structural organ are required. Examples would be ligaments, the aorta of mammals, and skin. Previous investigations have revealed little of ordered secondary structures in elastin. The Raman spectrum of bovine ligamentum nuchae elastin analyzed by Frushour and Koenig[103] agrees well with this

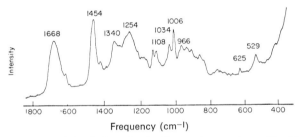

Fig. 33. Raman spectrum of elastin from bovine ligamentum nuchae: slit width 4 cm^{-1}, scan rate 10 cm^{-1} min^{-1}, laser power at sample 300 mW at 5145.3 Å, time constant 10 s (Ref. 103).

assertion. The amide I and III lines appear at 1668 and 1254 cm^{-1}, respectively. Similar frequencies were observed for other disordered proteins including alpha casein, denatured tropomyosin and denatured lysozyme, and discussed previously. In the latter case amide III was observed close to 1245 cm^{-1}. This difference might again reflect the sensitivity of the amide III to the distribution of ϕ, ψ angles of the disordered chain. There appears to be no unique amide III line for the random coil (disordered) conformation in proteins.

4.4 Comparison of Protein Structure in the Crystalline and Solution States

The most elegant and informative protein structural models have resulted from X-ray diffraction analysis of protein crystals in contact with the mother liquor. These crystals often contain over 50 percent by weight water and could almost be considered very concentrated solutions. Still there is much disagreement over the magnitude of difference between the solution and the crystalline conformations. McKenzie[83] has discussed the 'motility' flexibility in the conformation of native state proteins. In view of the small free energy minima for native state protein conformations McKenzie suggests that significant conformational change could occur upon dissolution of a crystalline protein. Other people (see review article by Rupley)[111] put forth a contrary opinion and cite similar reaction rates for biological activity of protein molecules in the solution and crystalline states. Raman spectroscopy has been recently applied to this problem because of its sensitivity to conformation of the polypeptide chain and local environment of the aromatic amino acid residues. The technique can be applied equally well to solutions, crystals and powders under different relative humidities.

The first comparison of Raman spectra of a protein in the solution and solid states was done with chymotrypsinogen and ribonuclease by Koenig and Frushour.[1,68] The solid was not composed of true single crystals but rather of the freeze-dried powder that may have been partially dehydrated. Definite spectral differences between the solution and solid state could be detected. Upon dissolution of chymotrypsinogen the intensity of the amide III line increased approximately 40 percent. When the freeze-dried ribonuclease was dissolved in water a curious result was observed; the Raman lines became sharper, suggesting more regular hydrogen bonding in the solution than freeze-dried solid state. Yu, Jo and Liu[57] re-examined the ribonuclease spectra under higher resolution. The sharpening of the Raman lines was confirmed and a significant change in the intensity of several tyrosine lines was detected. Both effects are apparent in Fig. 10. In the spectrum of the powder (zero and 100 percent relative humidity) the 852 cm^{-1} tyrosine line is slightly more intense than the adjacent tyrosine line at 832 cm^{-1} but upon dissolution the intensity ratio reverses. Also the intensity of the 644 cm^{-1} tyrosine line decreases. The authors suggest that changes in the environment of tyrosine buried in the protein interior are responsible for the intensity change and cannot be attributed to simple solvation effects of the single surface tyrosine.

In order to test this hypothesis they examined the effect of freeze-drying on several other proteins having 'buried' tyrosines. The snake venom proteins neutrotoxin alpha, cobramine A and cobramine B have respectively one, three and four tyrosines, all thought to be 'buried' in the protein interior.[92,112] Upon freeze-drying, the intensity of the 644 cm^{-1} tyrosine line increases though the ratio of the 852 and 832 cm^{-1} line intensities does not change significantly. Presumably the partial dehydration induces conformational changes in the polypeptide backbone that in turn perturb the local environments of the tyrosine residues. A more specific mechanism for the intensity changes has not been proposed by the authors.

In order to determine whether all of the spectral changes observed upon freeze-drying were characteristic of the solid state or were caused entirely by dehydration, Yu and Jo[112] obtained the spectrum of a single crystal of ribonuclease immersed in ethanol. The spectra of the crystal and ethanol are compared in Fig. 34 and in Fig. 35 the 644 cm^{-1} tyrosine line and amide III region for the

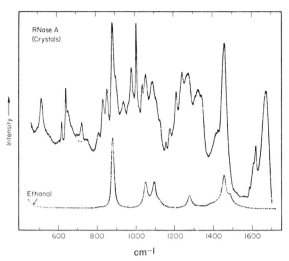

Fig. 34. Raman spectra of RNase A crystalline powder (upper) and 88% ethanol solution (lower). Conditions for the upper curve: crystalline powder in equilibrium with the vapor of 88% ethanol solution; $\Delta\sigma$, 4 cm^{-1}; signal strength, s = 2500 cps; scan rate, γ = 10 cm^{-1} min^{-1}; standard deviation in frequency determination, s.d. = 1%; laser power at sample, p = 95 mW. The strong line at 983 cm^{-1} is due to the SO_4^{2-} ions in the sample (Ref. 112).

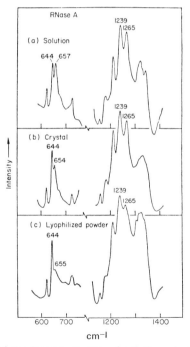

Fig. 35. A comparison of the Raman spectra of solution, single crystal, and lyophilized RNase A in the 600–750 and the 1180–1400 cm^{-1} regions (Ref. 112).

crystalline, freeze-dried and solution states are compared. The only significant spectral change observed upon crystallization (compared to the solution state) is the increase in the intensity of the 644 cm^{-1} line. The similarity of the amide III regions suggests that no detectable change in backbone conformation occurs upon crystallization but freeze-drying definitely broadens the separate components in the amide III region and this may arise from a slight perturbation of the backbone conformation and/or hydrogen bonding network.

Yu *et al.* have also examined single crystal Raman spectra of lysozyme,[58] carboxypeptidase,[112] insulin[93] and alpha lactoglobulin.[113] In the cases of lysozyme and alpha lactoglobulin the authors concluded that freeze-drying altered the backbone conformation but there was no conformational change in the backbone upon crystallization from solution. The spectra of alpha lactoglobulin in the crystalline and freeze-dried states are compared in Fig. 36 and the changes attributed to freeze-drying are denoted. Freeze-drying causes the three components of the amide III line (1274, 1260 and 1238 cm^{-1}) to coalesce into a broad band centered at 1260 cm^{-1}. The broadening of the disulfide S—S stretching line at 508 cm^{-1} also indicates a change in the backbone conformation.

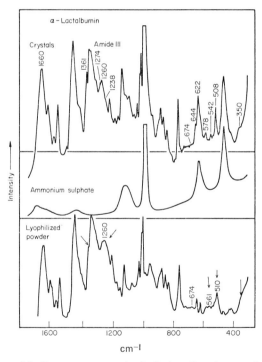

Fig. 36. Comparison of the Raman spectrum (top) of α-lactalbumin crystals with that (bottom) of lyophilized powder. The middle curve shows the contribution of ammonium sulfate solution to the top spectrum. All the spectra were taken with the same spectral slit width of 3 cm^{-1} and a scanning speed of 30 cm^{-1} min^{-1}. Laser power at the samples is about 70 mW (Ref. 113).

Definite differences are observed between the crystalline and solution state spectra of carboxypeptidase A[112] and beta lactoglobulin.[91] Carboxypeptidase has the distinction of having the only section of parallel chain beta sheet found in nature.[9] The amide III regions of the solution and crystalline spectra of carboxypeptidase are compared in Fig. 37. Two amide III components are

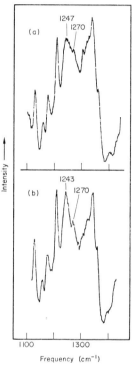

Fig. 37. Raman spectra of carboxypeptidase A in the crystalline state (a) and in solution (b). Conditions for a: $\Delta\sigma$, 4 cm^{-1}; s, 2500 cps; scan rate, 25 cm^{-1} min^{-1}; s.d., 1%; power, 80 mW. Conditions for b: pH 7.00; c, 60 mg ml^{-1}; ionic strength 3 M NaCl; $\Delta\sigma$, 4 cm^{-1}; s, 1000 cps; scan rate, 10 cm^{-1} min^{-1}; s.d., 0.7%; power, 150 mW (Ref. 112).

observed in the latter spectrum, one at 1247 cm^{-1} assigned to both the disordered and beta sheet regions and one at 1270 cm^{-1} assigned to the alpha helices. Upon dissolution the broader component at 1247 cm^{-1} has become sharpened and the peak maximum has shifted to 1243 cm^{-1}. The frequency shift suggests a conformational change upon dissolution, perhaps brought about by regularization of hydrogen bond strengths that would also account for the sharpening of the amide III lines. Frushour and Koenig[91] observed an increase in intensity and sharpening of the amide III line upon dissolution of both freeze-dried and crystalline beta lactoglobulin. A 30 percent increase in the amide III line intensity relative to the methylene deformation line at 1450 cm^{-1} upon dissolution can be observed in Fig. 23. The causes of these spectral changes are unknown at this point but may relate to the concept of motility as discussed by McKenzie.[83] The hydration upon dissolution might increase the tendency towards hydrophobic bonding in the center of the molecule and result in conformational changes of the interior segments of the polypeptide chain.

5 CONCLUSION

The Raman spectra of proteins follow the variations in the secondary structure in a consistent and predictable manner. Three regions of the spectrum, the amide I, III modes and a skeletal stretching mode near 900 cm^{-1}, are sufficiently sensitive for formulating structure–frequency correlations that are in good agreement with the spectra of homopolypeptides. Some discrepancies in assignments must be resolved, however, before quantitative use of these spectral regions can become practical. The contribution of the relatively weak amide III line of the alpha helix to the overall amide III contour in the protein spectra must be determined. Also, there appear to be significant frequency variations for the amide III mode of a single conformation. This should not be surprising in view of the possibilities of defects, as these conformations are found in globular proteins. There is a tendency to group together those regions of the protein tertiary structures that do not resemble the known ordered conformations (alpha

TABLE 10
Summary of conformationally sensitive protein Raman lines

	Amide I	Amide III	Skeletal mode
Alpha helix	1655–1659 cm^{-1} (s)	1250–1280 cm^{-1} (w)	900–960 cm^{-1} (s)
Antiparallel beta sheet	1667–1672 cm^{-1} (s)	1229–1240 cm^{-1} (s)	900–960 cm^{-1} (w)
Random coil (disordered)	1665–1675 cm^{-1} (s)	1243–1265 cm^{-1} (m)	900–960 cm^{-1} (w)

helix, beta sheet and 3_1) as random and equate them to the random coil conformations of polypeptides. This simplification may lead one astray; one must recognize that the frequencies for the disordered regions of all proteins are not identical. When the X-ray structures of more proteins become available one should be able to determine the distribution of the ϕ and ψ angles in the disordered regions of different proteins and correlate this distribution with the amide I and III line intensity and half width.

Still, the frequency variations are not so severe as to prohibit qualitative analysis of the protein structure from the Raman spectrum. A summary of the important structure–frequency correlations is given in Table 10. Regions of the spectrum not included in the table are sensitive to conformation, though in an indirect way. The intensities of the tyrosine and tryptophan lines reflect their local environment. Of course, similar effects are noted in the circular dichroic, ultraviolet and fluorescence spectra of these residues. The Raman information is potentially much more useful, however, because the normal mode associated with the intensity change can be calculated and used to determine the interaction responsible for the change.

ACKNOWLEDGMENT

The authors gratefully acknowledge the financial support provided by the National Science Foundation under grant WM 52976 and the National Institutes of Health under grant GM 19975-01.

REFERENCES

(1) J. L. Koenig, *Raman Spectroscopy of Biological Macromolecules: A Review*, *J. Polymer Sci. Part D*, 1972.

(2) J. L. Koenig, *Raman Spectra of Chain Conformations* in *Advances in Raman Spectroscopy*, Vol. 1 (J. P. Mathieu, Ed.), Heyden, London, 1973.

(3) B. G. Frushour and J. L. Koenig, *Raman Spectroscopy of Polypeptides: A Review*, to be published.

(4) D. L. D. Caspar, *Design Principles in Organized Biological Structures* in *Principles of Biomolecular Organization* (G. E. W. Wolstenholme and M. O'Connor, Eds.), Little, Brown, Boston, 1966.

(5) R. L. Lumry and R. Bittonen in *Structure and Stability of Biological Macromolecules* (S. N. Timasheff and G. D. Fasman, Eds.), Marcel Dekker, New York, 1969.

(6) H. A. Scheraga, *Protein Structure*, Academic Press, New York, 1961.

(7) J. T. Edsall, *Adv. Protein Chem.* **14**, xii, 1959.

(8) A. M. Liquori, *Minimum Energy Conformations of Biological Polymers*, in *Principles of Biomolecular Organization* (G. E. W. Wolstenholme and M. O'Connor, Eds.), Little, Brown, Boston, 1966.

(9) E. R. Dickerson and I. Geis, *The Structure and Action of Proteins*, Harper and Row, New York, 1969.

(10) W. A. Hiltner and A. J. Hopfinger, *Biopolymers* **12**, 1197 (1973).

(11) A. G. Walton and J. Blackwell, *Biopolymers*, Academic Press, New York, 1973.

(12) G. D. Fasman, in *Poly-α-Amino Acids*, Chapter 10 (G. D. Fasman, Ed.), Marcel Dekker, New York, 1967.

(13) T. Ooi, R. A. Scott, G. Vanderkooi and H. A. Scheraga, *J. Chem. Phys.* **46**, 4410 (1967).

(14) M. Joly, *A Physico-Chemical Approach to the Denaturation of Proteins*, Academic Press, London, 1965.

(15) F. J. Boerio and J. L. Koenig, *J. Macromol. Sci.; Rev. Macromol. Chem.* **C7**, 209 (1972).

(16) S. Bhagavantam and T. Venkatarayuda, *Theory of Groups and Its Applications to Physical Problems*, Academic Press, New York, 1969.

(17) S. Krimm, *Adv. Polymer Sci.* **1**, 103 (1960).

(18) R. Zbinden, *Infrared Spectroscopy of High Polymers*, Academic Press, New York, 1964.

(19) A. Elliott, *Infrared Spectra and Structure of Organic Long-chain Polymers*, St. Martins Press, New York, 1969, p. 83.

(20) R. G. Snyder, *J. Mol. Spectrosc.* **37**, 353 (1971).

(21) B. Fanconi, B. Tomlinson, L. A. Nafie, W. Small and W. L. Peticolas, *J. Chem. Phys.* **51**, 3993 (1969).

(22) B. G. Frushour, unpublished results and with J. L. Koenig, *Biopolymers*, accepted for publication.

(23) P. P. Yaney, *J. Opt. Soc. Amer.* **62**, 1297 (1972).

(24) B. Bulkin, E. Cole and A. Noguerola, *J. Chem. Ed.* **51**, A273 (1974).

(25) T. Miyazawa, T. Shimanouchi and S. Mizushima, *J. Chem. Phys.* **29**, 611 (1968).

(26) B. Fanconi, E. Small and W. L. Peticolas, *Biopolymers* **10**, 1277 (1971).

(27) J. Jakes and S. Krimm, *Spectrochim. Acta* **27A**, 19 (1971).

(28) J. Jakes and S. Krimm, *Spectrochim. Acta* **27A**, 35 (1971).

(29) Y. Abe and S. Krimm, *Biopolymers* **11**, 1817 (1972).

(30) Y. Abe and S. Krimm, *Biopolymers* **12**, 1841 (1972).

(31) T. Miyazawa, *J. Chem. Phys.* **32**, 1647 (1960).

(32) S. Krimm and Y. Abe, *Proc. Nat. Acad. Sci. USA* **69**, 2788 (1972).

(33) J. L. Koenig and B. G. Frushour, *Biopolymers* **11**, 1871 (1972).

(34) P. Painter and J. L. Koenig, to be published.

(35) E. W. Small and W. L. Peticolas, *Biopolymers* **10**, 69 (1971).

(36) G. J. Thomas, *Biochim. Biophys. Acta* **213**, 417 (1970).

(37) T. J. Yu, J. L. Lippert and W. L. Peticolas, *Biopolymers* **12**, 2161 (1973).

(38) B. G. Frushour and J. L. Koenig, *Biopolymers*, accepted for publication.

(39) A. Rubcic and G. Zerbi, *Macromolecules* **7**, 754 (1974).

(40) A. Rubcic and G. Zerbi, *Macromolecules* **7**, 759 (1974).

(41) H. Suzuki, *Structure and Stability of Biological Macromolecules* (S. N. Timasheff and G. D. Fasman, Eds.), Marcel Dekker, New York, 1969, p. 575.

(42) R. J. Bell, *Introductory Fourier Transform Spectroscopy*, Academic Press, New York, 1972.

(43) D. L. Tabb and J. L. Koenig, *Biopolymers*, submitted for publication.

(44) C. C. F. Blake, G. A. Mair, A. C. T. North, D. C. Phillips and V. R. Sharma, *Proc. Roy. Soc. London, Ser. B* **167**, 378 (1967).

(45) D. Garfinkel and J. T. Edsall, *J. Amer. Chem. Soc.* **80**, 3818 (1958).

(46) M. C. Tobin, *Science*, **161**, 68 (1968).

(47) R. C. Lord and N. T. Yu, *J. Mol. Biol.* **50**, 509 (1970).

(48) R. C. Lord and R. Mendelsohn, *J. Amer. Chem. Soc.* **94** 2133 (1972).

(49) M. C. Chen, R. C. Lord and R. Mendelsohn, *J. Amer. Chem. Soc.* **96**, 3038 (1974).

(50) M. C. Chen, R. C. Lord and R. Mendelsohn, *Biochim. Biophys. Acta* **328**, 252 (1973).

(51) H. Brunner and H. Sussner, *Biochim. Biophys. Acta* **271**, 16 (1972).

(52) J. S. Cohen and O. Jardetsky, *Proc. Nat. Acad. Sci. USA* **60**, 92 (1968).

(53) N. T. Yu and B. H. Jo, *Arch. Biochem. Biophys.* **156**, 469 (1973).

(54) G. Kartha, J. Bello and D. Harker, *Nature* **213**, 862 (1967).

(55) C. H. Carlisle, R. A. Palmer, S. K. Mazumdar, B. A. Gorinsky and D. G. R. Yeates, *J. Mol. Biol.* **85**, 1 (1974).

(56) R. C. Lord and N. T. Yu, *J. Mol. Biol.* **51**, 203 (1970).

(57) N. T. Yu, B. H. Jo and C. S. Liu, *J. Amer. Chem. Soc.* **94**, 7572 (1972).

(58) N. T. Yu and B. H. Jo, *J. Amer. Chem. Soc.* **95**, 5033 (1973).

(59) N. T. Yu and C. S. Liu, *J. Amer. Chem. Soc.* **94**, 5127 (1972).

(60) B. G. Frushour and J. L. Koenig, *Biopolymers* **13**, 1809 (1974).

(61) B. G. Frushour and J. L. Koenig, *Biopolymers* **13**, 455 (1974).

(62) N. Greenfield and G. D. Fasman, *Biochemistry* **8**, 4108 (1969).

(63) A. M. Bellocq, R. C. Lord and R. Mendelsohn, *Biochim. Biophys. Acta* **257**, 280 (1972).

(64) V. J. C. Lin and J. L. Koenig, *Biopolymers*, submitted for publication.

(65) V. J. C. Lin and J. L. Koenig, *Biopolymers*, submitted for publication.

(66) B. W. Matthews, P. B. Sigler, R. Henderson and D. M. Blow, *Nature* **214**, 652 (1967).

(67) P. B. Sigler, D. M. Blow, B. W. Matthews and R. Henderson, *J. Mol. Biol.* **35**, 143 (1967).

(68) J. L. Koenig and B. G. Frushour, *Biopolymers* **11**, 2505 (1972).

(69) N. T. Yu, C. S. Culver and D. C. O'Shea, *Biochim. Biophys. Acta* **263**, 1 (1972).

(70) N. T. Yu and C. S. Liu, *J. Amer. Chem. Soc.* **94**, 3250 (1972).

(71) N. T. Yu, C. S. Liu and D. C. O'Shea, *J. Mol. Biol.* **70**, 117 (1972).

(72) M. J. Adams, T. L. Blundell, E. J. Dodson, G. G. Dodson, M. Vijayan, E. N. Baker, M. M. Harding, D. C. Hodgkin, B. Rimmer and S. Sheat, *Nature* **224**, 491 (1969).

(73) N. T. Yu, B. H. Jo, R. C. C. Chang and J. D. Huber, *Arch. Biochem. Biophys.* **160**, 614 (1974).

(74) E. J. Ambrose and A. Elliott, *Proc. Roy. Soc. A* **208**, 75 (1951).

(75) M. J. Burke and M. A. Rouguie, *Biochemistry* **11**, 2435 (1972).

(76) D. Givol, F. DeLorenzo, R. F. Goldberger and C. B. Anfinsen, *Proc. Nat. Acad. Sci. USA* **53**, 676 (1965).

(77) D. G. Steiner and P. E. Oyer, *Proc. Nat. Acad. Sci. USA* **57**, 473 (1967).

(78) W. W. Fullerton, R. Potter and B. W. Low, *Proc. Nat. Acad. Sci. USA* **66**, 1213 (1970).

(79) B. H. Frank and A. J. Veros, *Biochem. Biophys. Res. Comm.* **32**, 155 (1968).

(80) S. Lowey, *J. Mol. Biol.* **240**, 2421 (1965).

(81) J. Sodek, R. S. Hodjes, L. B. Smillie and J. Jurasek, *Proc. Nat. Acad. Sci. USA* **68**, 3800 (1972).

(82) W. Moffit and J. T. Yang, *Proc. Nat. Acad. Sci. USA* **42**, 596 (1956).

(83) H. A. McKenzie, *Adv. Protein Chem.* **22**, 173 (1967).

(84) G. D. Fasman, H. Hoving and S. N. Timasheff, *Biochemistry* **9**, 3316 (1970).

(85) E. Schecter and E. R. Blout, *Proc. Nat. Acad. Sci. USA* **51**, 695 (1964).

(86) R. Aschaffenburg and J. Drewry, *Nature* **176**, 218 (1955).

(87) K. Bell, *Nature* **195**, 705 (1962).

(88) S. N. Timasheff, R. Townend and L. Mescanti, *J. Biol. Chem.* **241**, 1863 (1966).

(89) H. Roels, G. Preaux and R. Lontie, *Biochimie* **53**, 1085 (1971).

(90) R. Townend, T. F. Kumosinski and S. N. Timasheff, *J. Biol. Chem.* **242**, 4538 (1967).

(91) B. G. Frushour and J. L. Koenig, *Biopolymers*, accepted for publication.

(92) N. T. Yu, B. H. Jo and D. C. O'Shea, *Arch. Biochem. Biophys.* **156**, 71 (1973).

(93) C. Tanford, J. G. Buzzel, D. G. Rands and S. A. Swanson, *J. Amer. Chem. Soc.* **77**, 6421 (1955).

(94) W. G. Crewther, R. D. B. Fraser, F. G. Lennox and H. Hindley, *Adv. Protein Chem.* **20**, 191 (1965).

(95) M. F. Perutz, *Nature* 167, 1053 (1951).

(96) S. Krimm, *J. Mol. Biol.* **4**, 528 (1962).

(97) L. Pauling and R. B. Corey, *Proc. Nat. Acad. Sci. USA* **39**, 253 (1953).

(98) W. T. Astbury and H. J. Woods, *Philos. Trans. R. Soc. London Ser. A* **232**, 333 (1933).

(99) J. O. Warwicker, *J. Mol. Biol.* **2**, 350 (1960).

(100) S. Seifter and P. M. Gallop, *The Proteins*, Vol. IV (H. Neurath, Ed.), Chapter 20, 155 (1966).

(101) R. D. B. Fraser and T. P. MacRae, *J. Chem. Phys.* **31**, 122 (1959).

(102) E. J. Ambrose and A. Elliott, *Proc. R. Soc. London Ser. A* **206**, 206 (1951).

(103) B. G. Frushour and J. L. Koenig, *Biopolymers* **14**, 379 (1975).

(104) R. D. B. Fraser and E. Suzuki, *Spectrochim. Acta* **26A**, 423 (1969).

(105) F. Lucas, J. T. B. Shaw and S. G. Smith, *J. Mol. Biol.* **2**, 339 (1960).

(106) E. Suzuki, *Spectrochim. Acta* **23A**, 2203 (1967).

(107) W. Traub and K. A. Piez, *Adv. Protein Chem.* **25**, 243 (1971).

(108) A. G. Walton, M. J. Deveney and J. L. Koenig, *Calcif. Tissue Res.* **6**, 162 (1970).

(109) W. B. Rippon, J. L. Koenig and A. G. Walton, *Agric. Food Chem.* **19**, 692 (1971).

(110) E. W. Small, B. Fanconi and W. L. Peticolas, *J. Chem. Phys.* **52**, 4369 (1970).

(111) J. A. Rupley, *Protein Structure in Crystal and Solution*, in *Structure and Stability of Macromolecules* (S. N. Timasheff and G. D. Fasman, Eds.), Marcel Dekker, New York, 1969, Chapter 4.

(112) N. T. Yu and B. H. Jo, *J. Amer. Chem. Soc.* **95**, 5033 (1973).

(113) N. T. Yu, *J. Amer. Chem. Soc.* **96**, 4664 (1974).

(114) P. C. Painter and J. L. Koenig, unpublished data.

Chapter 3

RESONANCE RAMAN SPECTRA OF HEME PROTEINS AND OTHER BIOLOGICAL SYSTEMS

T. G. Spiro

Princeton University, Princeton, N.J., U.S.A.

T. M. Loehr

Oregon Graduate Center, Beaverton, Oregon, U.S.A.

1 INTRODUCTION

The advent of the laser has produced one revolution in Raman spectroscopy and is in the process of introducing another. In the pre-laser days, heroic measures were needed to concentrate enough light in the sample to obtain a Raman spectrum of good quality.[1] High-current mercury discharge lamps were generally used, with jackets of circulating dye solution to filter the exciting light. Illumination was coaxial about a cylindrical sample, with the scattered light being collected from the end of the cylinder. Relatively large amounts of sample, 5–20 ml, were required, and optical clarity was critical. For most biological materials these requirements put Raman spectroscopy quite beyond reach as an analytical technique. The high light power densities afforded by lasers changed the situation dramatically.[2] Scattering from very small volumes could now be measured with ease, and because lasers allow for front scattering as well as transmitted scattering, the optical quality of the sample became less important. Required sample volumes shrank from milliliters to microliters, and it became easy to record spectra of solutions, single crystals, powders, films, fibers, etc.[2-4] The introduction of spinning[5] or flowing[6] sample arrangements extended the utility of Raman spectroscopy still further by minimizing sample heating and consequent decomposition. This quantum jump in the applicability of the technique constituted the first laser Raman revolution.

In our view a second revolution is now under way, made possible by the availability of laser wavelengths over an increasingly wide range of the electromagnetic spectrum. One can now search systematically for resonance effects in the Raman spectrum by tuning the laser to an increasing variety of electronic transitions. Before lasers were available, Raman spectroscopists assiduously avoided colored samples, as a rule. There was a practical reason for this. With coaxial illumination most of the scattered light would be absorbed upon transit through the sample. This restriction also was greatly relaxed by the introduction of the laser, whose path through the sample could be made very short, especially with front scattering. Yet this virtue was little appreciated at first. In fact it was generally felt that an important advantage of the first commercially available laser, the He—Ne laser, was its long wavelength, 6328 Å, which would permit the taking of spectra of samples that absorbed light in the blue regions.

Another reason for avoiding absorbing materials was the preoccupation of Raman spectroscopists with Placzek's successful polarizability theory,[7] which allows one to think about the intensity of Raman bands as reflecting a change in molecular polarizability, *provided* that the excitation energy is far below that of the first electronically excited state. Placzek himself predicted interesting resonance effects, in the region where the polarizability theory does not apply. But although some rather pretty resonance Raman spectra were produced in the pre-laser days, especially by Shorygin[8,9] and by Behringer,[10] resonance effects were generally considered esoteric. With the laser, the experimental characterization of resonance effects has proceeded vigorously, and has stirred considerable interest among theorists. To the practising chemist or biochemist, interested in the applications of Raman spectroscopy,[11] resonance enhancement should be of equal import, because it offers a major increase in both sensitivity and selectivity,[12] sufficient to add a new dimension to the technique. The laser wavelength becomes a dynamic variable of the Raman experiment.

2   RESONANCE ENHANCEMENT

If resonance Raman (RR) spectroscopy is to be applied fruitfully to biostructural problems, it is important to try to anticipate what sorts of vibrational modes are enhanced by resonance with what sorts of electronic transitions. For this purpose theory provides a useful guide.

The phenomenon of resonance enhancement was predictable from the Kramers-Heisenberg-Dirac dispersion equation, used by Van Vleck[13] to describe the Raman scattering tensor.[14,15]

$$(\alpha_{ij})_{mn} = \frac{1}{h} \sum_e \left[\frac{(M_j)_{me}(M_i)_{en}}{\nu_e - \nu_0 + i\Gamma_e} + \frac{(M_i)_{me}(M_j)_{en}}{\nu_e + \nu_s + i\Gamma_e} \right] \tag{1}$$

Here m and n are the initial and final states of the molecule, while e is an excited state, and the summation is over all excited states. The quantities $(M_j)_{me}$ and $(M_i)_{en}$ are electric dipole transition moments, along the directions j and i, from m to e and from e to n, while ν_e is the frequency of the transition from m to e, and ν_0 and ν_s are the frequencies of the incident and scattered photons. When ν_0 approaches ν_e the left-hand term in the summation can become very large for an allowed electronic transition. It is prevented from reaching infinity by inclusion of the damping term $i\Gamma_e$, which is a measure of the bandwidth of the electronic transition. The magnitude of resonance enhancement is predicted to vary directly with the oscillator strength of the resonant electronic transition and inversely with its breadth. Maximal resonance effects are therefore expected for strong sharp absorption bands.

The dispersion equation does not specify the nature of the initial and final states of the scattering molecule, and therefore gives no information as to which vibrations are subject to resonance enhancement. To examine vibrational transitions it is customary to use the adiabatic approximation† (separability of electronic and vibrational wavefunctions) and to analyze the dependence of the electronic wavefunctions on the nuclear displacements, using the Herzberg-Teller formalism.[14] Alternatively one can apply third order time dependent perturbation theory to the scattering process.[17] The results of the two approaches are quite similar, and a convenient formulation,[18] neglecting what are considered to be less important terms, is

$$a_{ij} = \frac{1}{h} \sum_{e,f} \left[\frac{(M_j)_{ge} h_{ef}^{\Delta\nu} (M_i)_{fg}}{(\nu_e - \nu_0)(\nu_f - \nu_s)} + \frac{(M_i)_{ge} h_{ef}^{\Delta\nu} (M_j)_{fg}}{(\nu_e + \nu_s)(\nu_f + \nu_0)} \right] \tag{2}$$

where the summation is over all excited electronic states e and f, taken in pairs, and g is the electronic ground state. $h_{ef}^{\Delta\nu}$ is a vibronic coupling matrix element, connecting states e and f by a particular vibration $\Delta\nu$. Damping terms have been omitted from the frequency denominators for simplicity.

Two kinds of resonance process are expected to be important, corresponding to the conditions $e = f$ and $e \neq f$ in Eqn. (2), which Albrecht and Hutley[19] have called A and B terms, respectively. B terms involve vibronic mixing of two excited states, e and f. The active vibration may have any symmetry which is contained in the direct product of the two electronic transition representations. Its enhancement depends on the magnitude of both the vibronic mixing matrix element and the electronic transition moments. Thus efficient vibronic coupling of two allowed excited states is needed for B term enhancement.

A terms involve vibrational interactions with a single excited electronic state. This is the more common mechanism encountered in practice. The matrix element $h_{ef}^{\Delta\nu}$ is just the dependence of the energy of the state e on the vibrational coordinate, which is given by the Franck-Condon overlaps.[14] Only for totally symmetric vibrations are these overlaps non-zero for both the initial and final states of the scattering molecule. Moreover the overlaps are greatest for those vibrational modes which most directly lead to the molecular distortions experienced by the excited state. One therefore anticipate that not only are resonance enhanced vibrations localized on the chromophoric part of the molecule, but also the greatest enhancement will be found for those vibrations which are most strongly affected by excited state distortions. An understanding of the electronic transition will illuminate the nature of RR scattering and vice versa.

† Non-adiabatic effects may not be negligible at resonance, however.[16]

Equation (2) predicts scattering maxima at excitation wavelengths corresponding to resonance with both the incident ($\nu_e = \nu_0$) and scattered ($\nu_e = \nu_s = \nu_0 - \Delta\nu$) photons. This has indeed been observed in excitation profiles (plots of Raman intensity vs. excitation frequency) for ferrocytochrome c.[20] On the other hand closer examination of the equations suggests that maxima are also expected for resonance with *vibronic* levels,[21] i.e. $\nu_e + \Delta\nu_{excited} = \nu_0$. If the ground and excited state vibrational frequencies do not differ significantly, as is probably the case for cytochrome c,[20] then resonance of the scattered photon with ν_e occurs at approximately the same excitation frequency as resonance of the incident photons with $\nu_e + \Delta\nu_{excited}$. In that event, the two interpretations cannot be distinguished. Excitation profiles for the chromophoric iron-binding site of the protein transferrin have been interpreted[22] as revealing maxima at successive vibronic levels. The excited state vibrational frequency in this case is reduced substantially from that of the ground state.

The distinction between ν_s and ν_0 may be important in accounting for antisymmetric vibrational scattering, i.e. scattering from vibrations whose scattering tensors have the form $a_{ij} - a_{ji}$.[23] Such vibrations are forbidden in normal Raman scattering, since the molecular polarizability tensor must be symmetric.[7] This restriction breaks down under resonance conditions, as Placzek[7] foresaw over forty years ago. The first example of antisymmetric vibrational scattering was discovered only recently,[20] in the RR spectra of cytochrome c and hemoglobin, although antisymmetry had earlier been detected in the electronic Raman effect.[24] Equation (2) cannot account for antisymmetric scattering if ν_s is replaced by ν_0, since subtraction of a_{ji} from a_{ij} then leads to cancellation of all the terms. This cancellation is prevented, however, if ν_s is retained, and the resulting equation for $a_{ij} - a_{ji}$ appears to account for the observed frequency dependence of the antisymmetric scattering intensity of ferrocytochrome c.[18]

3   OBSERVED RESONANCE RAMAN SPECTRA OF BIOLOGICAL MATERIALS

3.1   Metal-Containing Chromophores

Incorporation of metal ions into enzymes and proteins imparts specific biological activity, such as metal transport or storage, electron transfer, respiratory function and catalysis. In the case of transition metals, their absorption characteristics in the visible region of the spectrum have permitted analysis of electronic and structural properties useful in elucidating the details of metalloprotein function. Both ligand field and charge transfer transitions appear in this region of the spectrum. Additional information can be obtained from RR spectra if the coupling of vibrational modes of a metal ion chromophore with either ligand field or charge transfer electronic transitions is understood.

3.1.1   Ligand field transitions

Ligand field or d-d transitions are parity forbidden since donor and acceptor
levels are both d-orbitals. The absorption intensities observed for typical
transition metal complexes are explained by mixing with allowed vibrational
transitions. It is predicted from theory, however, that forbidden electronic
transitions coupled with vibrational modes produce only negligible enhancement
of RR intensities.[25] This prediction is verified by the Raman studies of some
inorganic Co(III) and Pd(II) complexes,[25] Co(II) imidazole complexes[26] and
cobalt-substituted zinc enzymes.[26]

Excitation profiles for $Co(NH_3)_6^{3+}$ and PdX_4^{2-} (X = Cl, Br, I) actually
exhibit a *minimum* for the intensity of the metal-ligand breathing mode at the
ligand field λ_{max}, and this behavior is ascribed to interference between scattering
contributions from the d-d band and from higher energy allowed charge transfer
bands.[25]

Octahedral Co(II) imidazole- and tetrahedral Co(II) histidine complexes
show negligible dependence of Raman intensities upon excitation wavelength.[26]
Enhancement of the Raman intensities in tetrahedral Co(II) imidazole complexes,
especially the imidazole ring modes, appears to be via a preresonance effect with
a u.v. charge transfer state.[26]

The substitution of cobalt(II) for zinc(II) in carbonic anhydrase,[27] alkaline
phosphatase[28] and carboxypeptidase[29] occurs with little or no loss in catalytic
activity. Furthermore, such substitution permits these enzymes to be studied by
spectroscopic and magnetic techniques. However, excitation within the d-d
visible absorptions failed to give resonance enhanced Raman spectra for
Co-carbonic anhydrase and Co-carboxypeptidase.[26] The prospects for using
ligand field bands in RR studies appear dim.

3.1.2   Charge transfer transitions

Transitions in which an electron is effectively transferred from ligand to metal
orbitals can be fully allowed, and give rise to absorption spectra from near u.v.
to near infrared with molar extinction coefficients of the order of 10^3 $M^{-1}cm^{-1}$.
Furthermore, such transitions connect electronic states which can effectively
couple with vibrational modes of the metal-containing chromophores and
provide a mechanism for the resonance enhancement of the intensities of metal-
ligand or, in the case of greater degrees of electron delocalization, intra-ligand
vibrational modes. The following Cu- and non-heme Fe-proteins are discussed
in terms of RR effects arising principally from ligand → metal charge transfer
transitions.

COPPER-CONTAINING PROTEINS

i. *Hemocyanin*

Hemocyanins are high molecular weight, oxygen-carrying proteins found in
the phyla *Arthropoda* and *Mollusca*. The physiology, biochemistry and physical

characterization of hemocyanin have received considerable attention and are the subject of excellent reviews.[30,31]

Hemocyanin contains no prosthetic groups other than metal ions and it is known that two copper ions are involved in the binding of one oxygen molecule. Both colorless deoxyhemocyanin and the blue oxyhemocyanin are diamagnetic according to magnetic and electron paramagnetic resonance (EPR) studies.[32] Chemical and spectral evidence, however, has indicated that the metal ions in deoxyhemocyanin are in the Cu(I) state and are oxidized to the Cu(II) state in oxyhemocyanin.[31] In the latter case, antiferromagnetic coupling of the two Cu(II) ions at the O_2 binding site would account for the diamagnetism.[32]

The principal electronic absorption bands of *Cancer magister* oxyhemocyanin occur at 340 nm ($\varepsilon_{Cu} \simeq 10{,}000$ $M^{-1}cm^{-1}$)[32] and 575 nm ($\varepsilon_{Cu} \simeq 600$ M^{-1} cm^{-1}).[33] A RR study of oxyhemocyanin has been conducted by Loehr, Freedman and Loehr[34] in which the 340 and 575 nm absorption bands are invoked for the intensity enhancement of the 282 cm^{-1} and the 742 cm^{-1} Raman transitions, respectively. The intensity variation with excitation wavelength as well as the effect of isotopic substitution of molecular oxygen is shown in Fig. 1.

The 742 cm^{-1} peak is assigned to the symmetric stretching of bound dioxygen. Substitution with $^{18}O_2$ selectively shifts this peak to 704 cm^{-1}. The observed

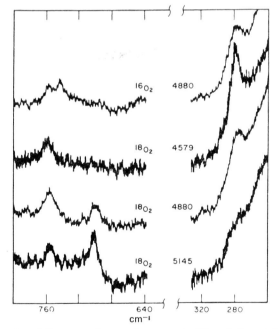

Fig. 1. Raman spectra of *Cancer magister* oxyhemocyanin (60 mg/ml) containing $^{18}O_2$ or $^{16}O_2$ showing the dependence of the intensity of the 742 and 282 cm^{-1} peaks on excitation wavelength at 457.9, 488.0 or 514.5 nm. The peak at 758 cm^{-1} is due to a non-resonant Raman vibration of tryptophan (Loehr *et al.*, Ref. 34).

vibrational energy and the magnitude of the isotope effect identify the hemo-
cyanin-bound dioxygen as a peroxide ion. Evans[35] has shown that anhydrous
Na_2O_2 exhibits a very strong band at 738 cm^{-1} which disappears on exposure to
the atmosphere and new frequencies near 830 and 860 cm^{-1} appear which are
assigned to hydrates of peroxide. Loehr et al.[34] use this evidence to postulate
that the peroxide ion in oxyhemocyanin may exist in a hydrophobic environment. Recent RR studies of μ-peroxo binuclear cobalt(III) complexes [Fig.
2A] by Freedman, Yoshida and Loehr[36] place the O—O stretching frequency
between 790 and 815 cm^{-1} (Table 1). However, the O—O frequency of symmetrically π-bonded dioxygen molecules in Group VIII transition metal complexes [Fig. 2B] are observed in the range 820–909 cm^{-1}.[37] The relatively low

TABLE 1
O—O stretching frequencies of some μ-peroxo and μ-hydroperoxo binuclear cobalt(III) complexes (Ref. 36)

Complex	νO—O, cm^{-1}
$[(NH_3)_5Co(O_2H)Co(NH_3)_5](SO_4)(HSO_4)_3$	815
$[(NH_3)_5Co(O_2)Co(NH_3)_5](SO_4)_2$	808
$[(his)_2Co(O_2)Co(his)_2]^a$	805
$[(NH_3)_4Co(O_2H, NH_2)Co(NH_3)_4](NO_3)_4$	795
$[(NH_3)_4Co(O_2, NH_2)Co(NH_3)_4](NO_3)_3$	793

^{a} his = histidinato ligand.

O—O stretching frequency at 742 cm^{-1} for oxyhemocyanin may be cited to
argue against a symmetrically π-bonded dioxygen in this protein. The presence
of bound oxygen in a form electronically equivalent to peroxide, O_2^{2-}, clearly
establishes that the electronic structure of both copper atoms at the binding site
may be formulated as Cu(II).

The intensity enhancement of the 742 cm^{-1} peak arises from vibronic coupling
with a single electronic state (Albrecht's A term)[38] observed as the 575 nm
absorption band of oxyhemocyanin. This strongly suggests that the 575 nm

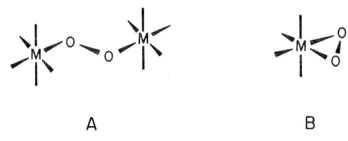

A   B

Fig. 2. Types of dioxygen coordination. A is commonly found in binuclear μ-peroxo- and μ-
superoxo-dicobalt complexes. B is known from mononuclear group VIII transition metal
complexes, involving π and π^* interaction of O_2 with the metal s, p and d orbitals (see text).

band involves $O_2^{2-} \to Cu(II)$ charge transfer rather than a copper-localized d-d transition. Similarly, the intensity enhancement of the ~ 800 cm^{-1} peak in the μ-peroxo cobalt(III) complexes[36] correlates with the near ultraviolet absorption band at 360 nm. This absorption band has been assigned to $O_2^{2-} \to Co(III)$ charge transfer[39] and is analogous to the $O_2^{2-} \to Cu(II)$ charge transfer band at 575 nm in hemocyanin. A similar pattern of energy shifts is seen in the ligand $\to$ metal charge transfer bands of stellacyanin, where the absorption bands in the cobalt-substituted protein are at 355 and 300 nm and the corresponding bands in the native copper protein are at 850 and 604 nm.[40]

Assignment of the 282 cm^{-1} peak in oxyhemocyanin[34] is less clear. This frequency appears to be unaffected by isotopic substitution with $^{18}O_2$. Resonance enhancement of the 282 cm^{-1} peak occurs via coupling with the oxyhemocyanin 340 nm transition. Absorption in the near u.v. (330–375 nm) has been cited as evidence for copper–copper interaction in binuclear systems exhibiting subnormal magnetic moments, whether they be simple copper carboxylate dimers[41] or copper proteins,[42] although other explanations are possible.[43,44] The 282 cm^{-1} peak probably arises from such a magnetically coupled copper dimer system. If the magnetic coupling is due to direct interaction, the low frequency Raman peak may correspond to Cu—Cu bond vibration. It is more likely, however, that the interaction is of the superexchange type and the observed band could arise from an enhanced symmetric stretching mode of a bridged species such as Cu—O—Cu. This explanation accounts for the lack of a vibrational isotope effect upon oxygenation with $^{18}O_2$. If the bridging oxygen atom arises from the incorporation of oxygen from solvent water upon oxygen binding, as has been suggested for hemerythrin[45] (see section 3.1.2, 'Hemerythrin'), the Cu—O—Cu stretching frequency should shift in $H_2^{18}O$ solution if some motion of the oxygen atom is involved in the vibration.

ii. 'Blue' copper proteins

The copper ions bound to proteins have been classified into three types according to their optical absorption spectra and magnetic properties, as reviewed by Malkin and Malmstrom.[42]

Type 1 or 'blue' copper is characterized by an intense absorption near 600 nm with extinction coefficients from 3000–11,000 M^{-1}cm^{-1}, whereas the extinction coefficients in common copper complexes are rarely greater than 100 M^{-1}cm^{-1}. It is this absorption which confers the distinctive intense blue color on the proteins. 'Blue' copper proteins also exhibit an absorption band near 450 nm and one in the near infrared. The EPR parameters for this type copper are unusual in that the hyperfine coupling constants of type 1 copper are abnormally small.

The EPR parameters for type 2 or 'non-blue' Cu(II) are similar to those of common copper complexes. While an absorption near 655 nm has been observed in one protein with only type 2 copper, the extinction coefficient is 400 M^{-1}cm^{-1} or lower.

The third type of Cu(II) shows no EPR signal and hence is termed 'EPR non-detectable'. These copper atoms apparently occur as spin paired Cu^{2+}—Cu^{2+} units. The optical absorption band at 330 nm in proteins containing type 3 copper has been associated with these EPR non-detectable copper pairs.

Copper ions of the three types can occur singly or together in metalloproteins: stellacyanin, azurin and plastocyanin contain only type 1 copper, erythrocuprein (superoxide dismutase) contains only type 2 copper, hemocyanin contains EPR nondetectable copper pairs, and the multicopper proteins laccase, ceruloplasmin and ascorbate oxidase contain all three types in a ratio of 1:1:2 (ascorbate oxidase may have a higher percentage of type 3 copper).[46]

The copper proteins containing only type 1 copper function as electron carriers. The blue multicopper proteins are oxidases, reducing oxygen to water in a four electron reduction. While the reduction mechanism and functions of the individual types of copper are not firmly established, it appears that substrate oxidation occurs at the type 1 copper, which accepts electrons most quickly. The type 3 copper pairs are probably the site of oxygen binding and reduction. The function of type 2 copper is less clear, but it may be necessary for the stabilization of intermediates in the O_2 reduction.[42]

Proteins containing type 1 copper are particularly good candidates for RR spectroscopy because of their intense absorption in the visible region. The high

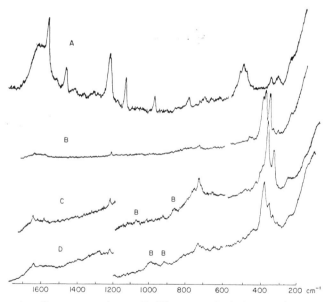

Fig. 3. RR spectra of copper proteins. A: Cu(II) ovotransferrin (44.6 mg/ml), pH 8, 488.0 nm excitation. B: *Rhus vernicifera* laccase (10.9 mg/ml). C: *Rhus vernicifera* stellacyanin (3.3 mg/ml below 600 cm⁻¹, 8.4 mg/cm³ above 600 cm⁻¹). D: human serum ceruloplasmin (11.6 mg/cm³) B, C and D are at pH 5.5. Spectra recorded with 647.1 nm Kr⁺ excitation. Bands marked B are due to buffer solution (Siiman *et al.*, Ref. 52).

extinction of the ~600 nm absorption is variously explained on the basis of distortion and low symmetry of the site,[47,48] charge transfer from oxidizable ligand atoms to Cu(II),[49] or a combination of such effects.

Ligand atoms of copper which have been implicated include nitrogen, oxygen and sulfur. For stellacyanin, EPR spectra were interpreted[50] in terms of the coordination of four nitrogen atoms, whereas recent chemical and spectroscopic studies by McMillin, Holwerda and Gray[40] indicate a cysteinyl-sulfur ligand. Fluorescence, absorption and EPR spectra of azurin suggest that copper interacts with tryptophan and sulfhydryl groups.[51]

Siiman, Young and Carey[52] published the first application of RR spectroscopy of the 'blue' copper proteins, stellacyanin, laccase and ceruloplasmin (Fig. 3). Similar experiments by Miskowski et al.[53] on ceruloplasmin, as well as on azurin and plastocyanin, have been reported (Fig. 4). The RR spectra are obtained by excitation within the 600 nm absorption band. Each of these proteins shows a spectrum characteristic of type 1 copper. Those proteins with mixed type 1, 2 and 3 coppers give essentially the same spectra as the proteins such as azurin which contain only a single copper atom per molecule. The spectra indicate that the type 1 site must be similar in these proteins, but that some structural differences must be present which affect the intensities and positions of the lines (Table 2).

By reference to the Raman spectra of model copper–peptide complexes with known structure, Siiman et al.[52] make the following vibrational assignments for atoms at the type 1 copper site.

frequency, cm^{-1} (intensity)	assignment
350–400 (s)	Cu—N (peptide N) and/or Cu—O (peptide CO oxygen)
260 (w)	Cu—S (cys) and/or Cu—N (imidazole)
330, 450 (w)	bending modes (CNC or CCN)
500, 750 (w)	bending modes (C=O)
1240 (w)	C—N (peptide)
1650 (w)	C=O (peptide CO with bond to Cu)

The authors conclude that the most likely copper ligands for either five-coordination or distorted tetrahedral coordination are carbonyl oxygen and amide nitrogen of a single peptide bond, one or two histidine nitrogens, and cysteine sulfur.

The spectral data and group vibrational assignment of Miskowski et al.[53] are in quite close agreement with those of Siiman, Young and Carey[52] for 'blue' proteins. The analysis of the electronic and RR spectra, however, is considerably more detailed by the former group of workers and leads to a model of the 'blue' copper site having approximately trigonal bipyramidal structure with two nitrogen atoms and one sulfur atom in the equatorial plane. Miskowski et al.[53] develop their argument as follows.

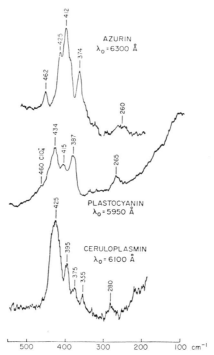

Fig. 4. RR spectra of 'blue' copper proteins. *Top: Pseudomonas aeruginosa* azurin, purity index (A_{280}/A_{600}) 4.1, pH 7.4, 0.6 mM. *Middle:* Spinach plastocyanin, purity index 1.4, pH 7.4, 1.4 mM. *Bottom:* Human serum ceruloplasmin, pH 5.8, 0.9 mM. Spectra are obtained with excitation from a tunable rhodamine 6G dye laser at the indicated wavelengths (Miskowski *et al.*, Ref. 53).

TABLE 2
Copper content and resonance Raman spectra of some 'blue' copper proteins by excitation within the 600 nm absorption band[c]

	Azurin[a]	Stellacyanin[b]	Plastocyanin[a]	Laccase[b]	Ceruloplasmin[a]
Cu_{tot}/molecule	1	1	2	4	8
$Cu_{'blue'}$/molecule	1	1	2	1	2
	260 w	260 w	265 w	260 w	280 w
				330 w	
		340 s		360 w	355 w
	374 s	385 s	387 s	380 s	375 w
		400 m		395 s	395 m
	412 s		415 m	405 s	
	425 s		434 s		425 s
	462 m	500 w		500 w	
		750 w	765 w	750 w	
		1240 w		1240 w	
		1650 w		1650 w	

[a] From Ref. 53.

[b] From Ref. 52; peak positions are approximate.

[c] Peak intensities are approximated by w, weak; m, medium; s, strong; frequencies given in cm^{-1}.

Approximately *four* RR peaks are observed in the 350–470 cm^{-1} region. Thus, about *four* ligand atoms must be attached to copper *if all these peaks can be attributed to fundamental Cu–ligand vibrations*. The atoms most likely to be bonded are N or O atoms which are not part of or attached to an aromatic ring, since the Cu–ligand vibrations are ~200 cm^{-1} for aromatic ligands.[54] Possible amino acids serving as ligands include lysine, arginine, glutamic acid, aspartic acid, glutamine, asparagine and serine, as well as the peptide group amide–N and carbonyl–O. The peptide nitrogen has been shown to be a particularly good Cu–ligand in studies of Cu(II)–peptide complexes.[55]

Although the ~270 cm^{-1} RR peak could be attributed to a Cu—N(imidazole) vibration,[56] much evidence supports the presence of a sulfur ligand[40,49,51] and hence its assignment to a Cu—S(cys) vibration. Furthermore, the observed 765 cm^{-1} peak is plausible for a C—S vibration resonance enhanced by proximity to the type 1 chromophore. The intensity of the 270 cm^{-1} Cu—S peak is, however, weak, a fact reasonably accounted for if the donor orbital for the S → Cu charge transfer transition, which is presumed to be in part responsible for the ~600 nm optical absorption, is S(π) rather than S(σ). The excitation profiles of the strongest RR peaks of ceruloplasmin and plastocyanin track the ~600 nm absorption band and suggest that these must arise from A terms in the Raman tensor expansion.[38,57] A terms involve coupling of vibrations to a single excited electronic state in which the molecular geometry is distorted along the normal coordinates of those vibrations.[58] The highest resonant intensities accompany those bond vibrations whose bonds are weakened in the electronically excited state. Such a situation is expected for Cu—N, but not for Cu—S, stretching vibrations, since the terminal orbital for the S → Cu transition would be directed toward N, rather than S, ligands.

The specific structural model that is proposed[53] is a trigonal bipyramid with a cysteine sulfur ligand sharing the equatorial plane with two strong field ligands, L, presumably N (but not imidazole). The axial positions are occupied by less strongly bound ligands, L', which might be N or O (Fig. 5). Although the 600 nm transition is ascribed to charge transfer between equatorial ligands and copper, the presumed appearance of Cu—L' vibrations in the RR spectra could be due to vibrational mixing between Cu—L and Cu—L' stretching modes. An attractive feature of the model relates to the electron transport function of these proteins. Reduction of the copper center to a stable trigonal Cu(I)SL$_2$ complex[59,60] may be accomplished by additional weakening of the Cu—L' bonds. This would support the idea[61] that unusual geometries around metal ions facilitate approach to the transition state in reactions catalyzed by metalloenzymes.

Additional work is needed to establish firmly the nature of the type 1 copper site in 'blue' copper proteins. Some of the proteins appear to have more than four RR peaks in the 350–470 cm^{-1} region of the spectrum. This makes it difficult to associate peaks with specific copper ligand vibrations and increases the attractiveness of proposed contributions[52] from peptide bond bending modes.

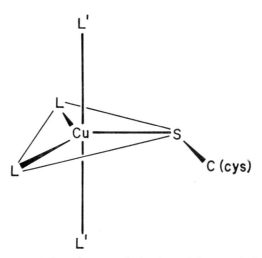

Fig. 5. Schematic representation of proposed structure of the type 1 site in 'blue' copper proteins. L and L′ may be N or O, σ-bonded to Cu; a cysteinyl S is π-bonded to Cu. The observed RR peaks in the 350–470 cm⁻¹ region are attributed to Cu—L and Cu—L′, 270 cm⁻¹ to Cu—S, and 765 cm⁻¹ to C—S vibrations (see text) (Miskowski *et al.* Ref. 53).

NON-HEME IRON-CONTAINING PROTEINS

i. *Iron–sulfur proteins*

These are non-heme iron proteins in which the iron atom is coordinated to four sulfur atoms of either cysteine or cysteine and inorganic (acid-labile) sulfur. Bacterial rubredoxins have only a single iron atom coordinated to four cysteine sulfur atoms; mammalian adrenodoxins and plant ferredoxins contain binuclear iron–sulfur clusters; and bacterial ferredoxins contain tetranuclear iron–sulfur clusters. The proteins serve as electron transfer agents for a variety of natural processes such as N_2-fixation and photosynthesis.[62–64]

Rubredoxin. Rubredoxin is the smallest of the iron–sulfur proteins with 54 amino acids and a molecular weight of 6100. The 3-dimensional structure of rubredoxin has been elucidated and refined to a resolution of 1.5 Å by Watenpaugh, Sieker, Herriott and Jensen[65] in one of the most detailed crystallographic analyses of a protein. The FeS_4 complex appears to be nearly tetrahedral, with bond angles in the range 101–115°. However, of the four Fe—S(cys) distances one is significantly shorter by an astonishing 0.25 Å from the value of 2.30 ± 0.05 Å for the other three.

The first report of a RR study of a metalloprotein was that of rubredoxin by Long and Loehr.[66] Oxidized *Clostridium pasteurianum* rubredoxin in a non-crystalline solid phase shows only two spectral lines at 365 and 311 cm⁻¹ when excited within the negative Cotton effect[67,68] band centered at 632 nm. In the aqueous solution the RR spectrum of *Clostridial* rubredoxin[68] obtained by excitation within the strong absorption band centered at ∼490 nm (Fig. 6), peaks at 368 and 314 cm⁻¹ are identical to those found for the solid. In addition,

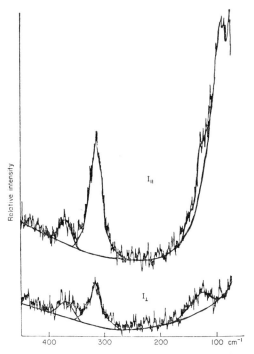

Fig. 6. RR spectrum of aqueous *Clostridium pasteurianum* rubredoxin, showing the presence of both polarized and depolarized vibrational modes. Excitation by 488.0 nm Ar$^+$ laser, power at sample ~ 5 mW (Long *et al.*, Ref. 68).

lower frequency vibrational modes are observed at *ca.* 150 and 126 cm^{-1}. Polarization measurements (see Fig. 6) indicate a depolarization ratio of 0.3 for the 314 cm^{-1} peak. A value of the order of 1/3 is expected for a resonance enhanced totally symmetric vibration.[38] The four observed peaks show an intensity and polarization pattern similar to that of the FeCl$_4^-$ tetrahedral anion and are assigned to the vibrational modes of the FeS$_4$ tetrahedron (Table 3).

TABLE 3
Resonance Raman spectrum of FeS$_4$
core of oxidized rubredoxin (Ref. 68)

Frequency, cm^{-1}	Assignment[a]
368	v_3 ⎱ tetrahedral stretching modes
314	v_1 ⎰
150	v_4 ⎱ tetrahedral bending modes
126	v_2 ⎰

[a] T_d nomenclature.

Recently, Yamamoto *et al.*[69] reported a similar set of RR bands for rubredoxin in the frequency range 100–400 cm^{-1}. Their normal coordinate analysis shows that these Raman lines can be identified as four of the six expected vibrations of an FeS$_4$ cluster having C_{3v} symmetry. This result is compatible with the finding of one 'short' Fe—S bond (*vide supra*) in the X-ray structural refinement.[65] The stretching frequencies of the FeS$_4$ core[66,68] do not change between solid and solution phases; hence, the coordination number and geometry about the iron atom are maintained.

Adrenodoxin. Mammalian adrenodoxin has a molecular weight of $\sim 12{,}000$ and contains two iron atoms, two acid-labile sulfides and four cysteinyl residues, involved in iron coordination.[70] Whereas exact structures are known for the simpler rubredoxin[65] and the more complex four- and eight-iron cluster bacterial ferredoxins,[71] much less is known about the two-iron cluster structure. It is believed that each iron atom in adrenodoxin is approximately tetrahedrally coordinated to two cysteinyl sulfur atoms and two inorganic sulfide-bridge atoms. Native, oxidized adrenodoxin possesses charge transfer bands at 414 and 455 nm, [70] and the Se-substituted protein has bands at 438 and 480 nm.[72]

Tang *et al.*[72] have obtained RR spectra of native adrenodoxin and the biologically active selenium-substituted adrenodoxin (Fig. 7). All Raman peaks occur in the 400–250 cm^{-1} region where iron–ligand vibrations are expected. Three polarized peaks are observed for adrenodoxin at 397, 350 and 297 cm^{-1}. Substitution of Se for the acid-labile S atoms gives rise to an RR spectrum in which the former 397 and 297 cm^{-1} peaks vanish, new peaks are observed at 355 and 263 cm^{-1}, and the 350 cm^{-1} peak is unaffected. Interpretation of these

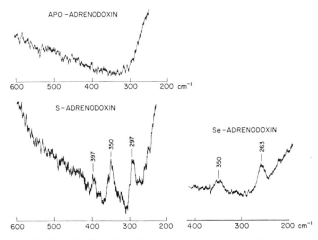

Fig. 7. RR spectra of adrenodoxins. *Top left:* apo-adrenodoxin, showing absence of vibrational modes. *Bottom left:* adrenodoxin with intact acid-labile sulfide bridges. *Right:* Selena-adrenodoxin, in which Se has replaced the acid-labile S bridges. Protein concentration is 1 mM, pH 7; spectra are obtained with 488.0 nm excitation (Tang *et al.*, Ref. 72).

observations is straightforward in that the $397/297 \, \text{cm}^{-1}$ and $355/263 \, \text{cm}^{-1}$ peaks may be associated with Fe—S (labile) and Fe—Se (labile) vibrations, respectively, and the $350 \, \text{cm}^{-1}$ peak with Fe—S (cysteine) vibration.

ii. *Hemerythrin*

Hemerythrin is the oxygen-carrying respiratory protein of certain marine invertebrates.[30] Hemerythrin is composed of eight subunits of ~ 7000 molecular weight, each of which contains two iron atoms at the oxygen binding site, with histidine and tyrosine as potential iron ligands.[45] Magnetic and spectroscopic studies have indicated that the two iron atoms are antiferromagnetically coupled and that the high spin Fe(II) in the deoxygenated protein is converted to high spin Fe(III) upon oxygenation.[45,73] This implies that oxygen is bound to hemerythrin as peroxide.

Definite proof for the presence of Fe-bound peroxide in oxyhemerythrin has been obtained by Dunn, Shriver and Klotz[74] using RR spectroscopy. Excitation within the 500 nm oxygen → iron charge transfer band results in the appearance of resonance enhanced Raman peaks at 844 and $500 \, \text{cm}^{-1}$ (Fig. 8). Replacement of $^{16}O_2$ by $^{18}O_2$ causes these peaks to shift to 798 and $478 \, \text{cm}^{-1}$, respectively (Fig. 8), indicating that both vibrational modes involve the bound oxygen molecule. The $798 \, \text{cm}^{-1}$ peak is assigned to the O—O stretch of peroxide by analogy to a peroxide stretching frequency of $836 \, \text{cm}^{-1}$ in $(NH_4)(HO_2)$.[74] A similar peroxide mode has been observed at $742 \, \text{cm}^{-1}$ in the copper-containing respiratory protein, hemocyanin[34] (see above). The $500 \, \text{cm}^{-1}$ peak in hemerythrin is assigned to an iron–peroxide stretching vibration.[74]

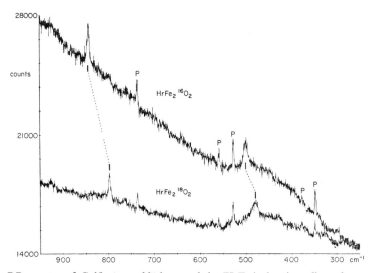

Fig. 8. RR spectra of *Golfingia gouldii* hemerythrin (HrFe₂) showing effect of oxygenation with $^{18}O_2$ or $^{16}O_2$. Protein concentration is 3.2 mM, pH 8. Spectra are obtained with 488.0 nm excitation and spinning sample cell in a back-scattering geometry. Lines marked p are laser plasma emissions (Dunn *et al.*, Ref. 74).

The similarity of the peroxide stretching frequencies of hemerythrin and hemocyanin as well as the stoichiometry of two metal ions per O_2 molecule and the antiferromagnetic coupling of the metal ions are all evidence in favor of similar modes of oxygen coordination in these two proteins. The oxy and oxidized (met) forms of hemerythrin are very similar to model systems containing oxo-bridged Fe(III) dimers in their electronic absorption and Mössbauer spectra.[75,76] A ligand can add to an oxo-bridged system such as oxy- or methemerythrin either as a second bridging ligand between the two iron atoms[75] or as a terminal ligand binding to only one of the iron atoms.[73] End-on binding is favored by the magnitude of the antiferromagnetic coupling constants[73] and from consideration of the amount of flexibility which would be required to accommodate such sterically diverse ligands as Cl^-, $O_2{}^{2-}$ and $N_3{}^-$ as bridging ligands.[77] The RR spectrum of metazidohemerythrin[77] gives further support for the concept of end-on binding of the adduct ligand. The asymmetric azide stretching frequency at 2049 cm^{-1}[77] is close to the frequency of the non-bridging azide in metazidohemoglobin at 2045 cm^{-1} in the infrared spectrum.[78]

iii. *Transferrin*

Transferrins are iron-binding proteins found in vertebrates. These proteins have exceptionally great affinities for iron and are thought to be involved in iron transport and in the inhibition of bacterial infection by iron chelation.[79] Chemical modifications and spectroscopic techniques indicate that each Fe(III) is bound to three tyrosines and two histidines,[80] with an equivalent of bicarbonate being required for development of the chromophore at 465 nm.[81] Excitation within this absorption band produces complex Raman spectra [Fig. 9; see also Fig. 3(a)] due to contributions from normal protein vibrational modes in addition to resonance enhanced peaks.[22,82,83]

The major RR peaks of Fe(III)-transferrin and their possible origins are listed in Table 4. The high frequencies of the peaks indicate that they are due to ligand vibrational modes rather than metal–ligand vibrations. Also, replacement of Fe(III) with Cu(II) has little effect on the spectrum.[22,52] Comparison with

TABLE 4
Resonance Raman spectra of Fe(III)-transferrins[a]

Frequency, cm^{-1}		
Transferrin[22,82] (human serum)	Ovotransferrin[82,83] (egg white)	Tentative assignment[22,82,83]
1609 s	1607 s	Tyrosine
1506 m	1505 m	Tyrosine or histidine
1285 s	1272 s	Tyrosine or histidine
1172 m	1172 m	Tyrosine

[a] Relative intensity denoted by s, strong; m, medium.

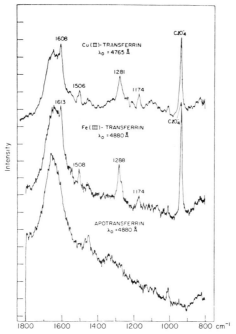

Fig. 9. Raman spectra of solutions of Cu(II)-, Fe(III)-, and apo-transferrin. Protein concentrations from top to bottom are 0.173, 0.148 and 0.150 mM respectively, all at pH 7.9. Cu(II)- and Fe(III)-transferrins also contain 0.05 M ClO_4^- as an internal standard. Excitation is with the indicated wavelengths (Gaber et al., Ref. 22).

the Raman spectra of tyrosine and histidine[83] indicates that the resonance enhanced peaks at 1609 and 1172 cm^{-1} can be attributed to phenol ring modes of tyrosine.[84] The peak at 1280 cm^{-1} could be due either to another tyrosine ring mode and CO stretch[85] or to an imidazole ring mode of histidine.[26,83] The peak at 1505 cm^{-1} correlates best with another imidazole ring mode of histidine.[83] It is the protonated form of the imidazole ring of histidine which exhibits peaks at these frequencies.[83]

Gaber et al.[22] have questioned the histidine assignments because they feel that all four resonance enhanced peaks can be accounted for by tyrosine ring vibrations. Their argument is based on the observation of strongly enhanced peaks at 1600, 1482, 1286, and 1168 cm^{-1} in the RR spectrum of Fe(III)-ethylenediamine di(o-hydroxyphenyl-acetate), Fe(EDDHA)$^-$. The Fe(III) in this compound appears to be coordinated to two phenolate oxygens, two amine nitrogens, and two carboxylate oxygens, and it is thought that the electronic transitions leading to resonance enhancement are due to phenolate $\rightarrow$ Fe(III) charge transfer. If this is the case, only phenolate vibrational modes are to be expected in the RR spectrum. The assignment of the $\sim$1500 cm^{-1} peak to tyrosine is still questionable, however, since there is no apparent ring mode at this position in the normal Raman spectrum of tyrosine.[83]

The visible absorption spectra of Fe(III)-transferrin and Fe(EDDHA)⁻ both show maxima at 465 nm which have been assigned to charge transfer from the $p\pi$-orbital of phenolate oxygen to the $d\pi^*$-orbital of Fe(III).[22] Consequently, other potential Fe(III) ligands in transferrin such as histidine or bicarbonate would not contribute enough to the electronic transitions in the visible region to undergo enhancement of their vibrational modes. This certainly appears to be the case with bicarbonate where replacement with oxalate[83] or ^{18}O-labeled bicarbonate[22] has no significant effect on the RR spectrum.

The failure to observe resonance enhanced peaks for Fe(III)-transferrin in the region of metal–ligand stretching vibrations (below 600 cm⁻¹) is also explained by the involvement of the $d\pi^*$-Fe(III) orbital in the charge transfer transition.[22] The weak antibonding nature of the Fe—O interaction would not cause much decrease in the strength of the Fe—O bond in the excited state (but the charge transfer transition would have a significant effect on the bond strengths in the phenolate moiety). Different behavior is expected for copper proteins where ligands interact with the σ^*–Cu(II) orbital and these Cu–ligand bonds are considerably weakened in the excited state. Thus, resonance enhanced peaks characteristic of metal–ligand vibrations have been observed in Cu(II)-trans-ferrin[52] as well as in the 'blue' copper proteins (see p. 105).

3.2   Heme Proteins

RR spectra were first reported for hemoglobin[86,87] and cytochrome c[20,88] in 1972. Since then RR studies of heme proteins[16,18,21,89–106] and porphy-rins[107–113] have been actively pursued in several laboratories. Considerable progress has been made in clarifying both the mechanism of RR scattering and the structural implications of the spectra.[114]

3.2.1   Heme electronic spectrum

The electronic properties of hemes, whose structure is illustrated in Fig. 10, are dominated by the aromatic system of the porphyrin ring. The visible and near-ultraviolet spectra contain two π–π^* transitions.[115] Both are of e_u symmetry under the D_{4h} point group, which applies approximately to metallopor-phyrins. Being nearly degenerate, they undergo considerable interaction, with the transition dipoles adding for the higher energy transition and nearly cancelling for the lower energy transition. The result is a very intense ($\varepsilon \approx 10^5$ M⁻¹cm⁻¹) band near 400 nm, called the Soret, γ or B, band, and a much weaker one (by a factor of 10 or more) near 500 nm, called the α, or Q_0, band. The lower energy transition, however, 'steals' back some 10% of the intensity of the higher energy transition through vibronic mixing, with the formation of a vibronic sideband, called the β or Q_v band, some 1300 cm⁻¹ above the center of the α band.

This basic α, β, Soret pattern, shown in Fig. 11 for ferrocytochrome c, is

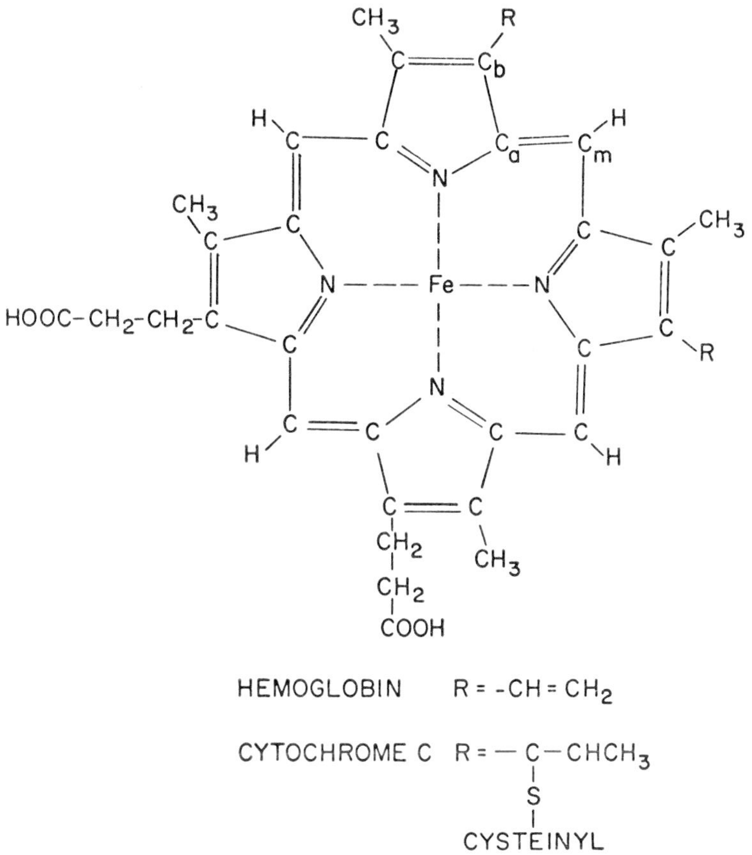

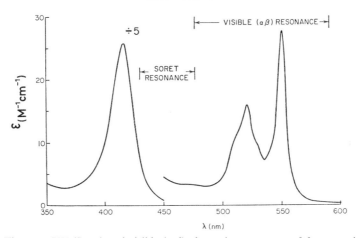

HEMOGLOBIN   R = -CH = CH₂

CYTOCHROME C R = — C—CHCH₃
                    |
                    S
                    |
              CYSTEINYL

Fig. 10. Structure of heme, indicating pyrrole substituents which occur in hemoglobin and cytochrome c.

Fig. 11. The near UV (Soret) and visible (α–β) absorption spectrum of ferrocytochrome c. Arrows span the approximate regions in which resonance with each of the two kinds of optical transitions dominates the Raman spectrum (Spiro and Strekas, Ref. 98).

subject to some variation.[116] The porphyrin orbital energies are influenced by the d orbital energies and occupation numbers of the central iron atom, which in turn are influenced by the electronic properties of the axial ligands. Generally, the greater the shift of π electrons from the iron atom to the porphyrin ring, the lower the energy of the electronic transitions. Also, electron withdrawing peripheral substituents, such as the vinyl groups of protoheme, or the formyl group of heme a, shift the electronic transitions to lower energy. The relative energies of the α and Soret transitions determine the degree of interaction, which in turn determines the intensity of the α band. Sometimes, as in deoxyhemoglobin, the α band is sufficiently weak, or the β band is sufficiently broad, to allow only one absorption envelope to be seen in the visible region. In the case of high-spin Fe(III) hemes, porphyrin $\rightarrow$ iron charge transfer transitions are also found in the visible region, and complicate the absorption spectrum considerably.[116,117]

3.2.2   Characteristics of heme scattering

When heme Raman spectra are obtained with excitation wavelengths below about 500 nm, the intense Soret band dominates the scattering mechanism[92,93,16] (see Fig. 11). The enhanced vibrational modes are all totally symmetric, and follow an A term wavelength dependence.[92] With excitation above about 500 nm, the α and β bands dominate the scattering.[20,21] Totally symmetric modes are not observably enhanced, presumably because A term scattering varies with oscillator strength, which is an order of magnitude smaller for the α band than the Soret band. Instead, non-totally symmetric modes are strongly enhanced. These must be due to B term scattering, attributable to the strong vibronic mixing of the α and Soret transition, which gives rise to the vibronic β band. The transition from the Soret to the α, β scattering region is illustrated in Fig. 12 for methemoglobin fluoride.

Excitation profiles show that all of the non-totally symmetric Raman bands reach maximum intensity at the center of the α band,[21] and that they reach second maxima within the β band, which, however, shift systematically to lower wavelength with increasing vibrational frequency of the Raman band.[20] Indeed these positions of the second maxima can be calculated by adding the individual vibrational frequencies to the frequency of the α transition (see Fig. 13). As discussed in section 2, they can therefore be attributed to resonance of the incident light with the individual vibronic levels that make up the β band ($\nu_0 = \nu_e + \Delta\nu_{\text{excited}}$) inasmuch as the ground and excited state vibrational frequencies are not expected to differ significantly for porphyrins.[118] In the case of ferrihemoglobin fluoride, this characteristic of the β band excitation profiles proved useful in confirming the assignment of the α band,[92] which is hidden among a group of charge-transfer transitions.[117]

Group theory determines that the symmetry of the vibronic modes are $E_u \times E_u = A_{1g} + A_{2g} + B_{1g} + B_{2g}$, but a_{1g} vibrational modes have been shown to be ineffective in mixing the two electronic transitions.[118] The b_{1g} modes (but

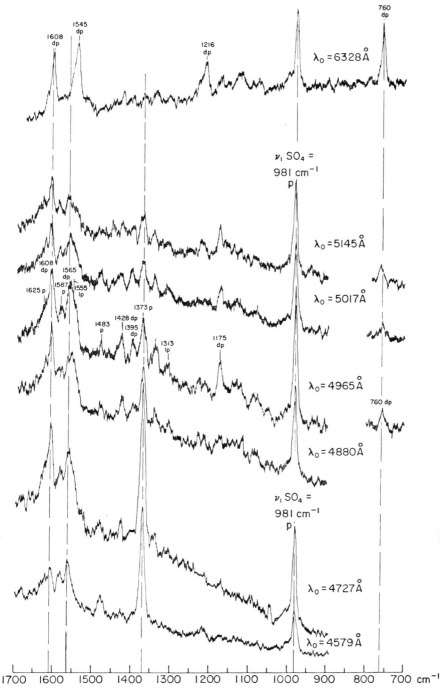

Fig. 12. RR spectra of ferri-hemoglobin fluoride (0.8 mM heme) containing 0.4 M $(NH_4)_2SO_4$ as internal standard, using He—Ne and a series of Ar^+ exciting wavelengths. Polarization of bands at $\lambda_0 = 4965$ Å are indicated (p = polarized, dp = depolarized, ip = inverse polarized). Typical instrumental conditions: slit width, 7 cm^{-1}; sensitivity, 10^{-9} A; time constant, 3s; scan rate, 50 cm^{-1} min^{-1} (Strekas and Spiro, Ref. 92).

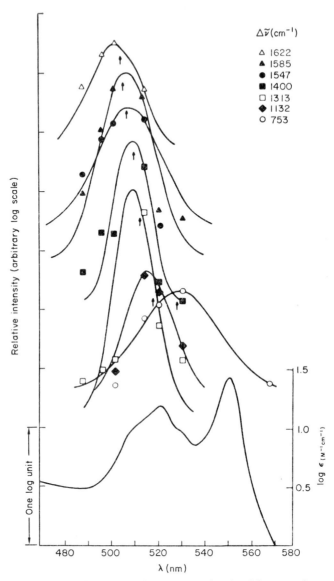

Fig. 13. Excitation profiles for the prominent Raman bands of ferrocytochrome c, and the electronic absorption spectrum, both on a logarithmic scale. The points represent intensities of the indicated Raman band, measured relative to the ν_1 sulfate peak from $(NH_4)_2SO_4$ internal standard, with the available Ar^+ and Kr^+ laser lines. The profiles are displaced for clarity on an arbitrary log intensity scale. The available points were fitted to the standard Gaussian curves displayed here by a DuPont 303 curve resolver (Spiro and Strekas, Ref. 20).

apparently not the b_{2g} modes[20,122]) are effective, and produce depolarized Raman bands (ρ, the ratio of scattered intensity polarized perpendicular and parallel, respectively, to the incident polarization, is 3/4). The a_{2g} modes are also effective. They are inactive in normal Raman scattering, because their scattering tensors are *anti*symmetric, i.e. $a_{xy} = -a_{yx}$. Antisymmetric scattering is allowed in the region of resonance, however, and gives rise to inverse polarization[20] ($\rho = \infty$, see Fig. 14).

While antisymmetric scattering requires that the scattering intensity vanish in parallel polarization, the cytochrome c Raman bands show significant residual parallel scattering, so that the polarization is anomalous ($\rho > 3/4$) but not entirely inverse.[20] There has been considerable discussion of the origin of the anomalous polarization.[18,20,96,97,109] It may reflect accidental degeneracy, i.e. modes of different symmetry coincident with the a_{2g} modes, or it may be due to lowering of the effective four-fold symmetry of the heme chromophore, with the result that the antisymmetric scattering tensors gain some symmetric character.[23] Positive evidence for accidental degeneracy for some of the a_{2g} modes comes from high resolution spectral scans in different polarizations.[96,109] On the other hand, direct determination of the magnitude of antisymmetric scattering, which can be achieved through the use of circularly polarized as well as linearly polarized light,[96,97] led to the observation[97] that some non-a_{2g} modes, with

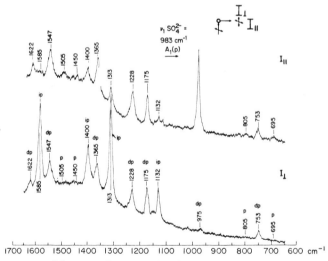

Fig. 14. RR spectra of ferrocytochrome c, showing anomalous polarization. The scattering geometry is shown schematically in the diagram at the top. Both the direction and the polarization vector of the incident laser radiation are perpendicular to the scattering direction. The scattered radiation is analyzed into components perpendicular ($I_\perp$) and parallel ($I_{\parallel}$) to the incident polarization vector. The excitation wavelength is 514.5 nm and the slit width is 10 cm^{-1}. The cytochrome c concentration is 0.5 mM, and the solution contains 0.5 M $(NH_4)_2SO_4$ as an internal standard giving rise to the strong, polarized band at 983 cm^{-1} from the ν_1 vibration of SO_4^{2-} (Spiro and Strekas, Ref. 20).

depolarization ratios in the normal range, have significant antisymmetric components, implying some reduction in the effective heme symmetry from D_{4h}.

3.2.3 Nature of the enhanced heme vibrational modes

The resonant α and Soret electronic transitions are polarized in the heme plane. Resonance enhancement should therefore be expected for those vibrations which alter the polarizability in the plane, i.e. those with tensor elements α'_{xx}, α'_{yy}, α'_{xy}, α'_{yx}, where x and y are the molecular coordinates defining the heme plane. Indeed the dominant RR bands are all between 1100 and 1650 cm^{-1}, where one expects in-plane porphyrin ring modes, involving the stretching of C—C and C—N partial double bonds and the bending of C—H bonds. Spectra observed with excitation above $\sim$500 nm, the region of vibronic scattering, contain only one other strong band, at $\sim$750 cm^{-1}, which is presumably an in-plane porphyrin deformation mode; below 750 cm^{-1}, where vibrations directly involving the iron atom are expected, the spectra are nearly blank. Apparently low frequency heme vibrations, including those involving the iron atom, are ineffective in mixing the α and Soret transitions. The dominance of the high frequency modes in vibronic scattering is consistent with the fact that the vibronic envelope in the absorption spectrum (Fig. 11) is the discrete β band, with a maximum $\sim$1300 cm^{-1} above the α transition.

With excitation below $\sim$500 nm, the region where totally symmetric modes are enhanced in resonance with the Soret transition, high frequency modes, especially the one at $\sim$1375 cm^{-1}, still dominate, but bands below 750 cm^{-1} now appear with moderate intensity.[90,94] Again this is consistent with the absorption spectrum, in which the Soret band is broader than the α band. The Soret band sometimes shows a shoulder $\sim$1300 cm^{-1} above its center, but only rarely is the shoulder resolved into a distinct band.[119] The low frequency vibrations which can be monitored with short wavelength excitation should include some which involve in plane Fe—N stretching, which should be sensitive to the movement of the iron atom out of the heme plane in high spin hemes (*vide infra*). Three such bands have been identified by Brunner and Sussner.[90]

It would be very useful to be able to monitor modes which involve the stretching of the axial ligand–iron bonds. Their frequencies could help to identify the axial ligands. The intensity of such modes, however, depends mainly on the tensor element α'_{zz}, whose enhancement requires resonance with a z-polarized (out of plane) electronic transition. A charge transfer transition involving the axial ligands should be particularly effective. Unfortunately such transitions are usually weak, at best, in heme visible spectra. The one confirmed axial ligand mode so far observed in heme Raman spectra is the Fe—O stretching mode of oxyhemoglobin, reported by Brunner[101] at 567 cm^{-1}, shifting to 540 cm^{-1} on $^{18}O_2$ substitution (Fig. 15). Appreciable out of plane absorption has been detected by Makinen and Eaton[120] in single crystal spectra of oxyhemoglobin, and was assigned to Fe $\rightarrow$ O_2 charge transfer. The relatively high frequency of the oxyhemoglobin Fe—O stretching mode (compare 500 cm^{-1}

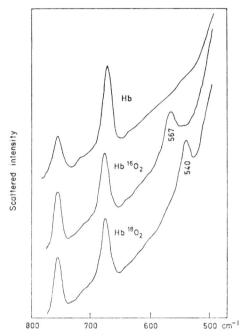

Fig. 15. RR spectra of 3×10^{-5} M hemoglobin (Hb), $Hb^{16}O_2$ and $Hb^{18}O_2$, respectively, at 15°C, pH 7.4. Abscissa: Raman frequency shift from exciting laser line at 488.0 nm; ordinate: scattered intensity in arbitrary units (Brunner, Ref. 101).

for ν_{Fe-O} in oxyhemerythrin, which contains bound peroxide)[94] suggests a substantial degree of multiple bond character in the Fe—O bond, in line with recent X-ray structural data for a protein free heme–O_2 complex.[121]

3.2.4 Assignment of the porphyrin ring modes

In the region of the dominant Raman bands, 1100–1650 cm^{-1}, porphyrin ring modes are expected, involving contributions from the stretching of C—C and C—N bonds and the bending of C—H bonds at the methine bridges. For a planar D_{4h} porphyrin, neglecting the peripheral pyrrole substituents, one expects these internal coordinates to contribute to $4a_{1g}$, $4a_{2g}$, $5b_{1g}$ and $4b_{2g}$ modes. The observed heme spectra are complex in this region, but overlapping bands can be resolved and catalogued via their differing polarization properties and excitation wavelength dependencies. When this is done for cytochrome c,[98] the spectra are found to contain three polarized modes (a_{1g}), four anomalously polarized modes (a_{2g}) and four depolarized modes (b_{1g} or b_{2g}). All of these can be attributed to porphyrin ring modes, without reference to the peripheral substituents, whose vibrations are presumably not resonance enhanced. The missing ring modes are either too weak to be observed, or else are involved in accidental degeneracies which have not been resolved (a_{2g}, b_{1g} and b_{2g} modes are expected to have nearly the same excitation wavelength dependence).

The frequencies of the cytochrome c ring modes are calculated with reasonable accuracy by a normal coordinate analysis[122] for planar octamethyl porphyrin, the results of which are shown in Table 5. This calculation employed the Urey-Bradley force field proposed by Ogoshi et al.,[123] who transferred force constants from related small molecules. Two non-Urey-Bradley force constants were added,[122] involving interaction of C—C stretches at the methine bridges, and of C—N stretches at the pyrrole nitrogen atoms. The potential energy distribution (Table 5) shows the internal coordinates to be quite highly mixed in the ring modes. The ~ 1585 cm^{-1} a_{2g} mode, however, which is sensitive to spin-state changes (vide infra) is mainly a methine bridge mode, as was earlier suggested.[98] On the other hand, the intense ~ 1375 cm^{-1} a_{1g} mode, which has been proposed as a marker of either iron atom out of plane displacement[87,90] or of oxidation state,[91,98] is not primarily a C—N breathing mode, as had been proposed,[90] but mainly involves breathing of the outer porphyrin ring. While b_{1g} and b_{2g} modes should both give rise to depolarized bands, the comparison of observed and calculated frequencies confirms an earlier inference,[20] based on correlations with sharp line fluorescence spectra of free-base porphyrins, that b_{2g} modes are not observed in the RR spectra.

TABLE 5
Porphyrin normal coordinate calculations[a]

Ferrocytochrome c frequency (cm^{-1})		Potential energy distribution[d] (contributions above 10%)	High → low spin frequency shift	
Exptl.[b]	Calc.[c]		Exptl.[e]	Calc.[f]
1626	1626(b_{1g})	51% C_aC_m, 20% C_bC_b, 16% C_aC_b	14	24[g]
1592	1598(a_{1g})	54% C_bC_b, 16% C_aC_m	4	7
1585	1587(a_{2g})	73% C_aC_m, 16% δCH, 14% C_aC_b	33	33[g]
1547	1534(b_{1g})	57% C_bC_b, 16% C_aC_m	1	20
1497	1493(a_{1g})	46% C_aN, 17% C_aC_b, 13% C_bC_b	23	18[g]
h	1476(b_{2g})	45% C_aC_m, 37% C_aN	h	4
1405	1424(b_{1g})	60% C_aN, 16% C_aC_m	0	8
1400	1399(a_{2g})	40% C_aN, 28% δCH, 13% C_aC_b	0	1
h	1387(b_{2g})	75% C_aC_b	h	2
1362	1376(a_{1g})	38% C_aC_b, 31% C_aC_m, 17% C_bC_b	4	3
1310	1284(a_{2g})	45% C_aC_b, 37% δCH	0	−8
1228	1257(b_{1g})	29% C_aC_b, 58% δCH	0	−7

[a] From Ref. 122.

[b] From Ref. 98.

[c] Planar octamethylporphyrin with modified Urey–Bradley force field (Ref. 122).

[d] Symbols: XY-force field element for stretching of the X—Y bond, see Fig. 10 for atom labelling; δCH-force field element for in plane bending of the methine bridge C—H bond.

[e] Ferrocytochrome c minus deoxyhemoglobin (Ref. 98).

[f] Planar minus 19° domed porphyrin, with force field altered to reflect decreased π conjugation at the methine bridge (Ref. 122).

[g] Spin-state marker bands (Ref. 98).

[h] These modes have not been detected.

Hemoglobin derivatives show several extra Raman bands in the porphyrin ring region.[98] These have also been observed for b-type cytochromes,[102] and are logically attributed to the influence of the vinyl substituents of protoheme, which are known to conjugate with the porphyrin π system.[116] (In c-type cytochromes, the vinyl groups are replaced by saturated thioether links, see Fig. 10.) Vinyl C=C stretching and C—H bending modes might be resonance enhanced. In addition, the asymmetric disposition of the vinyl groups destroys the symmetry center of the chromophore, and may induce Raman activity into infrared-active (e_u) porphyrin ring modes, of which there should be nine.[98]

3.2.5   Correlations with oxidation and spin states

Because the heme Raman bands can be identified via their three different states of polarization, their frequencies can reliably be correlated for different heme proteins. When this is done for cytochrome c and various derivatives of hemoglobin[95,98] (see Fig. 16) it is found that some of the porphyrin ring frequencies are nearly invariant, but others show appreciable shifts which correlate with the known spin and oxidation states.

Reduction of Fe(III) to Fe(II), without change in spin state, lowers several frequencies by a few cm^{-1}. This can be interpreted in terms of increased back

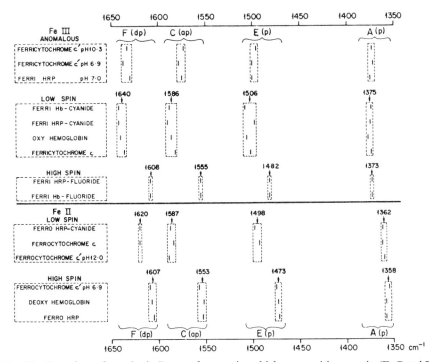

Fig. 16. Groupings of porphyrin Raman frequencies which are sensitive to spin (F, C and E) and oxidation (F and A) state of the heme (Rakshit and Spiro, Ref. 104).

donation of electrons from Fe(II) into π^* antibonding porphyrin orbitals, with a slight reduction in the strength of porphyrin bonds. Particularly noticeable in this regard is the polarized band at $\sim 1375\ cm^{-1}$, which is the most intense feature of spectra excited in the Soret region.[90] Essentially insensitive to spin state changes, this band decreases from $\sim 1375\ cm^{-1}$ for Fe(III) hemes to $\sim 1360\ cm^{-1}$ for Fe(II) hemes.[91]

Larger frequency reductions accompany a change from low to high spin iron.[98] In particular, bands at ~ 1640 (dp), [~ 1620 for Fe(II)], ~ 1590 (ip) and ~ 1500 (p) cm^{-1}, shift to ~ 1607, ~ 1555 and $\sim 1478\ cm^{-1}$, respectively. These shifts can be associated with known alterations in heme structure.[124] Low spin hemes are six coordinate, and the iron atom lies in the heme plane. High spin hemes are five coordinate, or if six coordinate, one axial ligand is more weakly bound than the other. The iron atom lies out of the mean heme plane, and the porphyrin ring domes slightly, in the same direction (Fig. 17). This change in

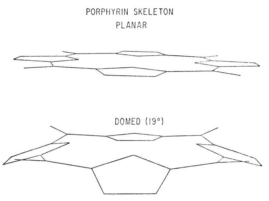

PORPHYRIN SKELETON
PLANAR

DOMED (19°)

Fig. 17. Perspective drawing of the porphyrin skeleton in planar and domed (19° angle between the pyrrole groups and the mean heme plane) conformations.

porphyrin ring conformation is the likely cause of the observed vibrational frequency changes.

A dome angle (the angle of the pyrrole rings with respect to the mean heme plane) of 19° is calculated for deoxyhemoglobin, on the basis of Perutz' estimate,[125] 0.75 Å, for the iron to mean heme plane displacement, and a Fe—N bond length of 2.086 Å, as suggested[126] for high spin Fe(II). Normal coordinate calculations on domed octamethyl porphyrin[122] reproduce the observed frequency shifts (see Table 5) if the force constants are altered slightly in the directions expected, on the basis of simple molecular orbital considerations, for the disruption of porphyrin π-conjugation engendered by the doming. The calculations suggest that the spin state frequency shifts arise primarily from alteration in π-bond order, due to less effective overlap of the π-orbitals in the domed high spin structure.

Initially six bands, labelled A to F, were proposed as markers of spin and/or oxidation state.[98] Subsequent experience indicates, however, that bands B (~ 1560 cm^{-1}, dp) and D (~ 1580 cm^{-1}, p) are unreliable because of interferences from nearby bands, especially in protoheme derivatives.

3.2.6   Applications

COOPERATIVITY IN O$_2$ BINDING TO FE AND CO HEMOGLOBIN

Oxygenation of deoxyhemoglobin produces a change in spin state and a substantial movement of the proximal histidine, bound to the iron atom, as the latter moves into the heme plane. This movement has been proposed by Perutz[125] as the 'trigger' for the quaternary structure change which is responsible for the cooperative binding of oxygen by hemoglobin.

The iron atoms in hemoglobin can be chemically replaced by cobalt, and the resulting protein also binds oxygen reversibly and cooperatively.[127–129] Yet deoxycobalt hemoglobin (CoHb) contains low spin Co(II). Binding of O$_2$ leads to electron transfer (as shown by EPR[130]), but no change in spin state. Since low spin Co(II) can fit into the heme plane, the cooperative binding of O$_2$ by deoxy-CoHb would seem to raise some doubt about the 'trigger' mechanism. It has been argued[126] that the protein, imposing its own constraints, could pull the cobalt atom substantially out of the heme plane in the deoxy form. Yet the doming of the heme group which should accompany such a displacement is not evident in the RR spectra of CoHb.[100] Table 6 shows that the large frequency lowerings of indicator bands,[98] F, C and E, upon removing oxygen from O$_2$FeHb are not observed for O$_2$CoHb, whose small shifts are similar to those which accompany the reduction of ferricytochrome c. Similar small shifts are observed on converting the bis-pyridine complex of Co(III) protoporphyrin-IX to the five coordinate mono-pyridine complex of the Co(II) porphyrin (low spin). Moreover, the cobalt porphyrin frequencies themselves are essentially the same in solution and in the globin matrix.[131] Also, the EPR spectrum of deoxy CoHb is essentially that of five coordinate Co(II) porphyrin.[128,132]

It appears, therefore, that the heme structures in O$_2$CoHb and deoxy CoHb are essentially the same as in six coordinate Co(III) and five coordinate Co(II) porphyrins, respectively. Crystal structures of the latter reveal the cobalt atom to be only 0.15 Å out of the porphyrin plane.[133] The bond between the cobalt atom and the fifth ligand (imidazole) is quite long, however, and its contraction on oxidation to Co(III) porphyrin amounts to some 0.2–0.3 Å. Therefore, the displacement of the bound histidine on oxygenation of CoHb could be as much as 0.35–0.45 Å,[133] without any distortion by the protein. This compares with about 0.9 Å estimated for the corresponding displacement in FeHb. Hopfield[134] has proposed a distributed energy model for the cooperative binding of O$_2$ by hemoglobin, which predicts proportionality between the histidine displacement and the free energy of cooperativity, defined as $-RT\ln K_4/K_1$, where K_1 and K_4 are the first and fourth equilibrium constants for O$_2$ binding. The binding curve for CoHb indicates that the free energy of cooperativity is only about

TABLE 6
Structure-sensitive Raman band positions (cm⁻¹) and shifts in iron and cobalt hemes

Molecule (form)		Band		
		F(dp)	C(ap)	E(p)
Iron hemoglobin[a]				
(Deoxy)		1607	1552	1473
(Oxy)		1640	1586	1506
	Δ	33	34	33
Cytochrome c[a]				
(Ferro)		1620	1584	1493
(Ferri)		1636	1582	1502
	Δ	16	−2	9
Cobalt hemoglobin[b]				
(Deoxy)		1643	1593	1508
(Oxy)		1649	1596	1512
	Δ	6	3	4
Cobalt protoporphyrin IX[c]				
(Co(II)pip)[d]		1642	1593	1508
(Co(III)pip₂)		1645	1596	1509
	Δ	3	3	1

[a] Ref. 98.   [b] Ref. 100.   [c] Ref. 131.   [d] pip = piperidine.

one-third that of FeHb,[100] which is less than the ratio of estimated histidine displacements. On this model, then, the 'trigger' mechanism could still be operative for deoxy-CoHb, despite the cobalt atom lying nearly in the heme plane.

HEMOGLOBIN QUATERNARY STRUCTURE

Perutz[135] reasoned that if the change in heme stereochemistry accompanying the high to low spin transition triggers the hemoglobin quaternary structure change that produces cooperative oxygen binding, then the reverse situation should also be true, i.e. a change in quaternary structure should influence the heme stereochemistry. Evidence for such a relationship comes from the observation that high spin character increases when IHP (inositol hexaphosphate) is added to aquo-methemoglobin, switching it from the oxy- to the deoxy-quaternary structure.[135–137] Consistent with this, the RR spectrum of aquo-methemoglobin shows a slight shift in intensity from the bands characteristic of low spin to those characteristic of high spin Fe(III) heme.[138] No change in the frequencies themselves can be observed, however, indicating that the switch in quaternary structure produces no distortion of the porphyrin ring, but only a shift in the equilibrium between the low and high spin structures. Similarly deoxy-hemoglobin shows no change in its Raman spectrum when it is modified with (N-maleimidoethyl)ether or 2,2'-dicarboxy-4,4' diiodoacetamidoazobenzol, which switch it to the oxyquaternary structure.[99] On the other hand, nitrosyl-hemoglobin does show one extra band, on addition of IHP, which switches it from oxy- to the deoxy-quaternary structure.[106] The significance of this new band is unclear.

ANOMALOUS HEME STRUCTURES IN CYTOCHROME C' AND HORSERADISH PEROXIDASE

Figure 16 shows a correlation diagram for those heme Raman bands, which have been identified as being characteristic of spin state (F, C and E) or oxidation state (F and A), using an expanded set of heme protein derivatives. The frequencies group themselves into narrow ranges in the four different classifications. The three heme derivatives at the top of the figure, ferricytochrome c' (from *Rhodopseudomonas palustris*) at pH 10.3 and 6.9[103] and Fe(III) HRP (horseradish peroxidase) at pH 7.0[104] are listed as anomalous, since they do not fit the correlation. Their frequencies are close to the expected low spin Fe(III) positions but they are definitely not low spin. Fe(III) horseradish peroxidase is essentially high spin on the basis of magnetic susceptibility,[139] Mössbauer[140] and EPR[141] measurements. At pH values less than 11.0 ferricytochrome c' has a magnetic susceptibility intermediate between the values expected for low and high spin Fe(III) heme, which increases on going from pH 6.9 to pH 10.3.[142] This finding was interpreted in terms of a spin state mixture, as in hydroxy-methemoglobin.[143] But whereas the latter shows Raman bands characteristic of both low and high spin heme, cytochrome c' shows only a single set of bands.[103] EPR measurements on ferricytochrome c' from *Chromatium*[144] have been interpreted in terms of an intermediate spin state, $S = 3/2$.

Inasmuch as the frequencies of bands F, C and E, and their shifts from low to high spin hemes, are associated with porphyrin conformation,[122] the position of the anomalous heme frequencies suggests that the heme groups in these proteins are closer to the low spin than the usual high spin structure, i.e. closer to being planar. For HRP this interpretation is plausibly related to the mechanism of the peroxidase reaction, which appears to involve intermediates with heme iron oxidized to Fe(IV).[140,141,145] In a less domed high spin heme, one electron would be in a relatively high energy orbital ($d_{x^2-y^2}$) and might be subject to facile removal.[104]

ELECTRON DISTRIBUTION IN OXYHEMOGLOBIN

While deoxyhemoglobin contains high spin Fe(II), oxyhemoglobin is diamagnetic. A considerable controversy has developed over the electron distribution in O_2Hb since Weiss' suggestion[146] that it is actually a superoxide complex of low spin Fe(III), $Fe^{3+}O_2^-$, with the diamagnetism explained by exchange coupling between the unpaired electrons on Fe^{3+} and O_2^-. Weiss pointed to the similarity of the absorption spectra of O_2Hb and low spin methemoglobin derivatives, and recently Barlow et al.[147] have reported the O_2Hb O—O stretching frequency at 1104 cm^{-1}, characteristic of superoxide ion, from $^{16}O_2$—$^{18}O_2$ difference infrared spectra of packed red cells. On the other hand Collman et al.[148] have reported an O—O stretching frequency at 1385 cm^{-1}, using low temperature infrared spectroscopy, for the model heme–O_2 complex mentioned above, whose short Fe—O distance, consistent with the relatively high Fe—O stretching frequency found in the O_2Hb Raman spectrum by Brunner,[101] supports Pauling's[149] suggestion of substantial multiple bonding (back-bonding) between low spin Fe^{2+} and bound O_2. Collman et al.[148] point out that the 1385 cm^{-1} O—O frequency is ~ 100 cm^{-1} lower than that of singlet O_2, a decrease similar to that observed for CO on binding to Hb.[150]

Yamamoto et al.[91] observed that the 1375 cm^{-1} heme resonance Raman frequency appears to be an indicator of oxidation state† and that O_2Hb classifies as Fe(III) in line with Weiss' model. This classification was confirmed by Spiro and Strekas,[98] who also found, however, that COHb classifies similarly and pointed out that back-bonding to O_2 or CO could account for the Raman frequencies without an actual superoxide (or CO^-) radical being involved. Recently Barron and Szabo[106] have found that NOHb also classifies as Fe(III). In this case EPR measurements show that the unpaired electron on NO is largely transferred to the iron atom on binding to heme, so that Fe(III) seems an unrealistic formulation. A reasonable interpretation is that the ~ 1375 cm^{-1} frequency is responsive to π back-donation from the iron atom to the porphyrin ring. This should be greater for Fe(II) than Fe(III), resulting in a lowering of the 1375 cm^{-1} frequency, as is observed for cytochrome c. But π acid ligands such as O_2, CO and NO selectively deplete the iron π orbitals. Their effect on back-donation to porphyrin appears to be about the same as actual removal of one electron from Fe(II). It is notable that the cyanide complex of reduced horseradish peroxidase[95,104] gives the same frequency, ~ 1362 cm^{-1}, as ferrocytochrome c. Although isoelectronic with CO, CN^- is a much weaker π acceptor, and apparently does not compete effectively with the porphyrin ring for the iron π electrons.

HEME COMPONENTS OF THE MITOCHONDRIAL ELECTRON TRANSPORT CHAIN

Numerous studies of cytochrome c have been carried out,[16,18,20,88,89,93,95–97] mostly for the purpose of elucidating the nature of heme RR spectra, since cytochrome c is easily handled and gives good quality spectra. Other components of the mitochrondrial electron transport chain have begun to be examined.

† Brunner and coworkers[87,90] consider that the decrease from 1378 cm^{-1} in O_2Hb to 1358 cm^{-1} in deoxy-Hb reflects the movement of the iron atom out of the heme plane. The association with out of plane displacement does not hold up when the comparison is extended to other derivatives. Thus the frequency is 1373 for fluorometHb (out of plane) and 1362 cm^{-1} for ferrocytochrome c′ (planar). The correlation with oxidation state is much more general, as shown in Fig. 16.

Adar and Erecinska[102] have reported Raman spectra of succinate-cytochrome c reductase, which contains one c-type cytochrome (c_1) and two b-type cytochromes (b_{561} and b_{566}). In the reduced form, the former has its α absorption band maximum at 552 nm, while the latter have α band maxima at 561 and 566 nm. Adar and Erecinska found that excitation with a 568.2 nm (Kr^+) laser line produced selective enhancement of Raman spectra due to the b cytochromes, while 514.5 nm (Ar^+) excitation gave selective enhancement of the cytochrome c_1 spectrum. Thus, absorption spectral differences between different classes of hemes in a complex aggregate can be exploited to bring out their Raman spectra selectively by appropriate tuning of the laser source.

Salmeen et al.[94] have examined Raman spectra of cytochrome oxidase, which contains two a-type cytochromes (a and a_3). Heme a contains a formyl peripheral substituent, which interacts strongly with the porphyrin π system and shifts the electronic transitions appreciably to lower energies from those of protoheme. Salmeen et al. used 441.6 nm excitation, from a He—Cd laser, which is almost exactly in resonance with the Soret band of reduced cytochrome oxidase (445 nm) and close to resonance with the Soret band of the oxidized form (420 nm). Sensitivity was therefore very high, and they were able to obtain spectra down to 5 μM in heme. Also, electron transport particles gave Raman spectra characteristic of cytochrome oxidase, since the b- and c-type cytochromes which are also present have Soret bands which are higher in energy, and therefore further from resonance with the 441.6 nm He—Cd laser line.

No one has yet reported Raman spectra of isolated heme a, which might be expected to be perturbed by the influence of the formyl group. The cytochrome oxidase spectra of Salmeen et al.[94] are basically similar to those of other heme proteins, excited in the region of the Soret band. There are some differences, however, particularly two new bands, at 215 and 1660 cm^{-1}, which appear in the spectrum of reduced cytochrome oxidase, but not in the spectrum of the oxidized form or in the spectra of other heme proteins. Assignment to the C=O stretch of the formyl group is suggested for the 1660 cm^{-1} band, which apparently goes out of resonance upon oxidation, conceivably through a structure change.

Like ferrocytochrome c, reduced cytochrome oxidase has a strong polarized band at 1358 cm^{-1}, as do reduced electron transport particles. Whereas this band (which is the oxidation state marker band A) shifts to 1374 cm^{-1} on oxidation of ferrocytochrome c, it does not move at all on oxidation of the electron transport particles, although other spectral changes are observed, including the disappearance of the 215 and 1660 cm^{-1} bands as mentioned above. Oxidation of reduced cytochrome oxidase gives a spectrum with two bands, one at 1358 cm^{-1} and the other at 1375 cm^{-1}, each with about half the intensity of the 1358 cm^{-1} band in reduced cytochrome oxidase. The reason for the curious behavior of the 1358 cm^{-1} band is not understood, but the appearance of two bands in the spectrum of oxidized cytochrome oxidase may be an indicator of inequivalence between heme a and heme a_3.

3.3 Other Delocalized π-Systems

3.3.1 Chlorophyll

The tetrapyrrole ring of chlorophyll is similar to that of heme, but the β carbon atoms of one of the pyrrole rings are saturated, and the symmetry is lowered by other modifications as well. Chlorophyll is highly photolabile and presents formidable experimental difficulties with respect to laser excitation. These have been surmounted by Lutz and his coworkers[151-154] who have reported RR spectra of plant chlorophyll both *in vitro* and *in vivo*. At wavelengths longer than 600 nm, the Raman spectra were obscured by fluorescence. Good quality spectra were obtainable with the various Ar^+ laser lines and with the He—Cd line at 441.6 nm.

The Raman spectra are similar to those of heme, but are more complex, presumably due to the lower symmetry. The two types of chlorophyll, a and b (the latter has a formyl group in place of a methyl group as one of the pyrrole β carbon substituents), give similar spectra, but intensities are generally higher for chlorophyll b, since its Soret band is at a somewhat longer wavelength (455 nm) than that of chlorophyll a (433 nm). Excitation profiles[153] confirm that the Soret transition is the dominant resonant electronic transition, although chlorophyll a shows resonance as well with a weaker transition at $\approx$ 498 nm, whose identity is not established. A lack of dependence of the excitation profile maxima on the Raman frequency indicates that this band is not of vibronic origin, as had been proposed. All of the Raman bands are polarized, with the exception of a chlorophyll a band at 1265 cm^{-1} which appears to be anomalously polarized. The depolarization ratios are wavelength dependent, and show complex behavior. In most cases they appear to approach 1/3 at wavelengths close to the Soret band, the value expected for a totally symmetric mode in resonance with a nondegenerate transition. Degeneracy of the Soret transition is lifted by the lack of axial symmetry in the chlorophyll plane.

Among the peripheral substituents which might be conjugated to the tetrapyrrole π system, the vinyl group does not appear to affect the Raman spectrum, and it was suggested that rotation out of the plane prevents conjugation.[154] The C=O stretch of a ketone substituent does show up strongly at $\approx$ 1700 cm^{-1}, however, identifiable by comparison with infrared studies.[155] The formyl C=O stretch of chlorophyll b can also be seen at 1664 cm^{-1}. When the chlorophyll molecules are allowed to aggregate, characteristic decreases in these C=O stretching frequencies are observed.[153] Some modifications of the skeletal modes are also observed.

Raman spectra of spinach chloroplasts and grana[152] show simultaneous contributions from chlorophyll a and b and from carotenoids which are present. These contributions can be selectively enhanced by appropriate adjustment of the laser wavelength. The chlorophyll spectra are seen to best advantage by excitation as close as possible to their Soret maxima, while at wavelengths longer than 475 nm, the carotenoid scattering is dominant. The chlorophyll spectra are similar to those observed *in vitro*, but considerable complexity in the C=O stretching region suggests a variety of states of aggregation.

3.3.2   Vitamin B_{12}

The corrin ring of vitamin B_{12} is also a conjugated tetrapyrrole, although all of the β carbon atoms are saturated, and one of the four methine bridges is replaced with a single bond between two pyrrole rings. Perhaps because of this break in the ring conjugation, the Raman spectra of vitamin B_{12} derivatives[156-159] resemble those of conjugated polyenes (*vide infra*) rather than those of porphyrins. They show a single very intense band near 1500 cm^{-1} and several weaker bands at lower frequencies. All of the bands are polarized. When excitation is changed from the visible region to 363.8 nm,[159] corresponding to a change from resonance with the α transition to resonance with the γ transition, the strong vitamin B_{12a} band at 1504 cm^{-1} drops in intensity while another band, at 1550 cm^{-1}, is greatly enhanced. This alteration in resonance enhancement pattern is reminiscent of that observed in heme Raman spectra on changing the excitation wavelength from the region of the α band to the region of the Soret band, and is similarly attributable to a change in the scattering mechanism.

Vitamin B_{12a}, aquo-cobalamin, contains Co(III) at the center of the corrin ring. A 5,6-dimethyl benzimidazole molecule, tied to the corrin ring through a ribose phosphate linkage, is coordinated to the cobalt atom as an axial ligand, and the other axial ligand is a water molecule. Replacement of the bound water molecule by other ligands, cyanide, methyl or 5'-deoxyadenosyl (B_{12} coenzyme) does not alter the RR spectrum observably. The cobalt atom in vitamin B_{12} can be reduced to Co(II) (B_{12r}) or Co(I) (B_{12s}). These reduction steps *are* accompanied by significant changes in the Raman spectra, new bands appearing with appreciable intensity.[156,158] It has been suggested[156,158] that these new bands reflect changes in corrin ring conformation which allow additional vibrational modes to become resonance enhanced. Wozniak and Spiro[156] considered these changes to result from dissociation of the benzimidazole ligand, since they found the same altered spectrum for methylcobalamin whose benzimidazole was dissociated by protonation at low pH. It seems likely, however, that they were in fact observing the spectrum of vitamin B_{12r}, since 'base off' methylcobalamin is efficiently photolyzed anaerobically to the Co(II) derivative.[158]

3.3.3   Polyolefins: carotenoids and visual pigments

Conjugated polyolefins have been known for some time[160] to give remarkable RR enhancements, particularly for a band at $\sim$1550 cm^{-1}, which is assigned to the in-plane stretching mode of the carbon–carbon double bonds. This mode would be expected to couple effectively to the resonant π–π^* transition, and Franck-Condon overlap calculations[161] bear this out. The excitation profiles for this band, as well as its first and second overtones, are well characterized by A-term scattering, in the case of β-carotene.[162]

The carotenoid pigments β-carotene[160,163,164] and lycopene[164] have been thoroughly studied, and Gill et al.[165] observed their spectra in situ in materials such as carrot and tomato tissue. Indeed the carotenoids are likely to dominate the Raman spectra of any biological sample in which they occur. As mentioned above, this is found to be the case for chloroplasts, unless the laser is carefully tuned to the immediate vicinity of the intense chlorophyll Soret bands.[152]

The visual pigment retinal has been studied by Rimai et al.[164] in its trans, 9-cis, and 13-cis stereoisomers. The enhanced Raman frequencies were found to be sensitive to isomerization and to substituents. Photoisomerism (to the trans form) was observed with blue-green exciting radiation in the chromophore absorption band, but not with red radiation. Retinal Schiff bases were also studied,[166] since it is well established that in rhodopsin, retinal is bound through Schiff base condensation of its carbonyl group with an amine group on the lipoprotein. The position of the –C=C– mode was found to correlate with the wavelength maximum of the chromophore absorption, suggesting that the vibrational frequency is a measure of π delocalization.

Rhodopsin itself has been the focus of recent studies. In 1970 Rimai et al.[167] obtained Raman spectra from frozen ($-70°C$) bovine retinas, which they assigned to a photoproduct, lumirhodopsin. The spectra were not of high quality, and only the –C=C– stretching frequency, 1555 cm^{-1}, could be obtained. Comparison with a model Schiff base, trans-retinylidene hexylamine, whose –C=C– stretching frequency, 1582 cm^{-1}, shifts to 1560 cm^{-1} on protonation, led to the suggestion that the Schiff base linkage in rhodopsin is protonated. Moreover, the frequency of the retina was found to be consistent with the above mentioned correlation with the absorption wavelength for protonated retinal Schiff bases.[166] Mendelsohn[168] examined the photoreceptor-like pigment of *Holobacterium halobium*, which has the advantage that, unlike mammalian rhodopsin, which is readily bleached, it undergoes only reversible photoreaction upon irradiation at room temperature. The –C=C– stretching frequency was found to be even lower, 1531 cm^{-1}, well below any model Schiff base. Upon addition of chloroform it decreased still further, to 1520 cm^{-1}, as did the absorption wavelength maximum. These characteristics suggested the possible involvement of a charge transfer complex between the bound retinal and a side chain of the protein. Mendelsohn located the –C=N– stretching mode at 1622 cm^{-1}, which indicated that the linkage was unprotonated, since for the hexylamine Schiff base of retinal, the –C=N– stretch is at 1623 cm^{-1}, shifting to 1645 cm^{-1} on protonation. Lewis et al.[169] suggest, however, that Mendelsohn's spectrum arose from a photo-intermediate, produced reversibly upon laser irradiation. Their own study of the bacterial pigment shows that the –C=N– linkage is protonated in the ground state ($\lambda_{max} = 570$ nm) giving a –C=N– stretching frequency of 1646 cm^{-1}. They observed a spectrum similar to Mendelsohn's, associated with a known photo-intermediate ($\lambda_{max} = 412$ nm).

The Raman evidence for photo-deprotonation of the bacterial pigment is plausibly connected with its ability to produce a proton flux across a membrane upon irradiation.[169]

Lewis et al.[170] had earlier obtained good quality spectra of digitonen extracts of bovine retina. They showed the $-C=N-$ stretch at 1645 cm^{-1}, indicating protonation, and the $-C=C-$ stretch at 1549 cm^{-1}, slightly lower than the value that Rimai et al.[167] had reported. Moreover, in addition to many weaker bands characteristic of retinal Schiff bases, the rhodopsin spectra showed resonance enhanced bands between 550 cm^{-1} and 950 cm^{-1}, at frequencies slightly lower than known ring vibrations of phenylalanine, tyrosine and tryptophan. It was suggested[170] that these bands do indeed arise from perturbed vibrations of aromatic side-chains, brought into resonance with the rhodopsin electronic transition via π-interactions with the retinylidene chromophore.

Lewis et al.[170] demonstrated that their Raman spectra were due to rhodopsin, and not an irreversibly formed photo-product, by monitoring the absorption spectrum before and after irradiation. Oseroff and Callender[171] have shown, however, that it is possible to set up a photostationary state, with significant concentrations of reversible photo-intermediates, under the conditions of laser illumination (5–50 mW) needed for Raman spectroscopy. They examined rhodopsin in bovine retinal rod outer segments at 80°K. At this temperature irradiation of rhodopsin, which contains 11-cis-retinal, reversibly converts it to isorhodopsin, which contains 9-cis-retinal, via a photo-intermediate, bathorhodopsin, whose conformation is unknown. To manipulate the composition of the sample, Oseroff and Callender employed a 'pump' laser beam, to establish the photo-stationary state, coaxial with a 'probe' beam, which produced the Raman spectrum. They were unable to separate out the rhodopsin spectrum, which however appeared to be quite similar to the spectrum of isorhodopsin. The latter contained the same major features reported by Lewis for rhodopsin, but apparently contained no bands below 960 cm^{-1}. The bathorhodopsin spectrum was appreciably altered. Its $-C=C-$ stretch was lowered to 1539 cm^{-1}, and three new bands appeared at 856, 877 and 920 cm^{-1}. (These do not correspond to the rhodopsin frequencies reported by Lewis et al.) Oseroff and Callender concluded that formation of bathorhodopsin is not a simple cis-trans isomerization, since the observed Raman frequencies are different from those of all-trans model compounds.

3.3.4   Nucleotide bases and aromatic protein side chains

When far ultraviolet Raman laser sources become generally available, new classes of biological chromophores will become accessible to RR studies. These include the very important nucleic acid purine and pyrimidine bases and the aromatic side chains of proteins. These aromatic ring chromophores absorb strongly at and below the wavelength region 260–280 nm. The electronic transitions involved are $\pi-\pi^*$, and in the case of nitrogen heterocycles, $n-\pi^*$ as well.[172] The former should provide strong resonance enhancement of in-plane stretching modes. This has recently been demonstrated for adenine by Tsuboi,[173] using 257.2 nm radiation obtained by doubling the frequency of the 514.5 nm Ar$^+$ laser line with a crystal doubler.

The absorption spectra of the biological aromatic residues are somewhat dependent on their macromolecular environment, and it is likely that their RR intensities, and possibly the frequencies as well, will show similar dependencies. Indeed, such effects have already been seen in ordinary Raman spectra and are attributable to preresonance enhancement. Thus some of the purine and pyrimidine ring modes are greatly intensified when helical polynucleotides change to random conformations. The ultraviolet absorption also increases, because of the decreased electronic interaction between the bases in the poly-nucleotide strand. Small and Peticolas[174] suggested that the two phenomena are directly linked. Although the conventional Ar$^+$ laser lines are far from resonance with the 260 nm electronic absorption, the relatively high intensity of the aromatic ring modes may nevertheless reflect preresonance enhancement. When the transition moment of the resonant electronic transition increases, the Raman enhancement increases correspondingly (see Eqn. 1). Aromatic ring modes also show up strongly in normal protein Raman spectra,[175] probably reflecting preresonance enhancement. There is evidence that tyrosine ring mode intensities are sensitive to the molecular environment.[176]

4   RESONANCE RAMAN LABELS

The chief limitation of RR spectroscopy as a biological structure probe is that not all sites of biological interest provide suitable chromophores for resonance enhancement of structurally sensitive vibrational modes. The scope of the technique can be extended, however, by using RR labels, analogous to spin or fluorescence labels. Small molecules with favorable RR characteristics can be attached to sites of biological function to report back structural information via their perturbed vibrational frequencies, or Raman intensities. Favorable characteristics include photostability, lack of fluorescence, and enhancement of vibrational modes which are structurally interpretable and which are influenced by structural features of the binding site.

The label technique was pioneered by Carey, Schneider and Bernstein,[177] who studied the RR spectrum of methyl orange bound to bovine serum albumin. The spectrum was similar to that shown by solid methyl orange. All the frequencies were decreased by 2–5 cm^{-1} from the values observed for methyl orange in aqueous solution, consistent with a lowering of the dielectric constant at the protein binding sites. There was also evidence from the disappearance of a band at 1168 cm^{-1} that the sulfonate group of methyl orange interacts at an unsymmetrical charge site on the protein. There was no evidence for conformation change on binding, however, contrary to a previous interpretation of visible spectral shifts.

Carey and coworkers have extended the use of RR labels to hapten–antibody interactions,[178] enzyme intermediates[179] and enzyme-inhibitor binding.[180] As labelled haptens,[178] they used 1-hydroxy-2-(2,4-dinitrophenylazo)-2,5-naphthalenedisulfonic acid, 1-hydroxy-2-(2,4-dinitrophenylazo)-3,6-naphthalenedisulphonic acid and 6-(2,4-dinitrophenyl)lysine. These were bound to bovine γ-globulin to produce rabbit antibodies, and RR spectra of the haptens bound to the antibodies were examined. Appreciable changes were found with respect to the unbound chromophores. The nitro group frequencies were in each case lowered by a few cm^{-1}, consistent with interaction with dipolar or charged sites on the antibodies, but not with twisting of the nitro groups. The azo group frequencies did show evidence of twisting about the azo bonds, induced by binding to the antibody. Two populations of differently twisted azo bonds were observed, and were interpreted in terms of structural heterogeneity in the antibody population.

The enzyme intermediates studied[179] were the 4-amino-3-*trans*cinnamoyl and 2-hydroxy-3-nitro-α-toluyl derivatives of α-chymotrypsin. The former was prepared as a stable intermediate at low pH, while the latter was studied with a continuous flow technique. A $5\ cm^{-1}$ increase in the $1350\ cm^{-1}$ symmetric $-NO_2$ stretch of the cinnamoyl derivative was interpreted in terms of a slight twisting about the $C—NO_2$ bond, while the constancy of the ethylenic stretch at $1625\ cm^{-1}$ was taken as an indication of preserved *trans* conformation at the $-C=C-$ bond and coplanarity of the phenyl ring. The toluyl derivative showed intensity changes in the $900–1200\ cm^{-1}$ region, suggesting interaction of the phenyl ring with the enzyme active site. For a study of enzyme-inhibitor interactions, Kumar *et al.*[180] chose 4-sulfonamido-4'-dimethylamino-azobenzene and 4-sulfonamido-4'-hydroxyazobenzene, which are powerful inhibitors of carbonic anhydrase, and which provide strongly enhanced Raman spectra. The vibrational signatures provided support for binding of anionic forms of the sulfonamides to the zinc ion at the active site, with loss of a proton from the sulfonamide group. No significant twisting of the azo bond was observed.

ACKNOWLEDGMENT

Cited work of the two authors has been supported by grants from the National Institutes of Health (GM 18865 to T.M.L. and GM 13498 and HL 12526 to T.G.S.) and the National Science Foundation (GP 41008X to T.G.S.).

REFERENCES

(1) G. R. Harrison, R. C. Lord and J. R. Loofbourow, *Practical Spectroscopy*, Chap. 18, Prentice-Hall, Englewood Cliffs, N.J., 1948.

(2) S. P. S. Porto and D. L. Wood, *J. Opt. Soc. Amer.* **52**, 251 (1962).

(3) T. R. Gilson and P. J. Hendra, *Laser Raman Spectroscopy*, Wiley (Interscience), New York, 1970.

(4) M. C. Tobin, *Laser Raman Spectroscopy*, Wiley (Interscience), New York, 1970.
(5) W. Kiefer and H. J. Bernstein, *Appl. Spectrosc.* **25**, 500 (1971).
(6) W. H. Woodruff and T. G. Spiro, *Appl. Spectrosc.* **28**, 74 (1974).
(7) G. Placzek, *Rayleigh and Raman Scattering*, UCRL Trans. No. 526L from *Handbuch der Radiologie* (E. Marx, Ed.), Vol. 2, Akademische Verlagsgesellschaft, Leipzig, 1934, p. 209.
(8) P. P. Shorygin and T. M. Ivanova, *Dokl. Akad. Nauk SSSR* **121**, 70 (1958).
(9) P. P. Shorygin and T. M. Ivanova, *Opt. Spectrosc.* **15**, 94 (1963).
(10) J. Behringer, *Z. Electrochem.* **62**, 544 (1958).
(11) T. G. Spiro, in *Chemical and Biochemical Applications of Lasers* (C. B. Moore, Ed.), Chap. 2, Academic Press, New York, 1974.
(12) T. G. Spiro, *Acc. Chem. Res.* **7**, 339 (1974).
(13) J. H. Van Vleck, *Proc. Nat. Acad. Sci. USA* **15**, 754 (1929).
(14) J. Tang and A. C. Albrecht, in *Raman Spectroscopy* (H. A. Szymanski, Ed.), Vol. 2, Plenum Press, New York, 1970, Chapter 2.
(15) J. A. Koningstein, *Introduction to the Theory of the Raman Effect*, D. Reidel Publishing Co., Dordrecht, Holland, 1972.
(16) J. M. Friedman and R. M. Hochstrasser, *Chem. Phys.* **1**, 457 (1973).
(17) W. L. Peticolas, L. Nafie, P. Stein and B. Fanconi, *J. Chem. Phys.* **52**, 1576 (1970).
(18) D. W. Collins, D. B. Fitchen and A. Lewis, *J. Chem. Phys.* **59**, 5714 (1973).
(19) A. C. Albrecht and M. C. Hutley, *J. Chem. Phys.* **55**, 4438 (1971).
(20) T. G. Spiro and T. C. Strekas, *Proc. Nat. Acad. Sci. USA* **69**, 2622 (1972).
(21) T. C. Strekas and T. G. Spiro, *J. Raman Spectrosc.* **1**, 387 (1973).
(22) B. P. Gaber, V. Miskowski and T. G. Spiro, *J. Amer. Chem. Soc.* **96**, 6868 (1974).
(23) W. M. McClain, *J. Chem. Phys.* **55**, 2789 (1971).
(24) O. S. Mortensen and J. A. Koningstein, *J. Chem. Phys.* **48**, 3971 (1968).
(25) P. Stein, V. Miskowski, W. H. Woodruff, J. P. Griffin, K. G. Werner, B. P. Gaber and T. G. Spiro, *J. Chem. Phys.* (submitted for publication).
(26) C. M. Yoshida, T. B. Freedman and T. M. Loehr, *J. Amer. Chem. Soc.* **97**, 1028 (1975).
(27) S. Lindskog, *Struct. Bonding (Berlin)* **8**, 153 (1970).
(28) C. Lazdunski, C. Petitclerc, D. Chappelet and M. Lazdunski, *Biochem. Biophys. Res. Commun.* 35, 744 (1969).
(29) G. H. Tait and B. L. Vallee, *Proc. Nat. Acad. Sci. USA* **56**, 1247 (1966).
(30) F. Ghiretti (Ed.), *Physiology and Biochemistry of Haemocyanins*, Academic Press, New York, 1968.
(31) K. E. VanHolde and E. F. J. van Bruggen, in *Subunits in Biological Systems* (S. N. Timasheff and G. D. Fasman, Eds.), Marcel Dekker, New York, 1971, pp. 1–53.
(32) T. H. Moss, D. C. Gould, A. Ehrenberg, J. S. Loehr and H. S. Mason, *Biochemistry* **12**, 2444 (1973).
(33) C. Simo, *Ph.D. Thesis*, University of Oregon Medical School, Portland, Oregon, 1966, p. 34.
(34) J. S. Loehr, T. B. Freedman and T. M. Loehr, *Biochem. Biophys. Res. Commun.* **56**, 510 (1974).
(35) J. C. Evans, *Chem. Commun.* 682 (1969).
(36) T. B. Freedman, C. M. Yoshida and T. M. Loehr, *J. C. S. Chem. Commun.* 1016 (1974).
(37) J. S. Valentine, *Chem. Rev.* **73**, 235 (1973).
(38) A. C. Albrecht and M. C. Hutley, *J. Chem. Phys.* **55**, 4438 (1971).
(39) A. G. Sykes and J. A. Weil, *Prog. Inorg. Chem.* **13**, 1 (1970).
(40) D. R. McMillin, R. A. Holwerda and H. B. Gray, *Proc. Nat. Acad. Sci. USA* **71**, 1339 (1974).
(41) W. E. Hatfield, C. S. Fountain and R. Whyman, *Inorg. Chem.* **5**, 1855 (1966).
(42) R. Malkin and B. G. Malmstrom, *Adv. Enzymol.* **33**, 177 (1970).
(43) M. Ettinger, *Biochemistry* **13**, 1242 (1974).

(44) M. Kato, H. B. Jonassen and J. C. Fanning, *Chem. Rev.* **64**, 99 (1964).

(45) I. M. Klotz, in *Subunits in Biological System* (S. N. Timasheff and G. D. Fasman, Eds.), Marcel Dekker, New York, 1971, pp. 55–103.

(46) M. H. Lee and C. R. Dawson, *J. Biol. Chem.* **248**, 6596 (1973).

(47) A. S. Brill and G. F. Bryce, *J. Chem. Phys.* **48**, 4398 (1968).

(48) W. E. Blumberg, in *The Biochemistry of Copper* (J. Peisach, P. Aisen and W. E. Blumberg, Eds.), Academic Press, New York, 1966, pp. 49–66.

(49) R. J. P. Williams, *Inorg. Chim. Acta Rev.* **5**, 137 (1971).

(50) J. Peisach, W. G. Levine and W. E. Blumberg, *J. Biol. Chem.* **242**, 2847 (1967).

(51) A. Finazzi-Agro, G. Rotilio, L. Avigliano, P. Guerrieri, V. Boffi and B. Mondovi, *Biochemistry* **9**, 2009 (1970).

(52) O. Siiman, N. M. Young and P. R. Carey, *J. Amer. Chem. Soc* **96**, 5583 (1974).

(53) V. Miskowski, S.-P. W. Tang, T. G. Spiro, E. Shapiro and T. H. Moss, *Biochemistry* **14**, 1244 (1975).

(54) K. Nakamoto, *Infrared Spectra of Inorganic and Coordination Compounds*, 2nd ed., John Wiley & Sons, New York, 1970, pp. 256–258.

(55) H. C. Freeman, in *Inorganic Biochemistry*, Vol. 1 (G. L. Eichorn, Ed.), Elsevier, New York, 1973, pp. 121–166.

(56) D. M. L. Goodgame, M. Goodgame and G. W. Rayner Canham, *J. Chem. Soc. A* 1923 (1971).

(57) J. Tang and A. C. Albrecht, in *Raman Spectroscopy*, Vol. 2 (H. A. Szymanski, Ed.), Plenum Press, New York, 1970, Chap. 2.

(58) A. Y. Hirakawa and M. Tsuboi, *Science* **188**, 359 (1975).

(59) P. G. Eller and P. W. R. Corfield, *Chem. Commun.* 105 (1971).

(60) D. Cooper and R. A. Plane, *Inorg. Chem.* **5**, 16 (1966).

(61) B. L. Vallee and R. J. P. Williams, *Proc. Nat. Acad. Sci. USA* **59**, 498 (1968).

(62) W. H. Orme-Johnson, *Ann. Rev. Biochem.* **42**, 159 (1973).

(63) S. J. Lippard, *Acc. Chem. Res.* **6**, 282 (1973).

(64) R. Mason and J. A. Zubieta, *Angew. Chem. Inter. Ed. Engl.* **12**, 390 (1973).

(65) K. D. Watenpaugh, L. C. Sieker, J. R. Herriott and L. H. Jensen, *Acta Cryst.* **B29**, 943 (1973).

(66) T. V. Long, II and T. M. Loehr, *J. Amer. Chem. Soc.* **92**, 6384 (1970).

(67) W. Lovenberg and W. A. Eaton, *Fed. Proc.* **29**, 1963 (1970); W. Lovenberg and B. E. Sobel, *ibid.* **24**, 233 (1965).

(68) T. V. Long, II, T. M. Loehr, J. R. Allkins and W. Lovenberg, *J. Amer. Chem. Soc.* **93**, 1809 (1971).

(69) T. Yamamoto, G. Palmer, L. Rimai, D. Gill and L. Salmeen, *Fed. Proc.* **33**, 1372 (1974).

(70) T. Kimura, *Struct. Bonding (Berlin)* **5**, 1 (1968).

(71) E. T. Adman, L. C. Sieker and L. H. Jensen, *J. Biol. Chem.* **248**, 3987 (1973).

(72) S.-P. W. Tang, T. G. Spiro, K. Mukai and T. Kimura, *Biochem. Biophys. Res. Commun.* **53**, 869 (1973).

(73) J. W. Dawson, H. B. Gray, H. E. Hoenig, G. R. Rossman, J. M. Schredder and R. H. Wang, *Biochemistry* **11**, 461 (1972).

(74) J. B. R. Dunn, D. F. Shriver and I. M. Klotz, *Proc. Nat. Acad. Sci. USA* **70**, 2582 (1973).

(75) K. Garbett, C. E. Johnson, I. M. Klotz, M. Y. Okamura and R. J. P. Williams, *Arch. Biochem. Biophys.* **142**, 574 (1971).

(76) H. B. Gray, *Adv. Chem. Ser.* **100**, 365 (1971).

(77) J. B. R. Dunn, *Ph.D. Thesis*, Northwestern University, Evanston, Illinois, 1974.

(78) S. McCoy and W. S. Caughey, *Biochemistry* **9**, 2387 (1970).

(79) R. E. Feeney and S. K. Komatsu, *Struct. Bonding (Berlin)* **1**, 149 (1966).

(80) B. L. Vallee and W. E. C. Wacker, in *The Proteins* (H. Neurath, Ed.), Vol. 5, Academic Press, New York, 1970.

(81) E. M. Price and J. F. Gibson, *Biochem. Biophys. Res. Commun.* **46**, 646 (1972).

(82) Y. Tomimatsu, S. Kint and J. R. Scherer, *Biochem. Biophys. Res. Commun.* **54**, 1067 (1973).

(83) P. R. Carey and N. M. Young, *Can. J. Biochem.* **52**, 273 (1974).

(84) R. J. Jakobsen, *Spectrochim. Acta* **21A**, 433 (1965).

(85) S. Pinchas, *Spectrochim. Acta* **28A**, 801 (1972).

(86) T. C. Strekas and T. G. Spiro, *Biochim. Biophys. Acta* **263**, 830 (1972).

(87) H. Brunner, A. Mayer and H. Sussner, *J. Mol. Biol.* **70**, 153 (1972).

(88) T. C. Strekas and T. G. Spiro, *Biochim. Biophys. Acta* **278**, 188 (1972).

(89) H. Brunner, *Biochem. Biophys. Res. Commun.* **51**, 888 (1973).

(90) H. Brunner and H. Sussner, *Biochim. Biophys. Acta* **310**, 20 (1973).

(91) T. Yamamoto, G. Palmer, D. Gill, I. T. Salmeen and L. Rimai, *J. Biol. Chem.* **248**, 5211 (1973).

(92) T. C. Strekas, A. J. Packer and T. G. Spiro, *J. Raman Spectrosc.* **1**, 197 (1973).

(93) L. A. Nafie, M. Pezolet and W. L. Peticolas, *Chem. Phys. Lett.* **20**, 563 (1973).

(94) I. Salmeen, L. Rimai, D. Gill, T. Yamamoto, G. Palmer, C. R. Hartzell and H. Beinert, *Biochem. Biophys. Res. Commun.* **52**, 1100 (1973).

(95) T. M. Loehr and J. S. Loehr, *Biochem. Biophys. Res. Commun.* **55**, 218 (1973).

(96) M. Pezolet, L. A. Nafie and W. L. Peticolas, *J. Raman Spectrosc.* **1**, 455 (1973).

(97) J. Nestor and T. G. Spiro, *J. Raman Spectrosc.* **1**, 539 (1973).

(98) T. G. Spiro and T. C. Strekas, *J. Amer. Chem. Soc.* **96**, 338 (1974).

(99) H. Sussner, A. Mayer, H. Brunner and H. Fasold, *Eur. J. Biochem.* **41**, 465 (1974).

(100) W. H. Woodruff, T. G. Spiro and T. Yonetani, *Proc. Nat. Acad. Sci. USA* **71**, 1065 (1974).

(101) H. Brunner, *Naturwissenschaften* **61**, 129 (1974).

(102) F. Adar and M. Erecinska, *Arch. Biochem. Biophys.* **165**, 570 (1974).

(103) T. C. Strekas and T. G. Spiro, *Biochim. Biophys. Acta* **351**, 237 (1974).

(104) G. Rakshit and T. G. Spiro, *Biochemistry* **13**, 5317 (1974).

(105) H. Brunner, A. Mayer, K. Gersonde and K. Winterhalter, *FEBS Lett.* (1974).

(106) L. D. Barron and A. Szabo, *J. Amer. Chem. Soc.* **97**, 660 (1975).

(107) K. N. Solovyov, N. M. Ksenfontova, S. F. Shkirman and T. F. Kachura, *Spectrosc. Lett.* **6**, 455 (1973).

(108) (a) A. L. Verma and H. J. Bernstein, *Biochem. Biophys. Res. Commun.* **57**, 255 (1974); (b) S. Sunder and H. J. Bernstein, *Can. J. Chem.* **52**, 2851 (1974).

(109) (a) A. L. Verma, R. Mendelsohn and H. J. Bernstein, *J. Chem. Phys.* **61**, 383 (1974); (b) A. L. Verma and H. J. Bernstein, *J. Chem. Phys.* **61**, 2560 (1974).

(110) (a) A. L. Verma and H. J. Bernstein, *J. Raman Spectrosc.* **2**, 163 (1974); (b) R. Mendelsohn, S. Sunder, A. L. Verma and H. J. Bernstein, *J. Chem. Phys.* **62**, 37 (1975).

(111) R. Plus and M. Lutz, *Spectrosc. Lett.* **7**, 7384, 133 (1974).

(112) R. H. Felton, N. T. Yu, D. C. O'Shea and J. A. Shelnutt, *J. Amer. Chem. Soc.* **96**, 3675 (1974).

(113) T. Kitagawa, H. Ogoshi, E. Watanabe and Z. Yoshida, *Chem. Phys. Lett.* **30**, 451 (1975).

(114) T. G. Spiro, *Biochim. Biophys. Acta Reviews on Bioenergetics* (in press).

(115) (a) M. Gouterman, *J. Chem. Phys.* **30**, 1139 (1959); (b) M. Gouterman, *J. Mol. Spectrosc.* **6**, 138 (1961).

(116) (a) D. W. Smith and R. J. P. Williams, *Struct. Bonding (Berlin)* **7**, 1 (1970); (b) P. S. Braterman, R. C. Davies and R. J. P. Williams, *Adv. Chem. Phys.* **3**, 359 (1964).

(117) W. A. Eaton and R. M. Hochstrasser, *J. Chem. Phys.* **49**, 985 (1968).

(118) M. H. Perrin, M. Gouterman and C. L. Perrin, *J. Chem. Phys.* **50**, 4137 (1969).

(119) M. Gouterman, F. P. Schwarz, P. D. Smith and D. Dolphin, *J. Chem. Phys.* **59**, 676 (1969).

(120) M. W. Makinen and W. A. Eaton, *Ann. N.Y. Acad. Sci.* **206**, 210 (1973).

(121) J. P. Collman, R. R. Gagne, C. A. Reed, W. T. Robinson and G. A. Rodley, *Proc. Nat. Acad. Sci. USA* **71**, 1326 (1974).

(122) P. Stein, M. J. Burke and T. G. Spiro, *J. Amer. Chem. Soc.* (in press).

(123) H. Ogoshi, Y. Saito and K. Nakamoto, *J. Chem. Phys.* **57**, 4194 (1972).

(124) J. L. Hoard, *Science* **175**, 1295 (1971).

(125) M. F. Perutz, *Nature* **228**, 726 (1970).

(126) J. L. Hoard and W. R. Scheidt, *Proc. Nat. Acad. Sci. USA* **70**, 3919 (1973).

(127) B. M. Hoffman and D. H. Petering, *Proc. Nat. Acad. Sci. USA* **67**, 637 (1970).

(128) G. C. Hsu, C. A. Spillburg, C. Brill and B. M. Hoffman, *Proc. Nat. Acad. Sci. USA* **69**, 2212 (1972).

(129) T. Yonetani, H. Yamamoto and G. V. Woodrow, *J. Biol. Chem.* **296**, 682 (1974).

(130) B. M. Hoffman, D. Diemente and F. Basolo, *J. Amer. Chem. Soc.* **92**, 61 (1970).

(131) W. H. Woodruff, D. M. Adams, T. G. Spiro and T. Yonetani, *J. Amer. Chem. Soc.* (1975) (in press).

(132) J.-C. W. Chien and L. C. Dickinson, *Proc. Nat. Acad. Sci. USA* **69**, 4452 (1974).

(133) R. G. Little and J. A. Ibers, *J. Amer. Chem. Soc.* **96**, 4452 (1974).

(134) J. J. Hopfield, *J. Mol. Biol.* **77**, 207 (1973).

(135) M. F. Perutz, *Nature* **237**, 495 (1972).

(136) J. V. Kilmartin, *Biochem. J.* **133**, 725 (1973).

(137) M. F. Perutz, A. R. Feisht, S. R. Simon and G. K. Roberts, *Biochemistry* **13**, 2174 (1974).

(138) W. C. Topp and T. G. Spiro (1975) (submitted for publication).

(139) D. Keilin and E. F. Hartree, *Biochem. J.* **49**, 88 (1951).

(140) T. H. Moss, A. Ehrenberg and A. J. Bearden, *Biochemistry* **8**, 4159 (1969).

(141) W. E. Blumberg, J. Peisach, B. A. Wittenberg and J. B. Wittenberg, *J. Biol. Chem.* **243**, 1854 (1968).

(142) A. Ehrenberg and M. D. Kamen, *Biochim. Biophys. Acta* **140**, 284 (1965).

(143) T. Iizuki and M. Kotani, *Biochim. Biophys. Acta* **194**, 351 (1969).

(144) M. Maltempo, T. H. Moss and M. A. Cusanovich, *Biochim. Biophys. Acta* **342**, 290 (1974).

(145) B. Chance, *Arch. Biochem. Biophys.* **21**, 416 (1949).

(146) J. Weiss, *Nature* **203**, 83 (1964).

(147) C. H. Barlow, J. C. Maxwell, W. J. Wallace and W. S. Caughey, *Biochem. Biophys. Res. Commun.* **55**, 91 (1973).

(148) J. P. Collman, R. R. Gagne, H. B. Gray and J. W. Hare, *J. Amer. Chem. Soc.* **96**, 6522 (1974).

(149) L. Pauling, *Nature* **203**, 182 (1964).

(150) J. O. Alben and W. S. Caughey, *Biochemistry* **7**, 175 (1968).

(151) M. Lutz, *C.R. Acad. Sci.* (*Paris*) **275B**, 497 (1972).

(152) M. Lutz and J. Breton, *Biochem. Biophys. Res. Commun.* **53**, 413 (1973).

(153) M. Lutz and J. Kleo, *C.R. Acad. Sci.* (*Paris*) **279D**, 1413 (1974).

(154) M. Lutz, *J. Raman Spectrosc.* (in press).

(155) K. Ballschmiter and J. J. Katz, *J. Amer. Chem. Soc.* **91**, 2661 (1969).

(156) W. T. Wozniak and T. G. Spiro, *J. Amer. Chem. Soc.* **95**, 3402 (1973).

(157) E. Mayer, D. J. Gardiner and R. E. Hester, *Biochim. Biophys. Acta* **297**, 568 (1973).

(158) E. Mayer, D. J. Gardiner and R. E. Hester, *J.C.S. Faraday Trans. II* **69**, 1350 (1973).

(159) E. Mayer, D. J. Gardiner and R. E. Hester, *Mol. Phys.* **26**, 783 (1973).

(160) J. Behringer and J. Brandmüller, *Ann. Phys.* (*Leipzig*) **4**, 234 (1959).

(161) A. Warshel and M. Karplus, *J. Amer. Chem. Soc.* **96**, 5677 (1974).

(162) F. Inagaki, M. Tasumi and T. Miyazawa, *J. Mol. Spectrosc.* **50**, 286 (1974).

(163) L. Rimai, R. G. Kilponen and D. Gill, *J. Amer. Chem. Soc.* **92**, 3824 (1970).

(164) L. Rimai, D. Gill and J. L. Parsons, *J. Amer. Chem. Soc.* **93**, 1353 (1971).

(165) D. Gill, R. G. Kilponen and L. Rimai, *Nature* **227**, 743 (1970).

(166) M. E. Heyde, D. Gill, R. G. Kilponen and L. Rimai, *J. Amer. Chem. Soc.* **93**, 6776 (1971).

(167) L. Rimai, R. G. Kilponen and D. Gill, *Biochem. Biophys. Res. Commun.* **41**, 422 (1970).

(168) R. Mendelsohn, *Nature* **243**, 22 (1973).

(169) A. Lewis, J. Spoonhower, R. A. Bogomolic, R. H. Lozier and W. Stockenius, *Proc. Nat. Acad. Sci. USA* **71**, 4462 (1971).

(170) (a) A. Lewis, R. S. Fager, E. W. Abrahamson, *J. Raman Spectrosc.* **1**, 465 (1973); (b) A. Lewis and J. Spoonhower, in *Neutron, X-Ray and Laser Spectroscopy in Biophysics* (S. Yip and S. Chen, Eds.), Academic Press, N.Y. (in press).

(171) A. R. Oseroff and R. H. Callender, *Biochemistry* **13**, 4243, (1974).

(172) (a) W. Hug and I. Tinoco, Jr., *J. Amer. Chem. Soc.* **95**, 2803 (1973); (b) *ibid.* **96**, 665 (1974).

(173) M. Tsuboi, reported at the Fourth International Conference on Raman Spectroscopy, Brunswick, Maine, Aug. 25–29, 1974.

(174) E. W. Small and W. L. Peticolas, *Biopolymers* **10**, 1377 (1971).

(175) (a) R. C. Lord and N-T. Yu, *J. Mol. Biol.* **50**, 609 (1970); (b) *ibid.* **51**, 203 (1970).

(176) N-T. Yu, B. H. To and D. C. O'Shea, *Arch. Biochem. Biophys.* **156**, 71 (1973).

(177) P. R. Carey, H. Schneider and H. J. Bernstein, *Biochem. Biophys. Res. Commun.* **47**, 588 (1972).

(178) P. R. Carey, A. Froese and H. Schneider, *Biochemistry* **12**, 2198 (1973).

(179) P. R. Carey and H. Schneider, *Biochem. Biophys. Res. Commun.* **57**, 831 (1974).

(180) K. Kumar, R. W. King and P. R. Carey, *FEBS Lett.* **48**, 283 (1974).

Chapter 4

RESONANCE RAMAN SPECTRA OF INORGANIC MOLECULES AND IONS

R. J. H. Clark

Christopher Ingold Laboratories, University College London, 20 Gordon Street, London WC1H 0AJ

1 INTRODUCTION

With the development of powerful laser light sources which now provide a large number of different excitation frequencies in well collimated beams, of double and triple monochromators, and of sensitive detection systems, it has proved possible to build up a considerable body of information on the frequency dependence of Raman band intensities. With the further development of dye lasers, whose frequencies are capable of being tuned over the entire visible region, and also into the ultraviolet region, major further advances in this area are to be expected. Of particular interest is the manner in which Raman band intensities increase as the exciting frequency (ν_0) approaches that of an electronic absorption band in the scattering molecule (the so-called preresonance Raman effect, abbreviated pre-RRE). Of even greater interest is what happens when the exciting frequency is made to coincide with that of an allowed (broad) absorption band of the molecule, leading to the so-called rigorous resonance Raman effect (abbreviated RRE). The RRE is characterized by an enormous increase in the intensity of a particular band, and an apparent breakdown in the harmonic oscillator selection rules such that overtones may appear with intensities comparable with that of the fundamental. Several important earlier reviews of the pre-RRE and RRE have appeared; [1-5] many of these (particularly those of Shorygin and coworkers)[5,6] relate to conjugated organic molecules for which RR spectra with certain bands having intensities as great as 10^6 times those of non-RR spectra have been reported. Two recent reviews have concentrated on theories of RR scattering[7] and on Raman band intensities.[8]

Several aspects of RR spectroscopy have contributed to the spectacular increase in interest in the subject in the last two or three years. In the first place it offers the possibility of investigating structural features of a specific site of biological activity within a macromolecule. This possibility arises where the atoms in the site give rise to an isolated electronic absorption band. On excitation within this band, most of the Raman bands are attenuated by absorption but some bands are greatly enhanced. The effect arises from coupling of vibrational and electronic transitions, the vibrational modes subject to the enhancement being localized on the chromophore, i.e. on the group of atoms which give rise to the electronic transition. RR spectroscopy thus offers a way to monitor vibrational frequencies of a chromophore. Because biological chromophores are often at sites of biological activity, the technique is considered to hold promise as a probe of biological structure. Studies of the use of dyes as labels involving biologically important molecules,[9,10] and detailed studies of haemoglobin and cytochrome c,[11-15] of vitamin B_{12} and some cobalt corrinoid derivatives,[16,17] and of other molecules[18-24] have been summarized in two important reviews on the subject.[25,26] The concentrations at which RR spectra were obtained for these materials in aqueous solution were sufficiently low (10^{-5}–10^{-3} M) so as to eliminate intermolecular interactions other than those with the solvent. This makes the data very relevant to the usual physiological situation.

RR spectroscopy also holds promise as a technique for the detection of small amounts of atmospheric pollutants, e.g. NO_2, O_3, i.e. it is considered possible that it could act as a quantitative remote probe for monitoring ambient pollutant content in the environment.

A further area which is attracting an increasing amount of attention is that of the RR spectra of structurally simple inorganic molecules, particularly those possessing allowed electronic transitions in the visible region. The present review is concerned primarily with this area, from which much unique spectroscopic information has already been obtained.

2   EXPERIMENTAL ASPECTS

A number of difficulties arise in attempting to measure quantitatively Raman band intensities in cases where ν_0, the exciting frequency, closely approaches ν_e, the frequency of the lowest allowed electronic transition of the molecule. These include those associated with (a) making allowance for the competition between scattering and absorption processes, (b) eliminating the thermal lens effect and the thermal decomposition of samples and (c) eliminating fluorescence and photolysis.

The major technical advance which has permitted Raman studies to be carried out using laser lines which fall within the contour of a strongly absorbing band has been the introduction of devices which permit relative motion between the sample and the focused laser beam. In this way localized overheating of the sample is avoided and the threat of sample decomposition is largely overcome. Both rotating sample and surface scanning techniques have been used.

2.1   Rotating Sample Techniques

The most widely used devices whereby samples may be rotated at speeds of *ca.* 2000 r.p.m. are illustrated in Fig. 1. These devices, whether they are for solids,[27] liquids[28,29] or gases,[30] are designed for spectrometers employing 90° viewing optics. Related items have been designed for spectrometers employing 180° viewing optics.[31] Much better signal to noise ratios are obtained for the Raman signals from rotating than from static liquids. This is a consequence of the fact that, in a strongly absorbing static medium, the thermal lens effect prevents the laser beam from coming to a sharp focus, and hence there is less Raman radiation collected for such an arrangement. Modifications of these devices permit the recording of RR spectra of liquids at low temperatures;[31] these usually operate by arranging a boil-off of liquid nitrogen past the sample, which has been suitably jacketed in order to prevent frosting.

2.2   Surface Scanning Techniques

In order to obtain the Raman spectrum of a highly absorbing solid (with the possibility of its being held at a range of different temperatures) a technique has been introduced whereby the laser beam, while remaining focused on the surface of the sample, is scanned rapidly over its surface in either a linear[32] or circular[33] manner. In the former case this is achieved by placing a rotating refractor plate (rotating at 200 r.p.m.) between the sample and the lens used to focus the laser beam on the crystal surface. In the latter case, it is accomplished by use of a rotating lens whose rotation axis coincides with the optic axis of the system and which refracts and focuses the laser beam into a circular path on the sample surface. Both procedures permit the recording of Raman spectra of highly coloured solids (crystals or powders) which would decompose instantly in a static beam.

2.3   Reduction of Fluorescence Background

Broad fluorescence bands may appear in addition to RR bands under resonance conditions and these distort and increase the background and so decrease the signal to noise ratio of the RR bands. The fluorescence emission mostly starts from the same excited state of the molecule as that which functions as the intermediate state in the RR effect.[34,35] The best method to reduce the fluorescence background is based on time discrimination against fluorescence.[36,37] The time taken for a transition between an initial and a final state is largely determined by the lifetime of the intermediate state, which may be *ca.* 10^{-9} s in the case of fluorescence but *ca.* 10^{-14} s (the order of magnitude of the vibrational period of a light wave) in the case of a Raman scattering process. The technique involves excitation of the sample with a pulsed laser, the pulse duration being of the order of nanoseconds or picoseconds, and time-adjusted

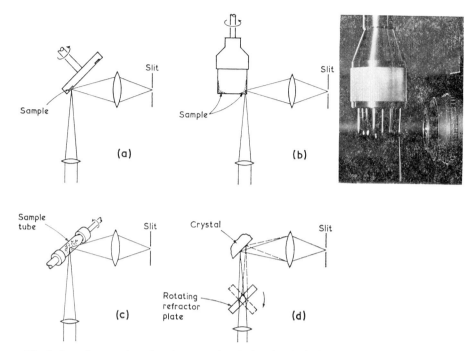

Fig. 1. Rotating sample and surface scanning devices for obviating thermal decomposition and the thermal lens effect. (a) Powdered solids, (b) liquids, (c) gases and (d) surface scanning technique for solids using a rotating refractor plate. In device (b) the liquid is flung to the walls of the cell by centrifugal force and the laser beam is brought to a focus within this thin film. In device (c), which can operate at temperatures of up to 300°C, it is essential that the (sealed) cells incorporate indentations of the Vigreux type in order to ensure rotation of the gas with the cell. The photograph shows device (b) in action.

gate electronics[38,39] to permit the separation of the RR from the fluorescence spectra. Difficulties remain if the rise and fall time of the fluorescence is less than *ca.* 1 ns (as it may be in the case of resonance fluorescence processes) owing to the very fast electronic systems then required. However, rejection of even short lived fluorescence by more than three orders of magnitude has been achieved by the use of picosecond gating of Raman scattering initiated by a mode locked laser.[37]

2.4 Corrections for Absorption

Most measurements of the intensities of RR bands have been carried out on liquids and as a function of the exciting frequency. As the main requirement for the RRE is high absorbance, corrections should be made for the subsequent attenuation of the Raman scattered radiation by the absorbing sample. Observed intensities in different regions of a Raman spectrum hence may be differently affected by self-absorption. Further, scattering is proportional to sample concentration whereas the attenuation of both the incident and the scattered radiation depends on this exponentially. Thus the relationship between the observed Raman band intensity and the concentration may be far from linear, and may in fact reach a maximum. For highly absorbing samples, the optimum concentration may be very low. Recent studies on spectrometers employing 90° collection optics have indicated that the optimum concentration corresponds to an absorbance per 1 cm path length of 13 (average of values for incident and scattered wavelengths).[40] Other analogous studies on spectrometers employing 180° collection optics, i.e. back-scattering geometry, have led to the conclusions that sample positioning and concentration are less critical than for 90° collection optics. This arises because it can be shown that, with this geometry, the intensity of the scattered light increases to an asymptotic limit with increasing concentration. With high sample absorbance, however, it becomes tedious to optimize the focusing of the sample image into the spectrometer because of the short effective penetration of the sample by the laser beam. A compromise for this geometry has been suggested of an absorbance of *ca.* 20 at the laser frequency for a 1 cm pathlength.[41]

For either geometry, it remains true that it is difficult to determine the effective absorption pathlength, and frequently the problem has been minimized by irradiation of the sample at grazing incidence. Another procedure is to use the technique of Raman difference spectroscopy with a rotating cell.[42] The cell is divided into two equal parts, one containing the absorbing solution, the other the non-absorbing solvent whose Raman bands are used as internal standards. By recording difference spectra of the solvent bands in solution and in the pure solvent as well as the optical spectrum of the same solution, it is possible to make the required corrections for absorption.

3   THE PRE-RESONANCE RAMAN EFFECT

When the exciting frequency (ν_0) is well removed from that of the lowest electronic transition of the molecule (ν_e) and the initial and final states of the molecule both involve the ground electronic state, Placzek[43] has shown that the intensity of molecular Raman scattering may be reduced to a mere dependence of the ground state polarizability on nuclear vibrations. The theory was thus intended to describe the off-resonance vibrational Raman effect only. This classical theory, the so-called bond polarizability theory, is widely used in the analysis of vibrational Raman spectra in terms of classical ground state properties of molecules.[3,44]

In this theory, the relationship between the total intensity of a Raman band scattered over a solid angle 4π by one randomly oriented molecule passing from initial state m to final state n is given by the expression[4]

$$I_{mn} = \frac{2^7\pi^5}{3^2c^4} I_0(\nu_0 \pm \nu_i)^4 \sum_{ij} |(\alpha_{ij})_{mn}|^2 \qquad (1)$$

where I_0 is the intensity of the incident light, v_i is the Raman shift of fundamental i, a_{ij} is the ijth element of the scattering tensor, and $i, j = x, y$, or z, the space-fixed coordinate axes. The tensor is viewed classically as the molecular polarizability in the case of Rayleigh scattering or as the first term in a Taylor series expansion involving its derivative in the case of Raman scattering. Thus the relationship between the scattered intensity and the scattered frequency is the familiar fourth power one. However, the simple polarizability theory completely breaks down under resonance conditions and this has necessitated the development of more general expressions to account for the intensity behaviour of Raman bands. These expressions are based on the Kramers-Heisenberg-Dirac dispersion equation, and they govern the entire frequency range, i.e. off-resonance scattering, pre-RR scattering, as well as RR scattering. The general form for the ijth component of the scattering tensor is given by the following sum over all vibronic states of the molecule[4]

$$(a_{ij})_{mn} = \sum_e \left[\frac{(M_j)_{me}(M_i)_{en}}{E_e - E_m - E_0 + i\Gamma_e} + \frac{(M_i)_{me}(M_j)_{en}}{E_e - E_n + E_0 + i\Gamma_e} \right] \qquad (2)$$

where m and n are the initial and final states of the molecule, e is an intermediate state, $E_0 (= hv_0)$ is the energy of the exciting radiation, and Γ_e is a damping constant which prevents the denominator at resonance from reaching zero and which represents the finite lifetime and 'sharpness' of the intermediate state. This damping of a state is due ultimately either to spontaneous radiative transitions from this state to other states, or, in addition, to non-radiative transitions to other states, thereby limiting the lifetime. The quantities $(M_j)_{me}$ and $(M_i)_{en}$ are the electric dipole transition moments along the directions j and i from m to e and from e to n. If v_e is the frequency of the transition from m to e, then in the off-resonance region $v_0 \ll v_e$ and a_{ij} is independent of v_0. On the other hand, when $v_e - v_0$ becomes very small, then one element in the summation, corresponding to the resonant electronic transition, becomes dominant (assuming finite transition moments). The magnitude of a_{ij} is thus clearly some function of the oscillator strength of this transition.

The above equation gives no information as to which vibrations of a molecule are subject to resonance enhancement. In order to obtain this information it is usual (following Albrecht)[4,45,46] to use the adiabatic approximation (separability of electronic and vibrational wave functions) and to expand the electronic wave function in a Taylor series in the nuclear displacements, using the Herzberg-Teller formalism.[4] If this is done, Albrecht has then argued that the vibrations subject to resonance enhancement are those which are responsible for mixing two relevant electronic states. A further approach, based on the application of third order time dependent perturbation theory to the scattering process, leads to the following simplified expression[47,48] for a_{ij}

$$a_{ij} = \frac{1}{h} \sum_{ef} \left[\frac{(M_j)_{ge} h_{ef}^{\Delta v} (M_i)_{fg}}{(v_e - v_0)(v_f - v_s)} + \frac{(M_i)_{ge} h_{ef}^{\Delta v} (M_j)_{fg}}{(v_e + v_s)(v_f + v_0)} \right] \qquad (3)$$

where the summation is taken over all excited electronic states e and f, taken in pairs, g is the electronic ground state, and v_e and v_s are the frequencies of the incident and scattered photons. The term $h_{ef}^{\Delta v}$ is a vibronic coupling matrix element connecting states e and f by a vibration designated Δv. The equation has been simplified by the omission of damping terms.

Resonance processes corresponding to the conditions $e = f$ and $e \neq f$ lead to expressions which Albrecht and Hutley[46] refer to as A and B terms respectively. A terms involve vibrational interaction with a single excited electronic state by way of Franck-Condon overlap integrals. Only in the case of totally symmetric vibrations are these integrals non-zero for both the initial and final states of the molecule. B terms involve vibronic mixing of two excited states, e and f, the active vibrations being permitted to have any symmetry which is contained in the direct product of the representations of the two electronic states. A and B terms may in general be distinguished by their different frequency dependences in pre-resonance Raman scattering. For an A term,

$$I_i \propto (v_0 - v_i)^4 \left[\frac{(v_e^2 + v_0^2)}{(v_e^2 - v_0^2)^2} \right]^2 = F_A^2 \tag{4}$$

where there is only a single active intermediate state e, with eigenfrequency y_e. For a B term,

$$I_i \propto 4(v_0 - v_i)^4 \left[\frac{(v_e v_s + v_0^2)}{(v_e^2 - v_0^2)(v_f^2 - v_0^2)} \right]^2 = F_B^2 \tag{5}$$

where two intermediate electronic states e and f must be taken into consideration $(v_f > v_e)$. The first expression is the same as that originally derived by Shorygin[5] and subsequently discussed by Savin, and (as indicated above) is intended to apply only to totally symmetric modes. The (dimensionless) factor F_A displays steeper dependence on v_0 than does F_B; if electronic state e is an active intermediate state for the scattering of fundamental mode i, then the first expression must eventually dominate as resonance is approached very closely. Several rather drastic approximations are inherent in the derivation of the above expressions, one in particular being that the active intermediate states f should lie at energies sufficiently high to permit the use of some average value for v_f.

The frequency dependence of an element of the scattering tensor, not only in the vibrational Raman effect, but also in the electronic and vibro-electronic Raman effects, has recently been discussed by Koningstein and Jakubinek.[49]

In presenting the results of studies on the pre-RRE it is common to plot the square root of the frequency corrected relative molar intensities, i.e. $[(1/f)(I_2 M_1/I_1 M_2)]^{\frac{1}{2}}$ values, which are quantities incorporating the $(v_0 - v_i)^4$ factor, against F_A or F_B, or some function closely related thereto, e.g. $[1 + (v_0/v_e)^2]/[1 - (v_0/v_e)^2]^2$. Two examples of a plot of $[(1/f)(I_2 M_1/I_1 M_2)]^{\frac{1}{2}}$ versus this last function are given in Fig. 2, from which it is clear that not only the totally symmetric but also the non-totally symmetric modes of titanium tetrachloride and titanium tetrabromide yield linear relationships.[50] Moreover, in agreement with the detailed

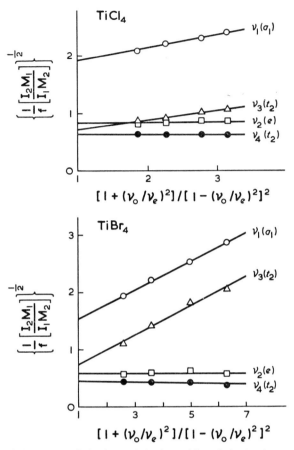

Fig. 2. Display of the pre-RR behaviour of the intensities of the fundamentals of titanium tetrachloride and titanium tetrabromide. The abscissa is the Shorygin function $[1 + (\nu_0/\nu_e)^2]/[1 - (\nu_0/\nu_e)^2]^2$, while the ordinate is the frequency-corrected molar intensity of the tetrahalide band relative to that of the standard (Ref. 50).

analysis of the pre-RRE by Behringer,[7] it is the two stretching modes rather than the bending modes which are subject to the enhancement. This appears commonly to be the case e.g. for the MnO_4^- and CrO_4^{2-} ions,[51] and other tetrahedral species, for octahedral anions,[52] square planar anions,[53] and other molecules.[54,55]

Many experimental results now available demonstrate that F_B^2 more commonly represents the experimental frequency dependence of Raman band intensities than does F_A^2, thus indicating that, with the exciting lines used, the Raman band intensity is derived from at least two electronic states vibronically mixed by the fundamental which displays the pre-RRE. This is the case for the chromate ion, for which ν_e and ν_f have been identified with the two lowest lying absorption bands (26,900 and 36,600 cm^{-1} respectively), corresponding to

$^1T_2 \leftarrow {}^1A_1$ type transitions.[50] Similar conclusions have been drawn for series of octahedral anions MX_6^{2-} (M = Pt or Sn, X = F, Cl, Br or I) and related ions,[52] square planar anions MX_4^- (M = Au, X = Cl or Br; M = I, X = Cl) and MX_4^{2-} (M = Pd or Pt, X = Cl or Br)[53] as well as for organic molecules such as p-nitroaniline.[46]

Experimental results on liquid pyrazine demonstrate[54] that, in accordance with Albrecht's theories, the normal mode most responsible for the forbidden intensity in an allowed electronic transition displays a striking pre-RRE.

A further consequence of the Albrecht theory is that when the frequency dependence of a Raman band intensity is governed by F_A^2, then $I \propto \varepsilon^2$ (where ε is the extinction coefficient of the absorption band associated with the lowest electronic transition) whereas when it is governed by F_B^2, then $I \propto \varepsilon$.[3,49,56]

One objective in studying Raman spectra of molecules with a variety of different exciting lines is to obtain data with which to correct both molecular and bond polarizability derivatives for the pre-RRE, i.e. to obtain $\bar{\alpha}'_{MX}$ values extrapolated to zero exciting frequency. This correction has become especially important as more and more deeply coloured compounds are studied. Earlier gas phase work in this area with mercury arc excitation has been summarized by Bernstein and coworkers,[57] and subsequently by Chantry[44] and Hester.[8] The most recent data have largely been gathered for simple inorganic molecules and ions, and these are summarized in Table 1. From these data it is evident that $\bar{\alpha}'_{MX}$ values depend both on p, the fractional covalent character of the MX bond,[58] and on some power function of the bond length, for example[59]

$$\bar{\alpha}'_{CX} \approx \bar{\alpha}'_{SiX} < \bar{\alpha}'_{GeX} < \bar{\alpha}'_{SnX}$$

and

$$\bar{\alpha}'_{MH} \approx \bar{\alpha}'_{MF} < \bar{\alpha}'_{MCl} < \bar{\alpha}'_{MBr} < \bar{\alpha}'_{MI}$$

The results are thus consistent with the simple delta function model of bond polarizability derivatives, whereby $\bar{\alpha}'_{MX}(\text{calc}) = (2/3)(\chi^{\frac{1}{2}}p/Z_{\text{eff}}a_0)(\frac{1}{2}n)r^3$, where χ and Z_{eff} are the geometric means of the electronegativities and effective nuclear charges, respectively, of M and X, Z_{eff} in each case being taken to be the atomic number of the atom minus the number of inner shell electrons, a_0 is the Bohr radius, p is the Pauling fractional covalent character, $\frac{1}{2}n$ is the MX bond order, and r is the equilibrium MX internuclear distance.

Data on $\bar{\alpha}'_{MO}$ values in various oxyanions[58] have been reviewed elsewhere.[8]

4   THE RESONANCE RAMAN EFFECT

4.1   General Introduction

Two different resonance effects are commonly regarded as being capable of taking place when a gas is excited within an absorption band, the resonance Raman effect (RRE) and resonance fluorescence (RF). RF from a gas arises

TABLE 1
Bond polarizability derivatives (Å^2) corrected for the pre-resonance Raman effect

Scattering species	$\bar{\alpha}'_{MX}$	Footnote ref.	Scattering species	$\bar{\alpha}'_{MX}$	Footnote ref.
			Solution data		
CCl_4	2.04	*a*	SnF_6^{2-}	0.81	*f*
$SiCl_4$	2.12	*a*	$SnCl_6^{2-}$	2.3	*f*
$GeCl_4$	2.86	*a*	$SnBr_6^{2-}$	3.9	*f*
$SnCl_4$	3.49	*a*	SnI_6^{2-}	*ca.* 5	*f*
$TiCl_4$	3.92	*a*	$PbCl_6^{2-}$	1.8	*f*
VCl_4	4.00	*b*	$PdCl_6^{2-}$	1.9_5	*f*
CBr_4	3.03	*a*	PtF_6^{2-}	1.5	*f*
$SiBr_4$	2.91	*a*	$PtCl_6^{2-}$	2.7	*f*
$GeBr_4$	3.61	*a*	$PtBr_6^{2-}$	3.6_5	*f*
$SnBr_4$	3.68	*a*	PtI_6^{2-}	6.6	*f*
$TiBr_4$	4.77	*a*	$RhCl_6^{3-}$	2.0	*f*
$SiBr_4$	4.21	*a*	$IrCl_6^{3-}$	1.4	*f*
GeI_4	5.02	*a*	$IrCl_6^{2-}$	3.0_5	*f*
SnI_4	6.75	*a*	$OsCl_6^{2-}$	1.7	*f*
TiI_4	*ca.* 13	*c*	$OsBr_6^{2-}$	3.6	*f*
BCl_3	1.89	*d*	$ReCl_6^{2-}$	1.3_5	*f*
BBr_3	2.57	*d*	$AuCl_4^-$	2.5_5	*g*
BI_3	3.75	*d*	$AuBr_4^-$	4.5	*g*
$ZnCl_4^{2-}$	0.87	*e*	$PdCl_4^{2-}$	1.4	*g*
$CdCl_4^{2-}$	1.04	*e*	$PdBr_4^{2-}$	2.2_5	*g*
$HgCl_4^{2-}$	2.10	*e*	$PtCl_4^{2-}$	1.3	*g*
$GaCl_4^-$	1.12	*e*	$PtBr_4^{2-}$	1.9	*g*
$ZnBr_4^{2-}$	1.80	*e*	ICl_4^-	2.6	*g*
$CdBr_4^{2-}$	2.46	*e*	ICl_2^-	2.3_5	*g*
$HgBr_4^{2-}$	5.11	*e*			
$GaBr_4^-$	3.08	*e*			
			Vapour phase data		
CF_4	1.14	*h*	CH_4	1.07	*h*
SiF_4	1.11	*h*	SiH_4	1.38	*h*
GeF_4	1.40	*h*	GeH_4	1.50	*h*
CCl_4	2.38	*i*	SnH_4	1.72	*h*
$SiCl_4$	2.68	*i*	$HgCl_2$	2.5	*j*
$GeCl_4$	3.57	*i*	$HgBr_2$	3.7	*j*
$SnCl_4$	3.71	*i*	HgI_2	5.3	*j*

a R. J. H. Clark and P. D. Mitchell, *J. Mol. Spectrosc.* **51**, 458 (1974).

b R. J. H. Clark and P. D. Mitchell, *J. Chem. Soc. Faraday Trans. II* **68**, 476 (1972).

c R. J. H. Clark and P. D. Mitchell, *J. Amer. Chem. Soc.* **95**, 8300 (1973).

d R. J. H. Clark and P. D. Mitchell, *Inorg. Chem.* **11**, 1439 (1972).

e T. V. Long and R. A. Plane, *J. Chem. Phys.* **43**, 457 (1965).

f Y. M. Bosworth and R. J. H. Clark, *J. Chem. Soc. Dalton Trans.* 1749 (1974).

g Y. M. Bosworth and R. J. H. Clark, *Inorg. Chem.* **14**, 170 (1975).

h R. S. Armstrong and R. J. H. Clark, *J. Chem. Soc., Faraday Trans.*, in press (1975).

i R. J. H. Clark and P. D. Mitchell, *J. Chem. Soc. Faraday Trans. II* **71**, 515 (1975).

j R. J. H. Clark and D. M. Rippon, *J. Chem. Soc. Faraday Trans. II* **69**, 1496 (1972).

from an absorption process[56] associated with an overlap between the exciting radiation and an electronic-vibrational-rotational absorption line of the scattering molecule (in practice, the exciting radiation is now usually a laser line which must, for RF to be observed, fall within the contour of the absorption line). Raman scattering from adjacent bands may also occur, but this is intrinsically much weaker than RF. If, on the other hand, a broad vibronic band is excited, only part of this band is in resonance, leading to a much reduced emission. The part of the band which is not in resonance, together with the other bands in the electronic manifold, will give rise to pre-RR scattering. In the limit in which the individual vibronic bands merge into a structureless continuum the entire phenomenon is known as the RRE. The RR intensity is thus the sum of the RF and pre-RR intensities (these effects may be of comparable magnitude for a structureless continuum). Moreover, these two contributions are such that the overall RR spectrum is simpler than its individual RF and pre-RR components. There is thus good practical reason for treating the RRE as a single phenomenon for very broad absorption bands. For very narrow absorption bands, the observed effect is either RF (as described above, in the case of coincidence with an electronic-vibrational-rotational line of the scattering molecule), or pre-RR scattering (where this is not the case).

The observed characteristics of the RR and RF spectra of the halogen gases have been summarized by Bernstein et al.[60] (Table 2). In both cases an overtone progression may be observed, but the band widths of RR bands alone increase progressively with increase in the vibrational quantum number, and in consequence fewer RR than RF overtones are usually observed. (At least 47 RF

TABLE 2
Observed differences between the RF and the RR spectra of the halogen gases[a]

Criterion	RF	RRE
Band envelope	Very sharp doublets ($\Delta J = \pm 1$) at low pressure	Broad Q branch with rotational wings
Overtone pattern	Irregular sequence of overtone intensities; some lines may be missing	Continuous band broadening and continuous decrease in peak intensity with vibrational quantum number
Depolarization ratio	Depolarized	Bands expected to be polarized in the normal Raman effect are polarized
Behaviour with increasing gas pressure	Quenching observed, with the doublets developing multiplet structure	Intensity increases, but shape of band does not alter much
Behaviour with foreign gases	As above	Neither intensity nor band shape alters much

[a] Ref. 60. The criteria given here are mainly external and are not generally applicable, particularly that relating to depolarization ratios (see Ref. 7).

overtones of the fundamental of iodine gas have been observed.) The effects differ with respect to the band contour and overtone structure, the value of the depolarization ratio, and the intermolecular quenching behaviour with increased pressure. These conclusions are, however, rather specific to the halogen gases, and as discussed by Behringer,[7] the experimental distinction between RF and RR spectra is not generally as clear cut as implied above. For gases the apparent transition from the one effect to the other occurs at the dissociation limit of the excited state, RF predominating if the exciting frequency is less than, and the RRE if it is more than, this limit. The different effects are illustrated in Fig. 3 with reference to the halogen gases.

By way of example, the electronic transition of relevance for iodine gas is $B^3\Pi_{0u}^+ \leftarrow X\Sigma_g^+$; for this transition the rotational angular momentum must change by $\Delta J = \pm 1$ in both absorption as well as emission. Accordingly, the

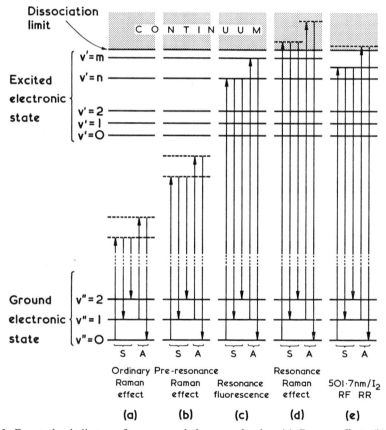

Fig. 3. Energy level diagram for gaseous halogen molecules. (a) Raman effect, (b) pre-resonance Raman effect, (c) resonance fluorescence (overwhelmingly, as $k_{RF} \approx 10^3 k_{RR}$), (d) resonance Raman effect, (e) Stokes resonance fluorescence and anti-Stokes resonance Raman effect (iodine vapour, $\lambda_0 = 501.7$ nm) (Ref. 9). S = Stokes, AS = anti-Stokes.

emission in RF will be a series of doublets corresponding to pairs of P and R lines.[61] Thus irradiation of iodine gas with 514.5 nm excitation produces a RF spectrum, characterized by sharp, depolarized RF doublets, with variable intensities throughout the progression (due to Franck-Condon factors). On irradiation into the continuum with 488.0 nm excitation RF is extremely weak, and hence a RR spectrum is now clearly revealed; it is characterized by broader bands (S band heads, Q and O branches) than in the RF case; the bands both increase in width and decrease monotonically in intensity as the vibrational quantum number increases, and in addition they are all polarized. On excitation with 501.7 nm radiation, the Stokes lines from iodine gas arise overwhelmingly from RF [excited through both the $R(39)$ line of the 64-0 band and the $R(26)$ line of the 62-0 band] whereas the anti-Stokes lines are due to the RRE (Fig. 4); this situation arises because excitation of molecules in the $v'' = 1$ state takes them into the continuum, from which RRE is expected to predominate, whereas excitation of those in the $v'' = 0$ state does not.

The differences between the RRE and RF may be attributed to the fact that the former occurs much faster than the latter.[7] RF may be regarded as a consequence of a succession of two independent well separated single photon

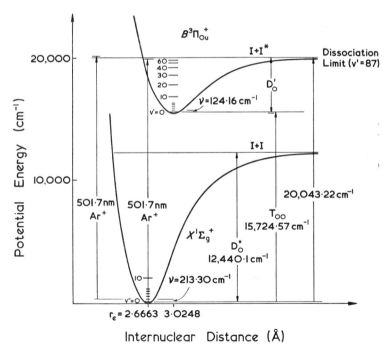

Fig. 4. Energy level diagram for molecular iodine, demonstrating that with 501.7 nm excitation, the Stokes lines arise primarily from RF whereas the anti-Stokes lines arise primarily from the RRE (spectral data from R. F. Barrow and K. K. Yee, *J. Chem. Soc. Faraday Trans. II* **69**, 684, 1973).

processes and thus the lifetimes of the intermediate states are very different from those for the RRE. RR scattering is then thought of as being a relatively instantaneous process, and thereby insensitive to collision induced quenching and redistribution of excitation energy. Hence, there is a progressive change from RF to RR scattering for iodine gas excited by 501.7 or 514.5 nm Ar$^+$ lines as the pressure inside the sample cell is increased from *ca.* 0.2 Torr to 30 atm by addition of an inert gas, e.g. argon. Such influences on fluorescence spectra are only possible if there is a sufficient time interval between absorption and re-emission so that additional processes redistributing energy may intervene. Under the conditions of high pressure, collisional broadening of the rotational-vibrational levels of molecular iodine takes place, and the levels effectively merge into a continuum. Associated with this pressure change is a change in the depolarization ratio from 0.75 (characteristic in the case of the halogen gases of RF) to 0.33 (characteristic similarly of the RRE).[62-64] (See footnote to Table 2, however.) Other related experimental studies have been made,[65,66] and theoretical aspects of RF have been studied in detail.[67]

The Stokes line intensity of the RR spectrum of gaseous iodine is found to have only about one-thousandth the intensity of the corresponding RF band. Hence a RR spectrum can be observed under RF conditions only if the RF is quenched, either as mentioned above with inert gas, or (usually) by the solvent when the molecule is taken into solution. Thus on excitation of a liquid, solution or solid within an absorption band, it is a RR spectrum which is generally observed, as the RF is normally quenched.

The mechanism of the energy transfer between photon and scatterer and the intensity distributions within the vibrational progressions seen in the RRE have been the subject of many intense but somewhat conflicting theoretical studies.[68-76] From the study by Peticolas *et al.*[71] it emerges that the selection rules for the appearance of overtones in the RRE are such that only totally symmetric fundamentals may show the effect. This and other related points are those most in need of detailed experimental verification, and simple molecules and ions are expected to provide the most satisfactory spectra to interpret in this respect. Many aspects of the theory of the RRE, however, remain to be worked out in detail.

4.2 Nature of the Overtone Progressions

A summary of the results obtained so far on the RR spectra of diatomic and triatomic species is given in Table 3 and of more complicated species in Table 4. Diatomic species have only one vibrational mode and this is, of course, totally symmetric. However, in the case of polyatomic species, the question of the symmetry of the fundamental displaying the RRE is important. In all the cases listed in Tables 3 and 4, polarization and other studies have identified the fundamental displaying the RRE as being a totally symmetric mode. This constitutes important support for the theoretical work of Peticolas *et al.*[71] More-

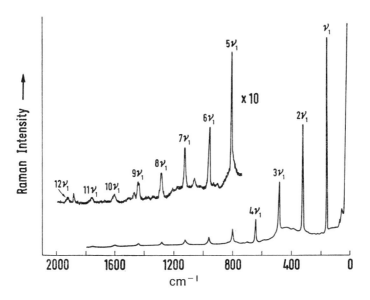

Fig. 5. RR spectrum of solid titanium tetraiodide (514.5 nm excitation) (Clark and Mitchell, Ref. 78).

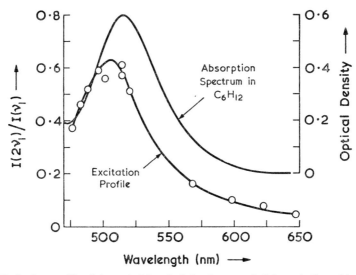

Fig. 6. Excitation profile of the ν_1 (a_1) band of titanium tetraiodide, as indicated by the plot of the ratio $I(2\nu_1)/I(\nu_1)$ versus excitation wavelength (R. J. H. Clark and N. D'Urso, unpublished work).

TABLE 3
Resonance Raman spectra of diatomic and triatomic species

Species	State	Exciting line (nm)	ν_0 cm^{-1}	ν_e cm^{-1}	$\nu_e-\nu_0$ cm^{-1}
I_2	gas	488.0 Ar$^+$	20,487	19,800	-690
	n-heptane soln.	514.5 Ar$^+$	19,430	19,200	-230
	CHCl$_3$ soln.	514.5 Ar$^+$	19,430	19,500	$+70$
	Ar matrix, 16K	514.5 Ar$^+$	19,430	19,600	$+170$
Br_2	gas	488.0 Ar$^+$	20,487	20,700	$+210$
Br^{35}Cl	gas	488.0 Ar$^+$	20,487	—	—
I^{35}Cl	gas	488.0 Ar$^+$	20,487	20,820	$+330$
IBr	gas	488.0 Ar$^+$	20,487	19,715	-770
I_2^+	HSO$_3$F soln.	632.8 He/Ne	15,800	15,625	-170
Br_2^+	SO$_3$F$^-$ salt in HSO$_3$F/SbF$_5$/SO$_3$ soln.	514.5 Ar$^+$	19,430	19,610	$+180$
^{35}Cl$_2^-$	Li$^+$ salt, Ar matrix	514.5 Ar$^+$	19,430	orange-red	—
S_2^-	Doped in NaI or KI	488.0 Ar$^+$	20,487	25,000	$\sim +4,500$
SSe$^-$	Doped in KI	488.0 Ar$^+$	20,487	—	—
Se$_2^-$	Doped in KI	488.0 Ar$^+$	20,487	—	—
NO_2	gas	488.0 Ar$^+$	20,487	22,000	$\sim +1,500$
I_3^-	K$^+$ salt in CH$_3$OH soln.	337.1 N$_2$	29,656	28,570	$\sim +1,090$
	K$^+$ salt in CH$_3$OH soln.	363.8 Ar$^+$	27,479	28,570	$\sim +2,000$
^{16}O$_3^-$	Na$^+$ salt, Ar matrix, 16K	488.0 Ar$^+$	20,487	$\sim 22,500$	$+2,000$
^{16}O$_3^-$	Cs$^+$ salt, Ar matrix, 16K	488.0 Ar$^+$	20,487	$\sim 22,500$	$+2,000$
^{18}O$_3^-$	Cs$^+$ salt, Ar matrix, 16K	488.0 Ar$^+$	20,487	$\sim 22,500$	$\sim +2,000$
S_3^-	Doped in NaCl	488.0 Ar$^+$	20,487	16,400	$\sim -4,000$
	Ultramarine	647.1 Kr$^+$	15,450	16,400	$+950$
ClO_2	Ar matrix, 16K	—	—	—	—

[a] W. Kiefer and H. J. Bernstein, *J. Mol. Spectrosc.* **43**, 366 (1972). The results are in good agreement with those obtained by fluorescence measurements, viz. $\omega_e = 214.51886$, $x_e\omega_e = 0.60738$ cm^{-1}, D. H. Rank and B. S. Rao, *J. Mol. Spectrosc.* **13**, 34 (1964). See also W. Kiefer and H. J. Bernstein, *Appl. Spectrosc.* **25**, 609 (1971), and R. F. Barrow and K. K. Yee, *J. Chem. Soc. Faraday Trans. II* **69**, 684 (1973) ($\omega_e = 214.5016 \pm 0.014$, $x_e\omega_e = 0.6147 \pm 0.002$).

[b] W. Holzer, W. F. Murphy and H. J. Bernstein, *J. Chem. Phys.* **52**, 399 (1970).

[c] W. Kiefer and H. J. Bernstein, *J. Raman Spectrosc.* **1**, 417 (1973). This paper also includes data on the overtone progressions of the iodine stretching vibration in carbon tetrachloride, cyclohexane, cyclopentane, n-pentane, n-hexane, carbon disulphide, benzene, toluene, p-xylene and p-dioxan. For the first five solutions, as well as for n-heptane and chloroform solutions, sufficient data were obtained to establish that the anharmonicity constant $\omega_e y_e$ is -0.006 ± 0.003.

[d] O. S. Mortensen, *J. Mol. Spectrosc.* **39**, 48 (1971); W. Kiefer and H. J. Bernstein, *Appl. Spectrosc.* **25**, 500 (1971).

[e] W. F. Howard and L. Andrews, *J. Raman Spectrosc.* **2**, 447 (1974).

[f] M. Berjot, M. Jacon and L. Bernard, *Compt. Rend.* **B274**, 1274 (1972).

[g] W. Kiefer and H. W. Schrötter, *J. Chem. Phys.* **53**, 1612 (1970).

[h] R. J. Gillespie and M. J. Morton, *J. Mol. Spectrosc.* **30**, 178 (1969).

[i] R. J. Gillespie and M. J. Morton, *Chem. Commun.* 1565 (1968).

[j] M. Booth, R. J. Gillespie and M. J. Morton, *Adv. Raman Spectrosc.* **1**, 364 (1973).

[k] W. F. Howard and L. Andrews, *J. Amer. Chem. Soc.* **95**, 2056 (1973). Similar but less extensive overtone progressions were observed with K, Rb or Cs in place of Li.

[l] W. Holzer, W. F. Murphy and H. J. Bernstein, *J. Mol. Spectrosc.* **32**, 13 (1969).

TABLE 3—*cont.*

ε_{max}	ω_e or ω_1 cm^{-1}	$x_e\omega_e$ or x_{11} cm^{-1}	Progression	$I(2\nu_1)/I(\nu_1)$†	Footnote ref.
680	214.534 ± 0.040	0.6070 ± 0.0085	$16\nu_1$	0.71	a,b
—	212.51 ± 0.10	0.56 ± 0.04	$20\nu_1$	—	c
—	212.18 ± 0.08	0.64 ± 0.03	$13\nu_1$	0.70	c,d
—	213.3 ± 0.3	0.52 ± 0.05	$11\nu_1$	—	e
75	318.5 ± 1	—	$5\nu_1$	1.07	b,f,g
—	440 ± 1	—	$2\nu_1$	~ 0.15	b
51	381 ± 1	—	$\geqslant 5\nu_1$	0.95	b
170	265 ± 1	—	$\geqslant 7\nu_1$	1.08	b
2,600	238	~ 0	$4\nu_1$	0.70	h
—	361.5 ± 1	0.9 ± 0.2	$6\nu_1$	1.0	i,j
—	249.2 ± 1.0	1.62 ± 0.1	$9\nu_1$	~ 1	k
—	594 ± 1	3.4 ± 0.4	$3\nu_1$	—	l,m
—	464 ± 1	2.0 ± 0.25	$3\nu_1$	—	l,m
—	325 ± 1	0.75 ± 0.25	$3\nu_1$	—	l,m
—	750.5 ± 0.5	0.45	$2\nu_2$	~ 1	n,o
25,500	112	—	$7\nu_1$	~ 0.6	p
—	~ 114	—	$8\nu_1$	0.59	q
—	$1{,}021.3 \pm 1$	5.33 ± 0.5	$3\nu_1$	~ 1.4	r
—	$1{,}028.2 \pm 1.0$	4.95 ± 0.25	$4\nu_1$	0.67	r
—	970.7 ± 0.3	4.42 ± 0.10	$5\nu_1$	0.50	r
—	532	1	$6\nu_1$	0.46	l,s,t
—	550.3 ± 0.3	0.75 ± 0.25	$6\nu_1$	0.51	u
—	949.1 ± 0.3	3.55 ± 0.05	$6\nu_1$	—	v

† Where the relative Raman scattering cross sections of the first overtone to the fundamental have not been quoted in the original paper, they have been taken to be the products of the relative peak heights and relative half-band widths.

[m] J. Rolfe, *J. Chem. Phys.* **49**, 4193 (1968). This paper relates to the resonance fluorescence spectra of the S_2^-, Se_2^- and SSe^- ions at 4.2 K; ω_e and $x_e\omega_e$ were found to be 595.7 and 2.5 cm^{-1} respectively (S_2^-), 327.8 and 0.75 cm^{-1} respectively (Se_2^-), and 462.3 and 1.6 cm^{-1} respectively (SSe^-).

[n] M. J. Marsden and G. R. Bird, *J. Chem. Phys.* **59**, 2766 (1973). The electronic spectrum of this molecule is very complicated and not completely analysed; see A. E. Douglas and K. P. Huber, *Can. J. Phys.* **43**, 74 (1965).

[o] D. E. Tevault and L. Andrews, *Spectrochim. Acta* **30A**, ⌒69 (1974) give details of the matrix-isolated species.

[p] K. Kaya, N. Mikami, Y. Udagawa and M. Ito, *Chem. Phys. Lett.* **16**, 151 (1972).

[q] W. Kiefer and H. J. Bernstein, *Chem. Phys. Lett.* **16**, 5 (1972).

[r] L. Andrews and R. C. Spiker, *J. Chem. Phys.* **59**, 1863 (1973). For the Na^+ salt and 514.5 nm excitation, three overtones of ν_1 were observed despite a lower value for $I(2\nu_1)/I(\nu_1)$ of 0.8. The electronic spectral data are those of M. E. Jacox and D. E. Milligan, *J. Mol. Spectrosc.* **43**, 148 (1972).

[s] W. Holzer, W. F. Murphy and H. J. Bernstein, *Chem. Phys. Lett.* **4**, 641 (1970).

[t] W. Holzer, S. Racine and J. Cipriani, *Adv. Raman Spectrosc.* **1**, 393 (1973).

[u] R. J. H. Clark and M. L. Franks, *Chem. Phys. Lett.* **34**, 69 (1975).

[v] F. K. Chi and L. Andrews, *J. Mol. Spectrosc.* **52**, 82 (1974).

TABLE 4
Resonance Raman spectra of polyatomic inorganic species

Species	State	Exciting line (nm)	ν_0 cm^{-1}	ν_e cm^{-1}	$\nu_e-\nu_0$ cm^{-1}
MnO$_4^-$	K$^+$ salt	514.5 Ar$^+$	19,430	$\sim$18,950	-480
	H$_2$O soln.^{a}	514.5 Ar$^+$	19,430	$\sim$18,950	-480
CrO$_4^{2-}$	K$^+$ salt	363.8 Ar$^+$	27,479	26,900	-580
SnI$_4$	C$_6$H$_{12}$ soln.	363.8 Ar$^+$	27,479	27,430	-40
TiI$_4$	solid	514.5 Ar$^+$	19,430	19,400	-30
	C$_6$H$_{12}$ soln.^{b}	488.0 or 514.5 Ar$^+$	19,430	19,400	-30
VCl$_4$	CCl$_4$ soln.	457.9 Ar$^+$	21,833	24,510	$+2,680$
MoS$_4^{2-}$	NH$_4^+$ salt in H$_2$O/NaOH^c	476.5 Ar$^+$	20,981	21,360	$+380$
	H$_2$O soln.	465.8 Ar$^+$	21,463	21,360	-100
PdBr$_4^{2-}$	K$^+$ salt in 2M HBr soln.	325.0 He/Cd	30,760	30,085	-670
	K$^+$ salt in KBr disc	325.0 He/Cd	30,760	—	—
AuBr$_4^-$	Et$_4$N$^+$ salt^e	457.9 Ar$^+$	21,833	$\sim$22,500	$\sim$700
	K$^+$ salt (2H$_2$O)	457.9 Ar$^+$	21,833	$\sim$23,000	$\sim$1,200
Te$_4^{2+}$	Conc. H$_2$SO$_4$ soln.	514.5 Ar$^+$	19,430	19,610	$+180$
Mo$_2$Cl$_8^{4-}$	K$^+$ salt	514.5 Ar$^+$	19,430	19,340	-90
	Rb$^+$ salt	514.5 Ar$^+$	19,430	18,750	-680
	enH$_2^{2+}$ salt	514.5 Ar$^+$	19,430	18,600	-830
	NH$_4^+$ salt	514.5 Ar$^+$	19,430	18,730	-700
	Cs$^+$ salt	514.5 Ar$^+$	19,430	18,780	-650
PtBr$_6^{2-}$	[(n-C$_4$H$_9$)$_4$N]$^+$ salt	325.0 He/Cd	30,760	31,800	$+1,000$
PtI$_6^{2-}$	H$_2$O soln.	488.0 Ar$^+$	20,487	20,250	-240
Re$_2$Cl$_8^{2-}$	[(n-C$_4$H$_9$)$_4$N]$^+$ salt	615.0 Ar$^+$/R6G^u	16,260	14,750	$-1,510$
Re$_2$Br$_8^{2-}$	[(n-C$_4$H$_9$)$_4$N]$^+$ salt	647.1 Kr$^+$	16,400	14,100	$-2,300$

^{a} Also observed, $n\nu_1(a_1)+\nu_1(t_2)$ as far as $n = 5$.

^{b} Also observed in carbon disulphide to $12\nu_1$, and in carbon tetrachloride to $11\nu_1$, and methyl cyclohexane to $13\nu_1$.

^{c} Also observed, $n\nu_1(a_1)+\nu_3(t_2)$ to $n\sim 2$.

^{d} Deduced from the original data; the overtone frequencies are not very accurate.

^{e} Also observed, $n\nu_1(a_{1g})+\nu_4(b_{2g})$ as far as $n = 6$, and $n\nu_1(a_{1g})+\nu_2(b_{1g})$ as far as $n = 5$.

^{f} Where relative Raman cross sections have not been quoted in the original paper, they have been taken to be the products of the relative peak heights and the relative half-band widths.

^{g} W. Kiefer and H. J. Bernstein, *Mol. Phys.* **23**, 835 (1972).

^{h} R. J. H. Clark and P. D. Mitchell, *Chem. Commun.* 762 (1973).

^{i} R. J. H. Clark and P. D. Mitchell, *J. Amer. Chem. Soc.* **95**, 8300 (1973). See also R. J. H. Clark and P. D. Mitchell, *J. Raman Spectrosc.* **2**, 399 (1974).

over, with but one apparent exception (NO$_2$), this mode is either principally or totally bond stretching in form, as expected, because during such a vibration the electron distribution in closed shell molecules is expected to vary more extensively than during a bending vibration.

Resonance is always most effective, as judged either by the extent of the overtone progression, or by the ratio of the Raman scattering cross-sections of $2\nu_1$ to ν_1, when $(\nu_e - \nu_0) \approx 0$. This situation is well illustrated for the case of titanium tetraiodide[77,78] (Figs. 5 and 6) and for the Mo$_2$Cl$_8^{4-}$ ion[79] in association with various cations (Fig. 7). The frequency and half band width of the lowest allowed electronic transition of the Mo$_2$Cl$_8^{4-}$ ion (*ca.* 19,000 cm^{-1})

TABLE 4—*cont.*

ε_{max}	ω_1 cm^{-1}	x_{11} cm^{-1}	Progression	$I(2\nu_1)/I(\nu_1)^f$	Footnote ref.
—	845.5 ± 0.5	1.1 ± 0.2	$8\nu_1$	~0.63	g
2,400	839.5 ± 0.5	1.0 ± 0.2	$8\nu_1$	~0.44	g
—	854.4 ± 0.5	0.71 ± 0.1	$10\nu_1$	0.67	g
—	150.1 ± 0.5	0.05 ± 0.05	$11\nu_1$	0.7–1.0 est.	h
—	161.0 ± 0.2	0.11 ± 0.03	$12\nu_1$	0.61	i
5,400	161.5 ± 0.2	0.11 ± 0.03	$13\nu_1$	0.61	i
2,800 (gas)	385	—	$3\nu_1$	~0.4	j
13,000	~465	—	$5\nu_1$	~0.45	k
13,000	454.0	~0.0	$6\nu_1$	0.50	l
11,500	—	—	$3\nu_1$	—	m
—	190 ± 2^d	0.6 ± 0.6^d	$5\nu_1$	~0.55	m
5,000	213.4 ± 0.5	0.29 ± 0.05	$9\nu_1$	0.84	n
5,000	210.5 ± 0.5	~0	$5\nu_1$	—	n
—	220.5 ± 1	1.0 ± 0.5	$3\nu_1$	~0.18	o
—	347.1 ± 0.5	0.50 ± 0.08	$5\nu_1$	~0.5	p
—	338.8 ± 0.9	0.43 ± 0.13	$8\nu_1$	~0.5	p,q
—	348.7 ± 0.8	0.52 ± 0.11	$8\nu_1$	~0.5	p,q
—	339.6 ± 0.9	0.76 ± 0.15	$9\nu_1$	~0.5	p,q
—	342.1 ± 0.8	0.66 ± 0.16	$11\nu_1$	~0.5	p,q
18,000	209.1 ± 0.5	0.1 ± 0.1^d	$7\nu_1$	~0.5	r,s
12,800	150.3 ± 0.5	~0	$3\nu_1$	~0.2	r,s
1,530	272.6 ± 0.4	0.35 ± 0.05	$4\nu_1$	0.22	t
—	276.2 ± 0.5	0.39 ± 0.06	$4\nu_1$	0.31	t

[j] T. Kamisuki and S. Maeda, *Chem. Phys. Lett.* **21**, 330 (1973).

[k] A. Ranade and M. Stockburger, *Chem. Phys. Lett.* **22**, 257 (1973).

[l] Unpublished work.

[m] H. Hamaguchi, I. Harada and T. Shimanouchi, *Chem. Lett.* 1049 (1973).

[n] Y. M. Bosworth and R. J. H. Clark, *Chem. Phys. Lett.* **28**, 611 (1974); *J. Chem. Soc. Dalton Trans.* 381 (1975).

[o] M. Booth, R. J. Gillespie and M. J. Morton, *Adv. Raman Spectrosc.* **1**, 364 (1973).

[p] R. J. H. Clark and M. L. Franks, *Chem. Commun.* 316 (1974).

[q] R. J. H. Clark and M. L. Franks, *J. Amer. Chem. Soc.* **92**, 2691 (1975).

[r] Y. M. Bosworth and R. J. H. Clark, *J. Chem. Soc. Dalton Trans.* 1749 (1974).

[s] H. Hamaguchi, I. Harada and T. Shimanouchi, *J. Raman Spectrosc.* **2**, 517 (1974).

[t] R. J. H. Clark and M. L. Franks, to be published.

[u] R6G = rhodamine 6G.

are slightly dependent on the counter-ion, and so by a variation of the latter, the effectiveness of the resonance may be studied using only a single exciting line (514.5 nm, in this instance). Further studies along these lines may be expected when the full potential of tunable dye lasers is realized. The relative positions of some argon ion laser lines to the vibrational structure of the lowest allowed electronic state of the $Mo_2Cl_8^{4-}$ ion is indicated in Fig. 8. The longest overtone progressions which have so far been observed are for iodine (to $v = 20$ at room temperature, and to $v = 25$ at *ca.* 20K) and titanium tetraiodide (to $v = 13$). The optimum concentration for solution studies on such compounds appears to be 10^{-4}–10^{-2} M.

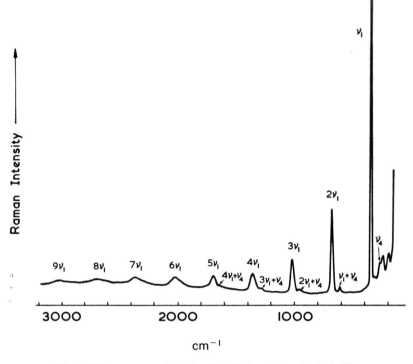

Fig. 7. RR spectrum of $Cs_4Mo_2Cl_8$ (Clark and Franks, Ref. 79).

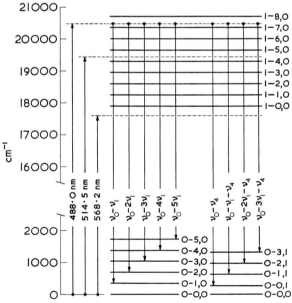

Fig. 8. Relative positions of some argon ion laser lines to the vibrational structure in the 19,000 cm⁻¹ electronic state of the $Mo_2Cl_8^{4-}$ ion. Transitions from the 514.5 nm intermediate state down to the ground electronic state indicate the two overtone progressions observed in the RRE.

Theoretical work by Kobinata,[74] and more recently by Koningstein and Jakubinek,[49,80] allows the prediction that the intensity of a given Raman band arising from an asymmetric vibration which mixes two excited electronic states should reach a maximum both at $\nu_0 \approx \nu_{00}$ as well as $\nu_0 \approx \nu_{01}$. The resonance Raman excitation profiles of certain Raman bands of oxyhaemoglobin are in agreement with these predictions.[81] Such also appears to be the case for the excitation profile of the t_z CoN stretching fundamental of the $Co(NCS)_4{}^{2-}$ ion[82] for which ν_{max} lies *ca.* 250 cm^{-1} above ν_{00}. However, for a totally symmetric vibration the ν_0 for maximum enhancement of the Raman intensity need not coincide with either ν_{00} or ν_{01}.[49] It is also concluded that, for the appearance of overtone progressions, it is necessary that the excited state has its potential surface displaced with respect to the ground state. The effect is more pronounced the steeper the slope of the upper potential curve above the lower minimum.[7] Moreover, it seems that, in agreement with theoretical predictions,[4] the electronic transition must be electric dipole allowed in order for overtone progressions to be observed. Certainly none has yet been reported for a compound for which ν_0 has been brought into coincidence with a ligand field transition. The latter, at least for centrosymmetric species, are electric dipole forbidden, and generally rather weak.

The Raman scattering cross-section for $2\nu_1$ relative to that of ν_1 rarely exceeds *ca.* 0.7 (Tables 3 and 4). Various attempts have been made to account for the Raman scattering cross-sections for the different members of the observed overtone progressions, the greatest success in this respect having been achieved by Kobinata.[74] Behringer[7] has concluded that overtones of high order may appear with intensities comparable with that of the fundamental in cases where the damping constant is small and where the frequency of the resonant transition is strongly dependent on the normal vibration displaying the RRE [*cf.* Eqns. (2) and (3)].

4.3 Calculation of Harmonic Frequencies and Anharmonicity Constants

The observation of large numbers of overtones of a totally symmetric fundamental of the species listed in Tables 3 and 4 makes it possible to determine accurately the appropriate harmonic frequency ω_1 and anharmonicity constant x_{11}. The observed wavenumber, $\nu(n)$, of any overtone of a polyatomic anharmonic oscillator is given by the expression[83]

$$\nu(n) = G(n) - G(0)$$
$$= n\omega_1 - (n^2 + n)x_{11} + \text{higher terms} \qquad (6)$$

where $G(n)$ is the term value of the nth vibrational level. Thus ω_1 and x_{11} may be determined from a plot of $\nu(n)/n$ versus n. Such a plot is shown in Fig. 9 for two of the three progressions observed in the RR spectrum of the $AuBr_4{}^-$ ion (see later). The values of ω_1 and x_{11} so determined for each species displaying the RRE are included in Tables 3 and 4. The technique of RR spectroscopy thus

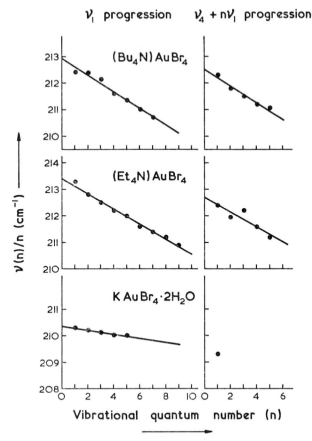

Fig. 9. Plot of $\nu(n)/n$ versus n for two of the overtone progressions observed in the R R spectrum of complexes of the $AuBr_4^-$ ion in the solid state (Bosworth and Clark, Ref. 86).

offers a powerful method for obtaining information relating to the anharmonicity of totally symmetric modes. It should, moreover, have particular relevance to the study of the effects of intermolecular interactions on the ground state properties of simple species by way of the influence of the solvent on ω_1 and x_{11} of the solute.[78] It should be borne in mind, however, that the higher terms in Eqn. (6) contribute to a change of intercept, and accordingly if these are significant this will restrict the accuracy with which ω_1 can be evaluated.

The half band widths of each member of the overtone progression invariably display a characteristic increase with increase in vibrational quantum number. This increase is partly due to environmental effects, and partly due to cross terms in the potential function which modify the frequencies of overlapping hot bands. Then, owing to the fact that the fundamentals of many of the molecular species, e.g. TiI_4, have such low frequencies, hot band contributions are significant, and a more accurate expression for $v(n)$ for a MX_4 tetrahedral molecule is as follows.[83]

$$G(n, m, p, q) - G(0, m, p, q)$$
$$= n\omega_1 - (n^2 + n)x_{11} - n[x_{12}(m+1) + x_{13}(p+\tfrac{3}{2}) + x_{14}(q+\tfrac{3}{2})] \quad (7)$$

Terms such as $x_{12}(m + d_i/2)$ therefore contribute to the breadth in a way proportional to n (d_i = degeneracy). For vapour state spectra, the increased broadening with increasing vibrational quantum number may be caused, in addition, by the rotational fine structure consisting of, in the case of linear molecules, overlapping S, Q and O branches to each member of the progression.[84] In particular, the broadening is brought about by the increasing separations of the S band heads with higher n values.

The only species listed in Tables 3 and 4 whose RR progressions do not display the characteristic feature of a regular increase in half band width with increasing vibrational quantum number are the matrix isolated species Cl_2^-, O_3^-, S_3^- and NO_2; evidently at the low sample temperatures at which the spectra were recorded, both rotational as well as hot band contributions to band broadening have been minimized.

For several of the polyatomic ions, e.g. the tetrahedral MnO_4^- and MoS_4^{2-} ions,[51,85] the square planar $AuBr_4^-$ ion,[86] and the binuclear $Mo_2Cl_8^{4-}$ ion[79] (which has D_{4h} symmetry), second progressions in the totally symmetric stretching fundamental which displays the main RRE progression are observed. This second progression in the $v_1(a_1)$ fundamental is based on one quantum of another fundamental [the $v_3(t_2)$ 'stretching' fundamental in the case of the two tetrahedral ions, and a Raman-active metal–halogen stretching fundamental in the cases of the $AuBr_4^-$ and the $Mo_2Cl_8^{4-}$ ions]. The values obtained for ω_1 and x_{11} from these second progressions are closely similar to, but, owing to the fewer observed members, less accurate than those obtained from the main progression. Cross terms, x_{ij}, may also be evaluated from these spectra.[86] Moreover, for the $AuBr_4^-$ ion, a third progression in the $v_1(a_{1g})$ fundamental has been observed, based on one quantum of the $v_4(b_{2g})$ bending fundamental.

4.4 Depolarization Ratios

Theoretical work by Mortensen[87,88] indicates that the measurement of depolarization ratios of vibrational Raman bands for totally symmetric modes of isolated, freely rotating molecules in regions of near resonance permits a determination of the symmetry of the electronic state with which the exciting line is in near resonance. The depolarization ratios (ρ_p or ρ_n) are defined as

$$\rho_p = \frac{I_\perp}{I_\parallel} = \frac{3\gamma'^2}{45\bar{\alpha}'^2 + 4\gamma'^2} \quad (8)$$

and

$$\rho_n = \frac{I_\perp}{I_\parallel} = \frac{6\gamma'^2}{45\bar{\alpha}'^2 + 7\gamma'^2} \quad (9)$$

where the p and n subscripts indicate whether the exciting radiation is polarized (p) or not (n, i.e. 'natural'). $I_\perp$ and $I_\parallel$ are the intensities of Raman scattered light observed in a direction perpendicular to the direction of propagation of the incident light and polarized perpendicular and parallel respectively to the direction of polarization of the incident light. The quantities $3\bar{\alpha}'$ and γ' are the trace and anisotropy of the derived polarizability tensor, and they are related to the cartesian tensor components as follows.

$$\bar{\alpha}' = \tfrac{1}{3}(a_{xx}' + a_{yy}' + a_{zz}') \tag{10}$$

$$\gamma'^2 = \tfrac{1}{2}[a_{xx}' - a_{yy}')^2 + (a_{yy}' - a_{zz}')^2 + (a_{zz}' - a_{xx}')^2 + 6(a_{xy}'^2 + a_{yz}'^2 + a_{zx}'^2)] \tag{11}$$

Mortensen's conclusions relate to axial molecules in non-degenerate (A_1) ground electronic states possessing at least one threefold or higher axis, but exclude cubic groups. The conclusions are that the depolarization ratio (ρ_p) is $\tfrac{1}{3}$ for a band involving an A_1 intermediate state but $\tfrac{1}{8}$ for one involving an E intermediate state (polarization of the electronic transition being 'σ' ($\parallel$) or 'π' ($\perp$) respectively). The depolarization ratios for the RR fundamentals listed in Tables 3 and 4, where available, are given in Table 5; the data relate only to

TABLE 5
Depolarization ratios of totally symmetric fundamentals of various molecular species under resonance Raman conditions

Species	State	ρ_p	Footnote Ref.
I_2[a]	Gas	0.35	[d]
	$CHCl_3$ soln.	0.26	[d]
Br_2[a]	Gas	0.25	[d]
BrCl	Gas	0.22	[d]
ICl	Gas	0.40	[d]
IBr	Gas	0.43	[d]
I_2^+[b]	HSO_3F soln.	0.30	[d]
Br_2^+	$HSO_3F/SbF_5/SO_3$ soln.	0.30	[d]
CS_2	Soln.	0.48[c]	[e]
I_3^-	K^+ salt in H_2O soln.	0.33	[d]
Te_4^{2+}	Conc. H_2SO_4 soln.	0.3	[d]

[a] Electronic transition in resonance has the assignment $^3\pi_{0u}^+ \leftarrow {}^1\Sigma_g^+$.

[b] Electronic transition is of the $\pi^* \leftarrow \pi$ type.

[c] This is a ρ_n value obtained with 253.7 nm Hg arc excitation. The corresponding value for ρ_p is calculated to be 0.32. The value of ρ_n obtained with off-resonance excitation (435.8 nm Hg arc) was 0.28.

[d] See Tables 3 and 4.

[e] L. Bernard and R. Dupeyrat, *J. Chim. Phys.* **65**, 410 (1968).

non-cubic species in view of the restriction mentioned above. The ρ_p values all lie in the range 0.33 ± 0.11 (including the datum for carbon disulphide, which is based on a ρ_n value), suggesting that the intermediate state has A_1 symmetry in each case, cf. Refs. 2, 56 and 89. Further results and discussion on the depolarization ratios of Raman bands of diatomic molecules under resonance conditions are given by Holzer and Le Duff,[90] Ito et al.[90] and Behringer.[7]

A depolarization ratio of $\rho_p = 0.12_5$ has been observed for a totally symmetric fundamental (at 1373 cm^{-1}) of ferri-haemoglobin fluoride[13] when the exciting frequency is brought into near coincidence with a planar (x, y) electronic transition of the molecule, i.e. $a_{xx}' \equiv a_{yy}'$, $a_{zz}' = 0$; this result is also in agreement with Mortensen's conclusions (see above).

Rea[91] has surveyed earlier work of Shorygin and coworkers on the resonance characteristics of several aromatic nitro compounds. He has also established[91] that for molecules with triply degenerate intermediate states, i.e. cubic molecules, the RR band will be completely polarized ($\rho_p = 0$), in agreement with observations (cf. Tables 3, 4) on ρ_p of the MnO_4^-, CrO_4^{2-}, MoS_4^{2-} ions, and TiI_4 and VCl_4 under resonance conditions, and the facts that, for example, the resonant state of the MnO_4^- and CrO_4^{2-} ions has the symmetry $^1T_2$.[92,93] The fact that ρ_p remains zero is a consequence of all three diagonal elements of the scattering tensor being enhanced by the same amount on decreasing $\nu_e - \nu_0$.[49]

The recent experimental discovery of inverse polarization ($\rho_p = \infty$) further indicates the importance of depolarization ratios. This ρ value is only possible under resonance conditions, and can only occur in the case of a non-totally symmetric vibration for which the molecular scattering tensor is antisymmetric, i.e. $a_{ij}' = -a_{ji}'$. Such modes are inactive in non-resonance Raman scattering. Specifically,[11] for haemoglobin and cytochrome c, which approximate to D_{4h} symmetry, the a_{2g} fundamentals display this effect. Deviations of ρ_p from ∞, within the range $\frac{3}{4} < \rho_p < \infty$, for such bands may provide sensitive tests for the loss of fourfold symmetry in the haeme chromophore by steric or other effects. This matter is discussed in detail elsewhere.[25,26,49]

4.5 Other Aspects of the Resonance Raman Effect

In the case of the MnO_4^- and CrO_4^{2-} ions, band structure in the lowest observed electronic transition corresponds to a progression in the totally symmetric fundamental, $\nu_1(a_1)$, in the excited state. The same fundamental is the one which is prominent in the RRE. This result is in agreement with the observations and conclusions of Behringer[2] and Shorygin[5,6] on the RR scattering of organic molecules.

The study of the $Mo_2Cl_8^{4-}$ ion[79] (one of the many metal–metal bonded species to display the RRE)[94,95] is particularly significant because it is known that the ca. 19,000 cm^{-1} transition involves excitation of a δ-electron of the so-called quadruple Mo—Mo bond.[96,97] Thus, by bringing the excitation frequency into coincidence with this electronic band it is not surprising that, of the three totally symmetric fundamentals of the ion [ν(MoMo), ν(MoCl) and δ(ClMoCl)], it is ν(MoMo) which is active in the RRE. A close relationship between absorption spectroscopy and RR spectroscopy is thus indicated.

In a somewhat related study,[16] the selective enhancement of band intensities in the RR spectra of cobalt corrinoids has been noted. The principal RR band of aquocobalamin (vitamin B_{12a}) excited by the 488.0 or 514.5 nm Ar^+ lines is at 1504 cm^{-1}, and this frequency is associated with the same corrin ring vibrational mode (a ring breathing mode of the small nitrogen heterocycles) which governs the vibrational spacing of the $\alpha\beta$-electronic band components of the molecule at *ca.* 500 nm. When resonance is switched to the so-called γ (or Soret) electronic band (centred at *ca.* 338 nm) by excitation with the 363.8 nm Ar^+ line, the 1504 cm^{-1} band loses most of its intensity, but a band at 1550 cm^{-1} is then enormously enhanced.

An intriguing application of the RRE is to the identification of the sulphur species present in the deep blue mineral ultramarine, which is essentially a sodium aluminosilicate with the idealized formula $(Na_8Al_6Si_6O_{24}S_4)_n$. One species is the S_3^- radical anion[98,99] (analogous to the ozonide ion), and it is trapped in holes in the open silicate lattice.[100] However, it is now established by both e.s.r. and RR studies that the S_2^- radical anion is present as well.[99] Two RR progressions based on the $\nu_1(a_1)$ band of the S_3^- ion, and one based on the fundamental of the S_2^- ion have been observed.[99] In general, negative ions may also be doped into alkali halide crystals, which prove to be convenient hosts for the study of otherwise unstable species in ordered matrices from liquid helium to room temperature.

Another class of compound for which RR studies are beginning to provide valuable new information is mixed-valence compounds, many properties of which have been reviewed by Robin and Day[101] and Hush.[102] For instance, the deep blue mixed-valence compound Cs_2SbCl_6, on irradiation with 488.0 or 514.5 nm excitation, i.e. within the contour of the mixed-valence transition centred at 17,900 cm^{-1}, yields a RR spectrum in which four progressions are evident, in all of which it is the $\nu_1(a_{1g})$ fundamental of the component $SbCl_6^-$ ion (326 cm^{-1}) which is the progressing fundamental.[103] Similar effects are displayed by the related complex $Cs_4SbBiCl_{12}$ in which the antimony(III) is replaced by bismuth(III).[103]

A further, very recent study of the RR band intensities of the $PtBr_6^{2-}$ and PtI_6^{2-} ions has indicated[104] that the a_{1g} component of overtones and combination tones involving non-totally symmetric fundamentals may also be enhanced under resonance conditions. However, their intensities fall off very rapidly with increase in the vibrational quantum number of the non-totally symmetric fundamental. The results suggest that long overtone progressions are characteristic of the RRE due to an *A* term whereas the absence thereof is characteristic of the RRE originating from a *B* term.[104]

4.6   Excitation within a Ligand Field Band

In view of the relationship between a_{ij}' and the electric dipole transition, it was hardly to be expected that excitation of a molecule in the vicinity of a (Laporte forbidden, albeit vibronically allowed) ligand field transition would influence the intensity of Raman scattering therefrom. However, both for square planar MX_4^{n-} ions[53,105] as well as for octahedral MX_6^{n-} ions,[52] the intensities of the a_{1g} fundamentals of the ions are significantly reduced in the region of a d–d transition. This reduction in intensity is clearly indicated in the excitation profiles for the Raman active fundamentals of the $PdBr_4^{2-}$ ion (Fig. 10). The

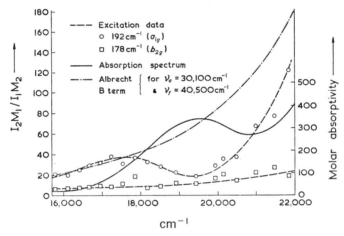

Fig. 10. Absorption spectrum and Raman band excitation profiles of fundamentals of the $PdBr_4^{2-}$ ion in aqueous solution in the vicinity of a ligand field transition (Miskowski et al., Ref. 105). The internal intensity standard is the a_1 band of the sulphate ion.

effect, which is not yet understood, appears to be specific to centrosymmetric ions. By contrast with the latter, excitation of the $Co(NCS)_4^{2-}$ ion within the contour of the d–d transition at ca. 16,300 cm^{-1} ($^4T_1(P) \leftarrow\, ^4A_2$) gives rise to an enhancement of the intensities of the CoN, NC and CS stretching fundamentals of the ion. This has been ascribed[82] to the partly allowed character of the resonant electronic transition (due to d–p mixing) in which respect it differs from d–d transitions of centrosymmetric ions.

5   CONCLUSION

Resonance Raman studies of simple inorganic species have demonstrated that (a) the extent of resonance is essentially a function of $v_e - v_0$; (b) it is a totally symmetric stretching fundamental of the species which exhibits the long progressions (almost invariably a stretching mode); and (c) to be most effective, it appears that ν_0 should become coincident in frequency with a sharp electric dipole allowed transition of high oscillator strength. Because the technique gives rise to such high intensities for fundamentals displaying the effect, it holds promise for the detection of trace amounts of materials possessing strong absorption bands in regions accessible by fixed or tunable laser excitation frequencies. Moreover, by virtue of the accuracy with which ω_1 and x_{11} may be determined from the RRE, very small changes in the strengths of solvent interactions with the scattering molecules may be detected, even when the latter are in very low concentrations. The prospect of obtaining cross terms in the potential function, albeit not very precisely, is also a matter of considerable interest. Detailed studies of excitation profiles of Raman bands are likely further to demonstrate the close relationship between Raman and electronic absorption spectroscopies.

REFERENCES

(1) J. Behringer, *Z. Elektrochem.* **62**, 906 (1958).
(2) J. Behringer, in *Raman Spectroscopy*, Vol. 1 (H. A. Szymanski, Ed.), Plenum Press, New York, 1967, p. 168.
(3) R. E. Hester, in *Raman Spectroscopy*, Vol. 1 (H. A. Szymanski, Ed.), Plenum Press, New York, 1967, p. 101.
(4) J. Tang and A. C. Albrecht, in *Raman Spectroscopy*, Vol. 2 (H. A. Szymanski, Ed.), Plenum Press, New York, 1970, p. 33.
(5) P. P. Shorygin, *Russ. Chem. Rev.* **40**, 367 (1971); *Soviet Phys.-Usp.* **16**, 99 (1973).
(6) T. M. Ivanova, L. A. Yanovskaya and P. P. Shorygin, *Opt. and Spectrosc.* **18**, 115 (1965).
(7) J. Behringer, *Chemical Society Specialist Reports on Molecular Spectroscopy*, Volume 2, (R. F. Barrow, D. A. Long and D. J. Millen, Eds.), 1974, p. 100; *J. Raman Spectrosc.* **2**, 275 (1974).
(8) R. E. Hester, *Chemical Society Specialist Reports on Molecular Spectroscopy*, Vol. 2, (R. F. Barrow, D. A. Long and D. J. Millen, Eds.), 1974, p. 439.
(9) H. J. Bernstein, *Adv. Raman Spectrosc.* **1**, 305 (1973).
(10) P. R. Carey, H. Schneider and H. J. Bernstein, *Biochem. Biophys. Res. Commun.* **47**, 588 (1972).
(11) T. G. Spiro and T. C. Strekas, *Proc. Nat. Acad. Sci. USA* **69**, 2622 (1972); *J. Raman Spectrosc.* **1**, 22 (1973).
(12) T. C. Strekas and T. G. Spiro, *Biochem. Biophys. Acta* **263**, 830 (1972).
(13) T. C. Strekas, A. J. Packer and T. G. Spiro, *J. Raman Spectrosc.* **1**, 197 (1973).
(14) W. H. Woodruff, T. G. Spiro and T. Yonetani, *Proc. Nat. Acad. Sci. USA* **71**, 1065 (1974).
(15) T. G. Spiro and T. C. Strekas, *J. Amer. Chem. Soc.* **96**, 338 (1974).
(16) E. Mayer, D. J. Gardiner and R. E. Hester, *J. Chem. Soc. Faraday Trans. II* **69**, 1350 (1973); *Mol. Phys.* **26**, 783 (1973).
(17) W. T. Wozniak and T. G. Spiro, *J. Amer. Chem. Soc.* **95**, 3402 (1973).
(18) H. Brunner, A. Meyer and H. Sussner, *J. Mol. Biol.* **70**, 153 (1972).
(19) H. Brunner, *Biochem. Biophys. Res. Commun.* **51**, 888 (1973).
(20) D. Gill, R. G. Kilponen and L. Rimai, *Nature* **227**, 743 (1970).
(21) L. Rimai, D. Gill and J. L. Parsons, *J. Amer. Chem. Soc.* **93**, 1353 (1971).
(22) T. V. Long, T. M. Loehr, J. R. Allkins and W. Lovenberg, *J. Amer. Chem. Soc.* **93**, 1809 (1971).
(23) T. V. Long and T. M. Loehr, *J. Amer. Chem. Soc.* **92**, 6384 (1970).

(24) S.-P. W. Tang, T. G. Spiro, K. Mukai and T. Kimura, *Biochem. Biophys. Res. Commun.* **53**, 869 (1973).

(25) T. G. Spiro, *Acc. Chem. Research* **7**, 339 (1974).

(26) T. G. Spiro and T. M. Loehr, *Adv. Infrared Raman Spectrosc.*, Vol. 1 (R. J. H. Clark and R. E. Hester, Eds.), Heyden, London, 1975, p. 98.

(27) W. Kiefer and H. J. Bernstein, *Appl. Spectrosc.* **25**, 609 (1971).

(28) W. Kiefer and H. J. Bernstein, *Appl. Spectrosc.* **25**, 501 (1971).

(29) R. J. Gillespie, personal communication; R. J. H. Clark, *Spex Speaker* **18**, 1 (1973).

(30) R. J. H. Clark, O. H. Ellestad and P. D. Mitchell, *Appl. Spectrosc.* **28**, 575 (1974).

(31) See, for example, W. Kiefer, *Appl. Spectrosc.* **28**, 115 (1974).

(32) J. A. Koningstein and B. F. Gächter, *J. Opt. Soc. Am.* **63**, 892 (1973).

(33) N. Zimmerer and W. Kiefer, *Appl. Spectrosc.* in press.

(34) P. P. Shorygin and T. M. Ivanova, *Dokl. Akad. Nauk SSSR* **121**, 70 (1958).

(35) P. P. Shorygin, *Pure Appl. Chem.* **4**, 87 (1962).

(36) P. P. Yaney, *J. Opt. Soc. Am.* **62**, 1297 (1972).

(37) T. Hirschfeld, *J. Opt. Soc. Am.* **63**, 1309 (1973).

(38) W. Kiefer and H. W. Schrötter, *Z. Angew. Phys.* **25**, 236 (1968).

(39) W. Kiefer, *Chem. Instr.* **3**, 21 (1971).

(40) T. C. Strekas, D. H. Adams, A. Packer and T. G. Spiro, *Appl. Spectrosc.* **28**, 324 (1974).

(41) D. F. Shriver and J. B. R. Dunn, *Appl. Spectrosc.* **28**, 319 (1974).

(42) W. Kiefer, *Appl. Spectrosc.* **27**, 253 (1973).

(43) G. Placzek, *Handbuch der Radiologie*, Vol. 6 (E. Marx, Ed.), Akademische Verlagsgesellschaft, Leipzig, 1934, p. 205. English translation by Ann Werbin, UCRL Trans. No. 526(L), U.S. Dept. of Commerce.

(44) G. W. Chantry, in *The Raman Effect*, Vol. 1 (A. Anderson, Ed.), Dekker, New York, 1971, p. 49.

(45) A. C. Albrecht, *J. Chem. Phys.* **34**, 1476 (1961).

(46) A. C. Albrecht and M. C. Hutley, *J. Chem. Phys.* **55**, 4438 (1971).

(47) W. L. Peticolas, L. Nafie, P. Stein and B. Fanconi, *J. Chem. Phys.* **53**, 1576 (1970).

(48) D. W. Collins, D. B. Fitchen and A. Lewis, *J. Chem. Phys.* **59**, 5714 (1973).

(49) J. A. Koningstein and B. G. Jakubinek, *J. Raman Spectrosc.* **2**, 317 (1974).

(50) R. J. H. Clark and P. D. Mitchell, *J. Mol. Spectrosc.* **51**, 458 (1974).

(51) W. Kiefer and H. J. Bernstein, *Mol. Phys.* **23**, 835 (1972).

(52) Y. M. Bosworth and R. J. H. Clark, *J. Chem. Soc. Dalton Trans.* 1749 (1974).

(53) Y. M. Bosworth and R. J. H. Clark, *Inorg. Chem.* **14**, 170 (1975).

(54) A. H. Kalantar, E. S. Franzosa and K. K. Innes, *Chem. Phys. Lett.* **17**, 335 (1972).

(55) K. Kaya, N. Mikami, Y. Udagawa and M. Ito, *Chem. Phys. Lett.* **13**, 221 (1972).

(56) O. S. Mortensen, *J. Mol. Spectrosc.* **39**, 48 (1971); *Mol. Phys.* **22**, 179 (1971).

(57) W. F. Murphy, W. Holzer and H. J. Bernstein, *Appl. Spectrosc.* **23**, 211 (1969).

(58) T. V. Long and R. A. Plane, *J. Chem. Phys.* **43**, 457 (1965).

(59) R. S. Armstrong and R. J. H. Clark, *J. Chem. Soc. Faraday Trans.* in press (1975).

(60) W. Holzer, W. F. Murphy and H. J. Bernstein, *J. Chem. Phys.* **52**, 399 (1970).

(61) S. D. Silverstein and R. L. St. Peters, *Chem. Phys. Lett.* **23**, 140 (1973); see also *Opt. Commun.* **7**, 193 (1973).

(62) I. R. Beattie, G. A. Ozin and R. O. Perry, *J. Chem. Soc.* (A) 2071 (1970).

(63) M. Berjot, M. Jacon and L. Bernard, *Can. J. Spectrosc.* **17**, 60 (1972).

(64) M. Berjot, L. Bernard and T. Theophanides, *Can. J. Spectrosc.* **18**, 128 (1973).

(65) D. G. Fouche and R. K. Chang, *Phys. Rev. Lett.* **29**, 536 (1972).

(66) R. L. St. Peters, S. D. Silverstein, M. Lapp and C. M. Penney, *Phys. Rev. Lett.* **30**, 191 (1973).

(67) A. Nitzan and J. Jortner, *J. Chem. Phys.* **57**, 2870 (1972).

(68) J. Behringer, *Z. Physik* **229**, 209 (1969).

(69) M. Jacon, M. Berjot and L. Bernard, *Compt. Rend.* **273B**, 596, 956 (1971).

(70) M. Berjot, M. Jacon and L. Bernard, *Opt. Comm.* **4**, 117 (1971).

(71) L. A. Nafie, P. Stein and W. L. Peticolas, *Chem. Phys. Lett.* **12**, 131 (1971).

(72) M. Jacon, *Adv. Raman Spectrosc.* **1**, 325 (1972).

(73) D. van Labeke and M. Jacon, *Opt. Comm.* **9**, 400 (1973).

(74) S. Kobinata, *Bull. Chem. Soc. Japan* **46**, 3636 (1973).

(75) P. F. Williams and D. L. Rousseau, *Phys. Rev. Lett.* **30**, 951 (1973).

(76) M. Mingardi and W. Siebrand, *Chem. Phys. Lett.* **23**, 1 (1973); *J. Chem. Phys.* **62**, 1074 (1975).

(77) R. J. H. Clark and P. D. Mitchell, *J. Amer. Chem. Soc.* **95**, 8300 (1973).

(78) R. J. H. Clark and P. D. Mitchell, *J. Raman Spectrosc.* **2**, 399 (1974).

(79) R. J. H. Clark and M. L. Franks, *Chem. Commun.* 316 (1974); *J. Amer. Chem. Soc.* **97**, 2691 (1975).

(80) O. S. Mortensen and J. A. Koningstein, *J. Chem. Phys.* **48**, 3911 (1968).

(81) T. C. Strekas and T. G. Spiro, *J. Raman Spectrosc.* **1**, 387 (1973).

(82) Y. M. Bosworth, R. J. H. Clark and P. C. Turtle, *J. Chem. Soc. Dalton Trans.* in press.

(83) G. Herzberg, *Infrared and Raman Spectra of Polyatomic Molecules*, Van Nostrand, Princeton, N.J., 1945, p. 205.

(84) W. Kiefer and H. J. Bernstein, *J. Mol. Spectrosc.* **43**, 366 (1972).

(85) A. Ranade and M. Stockburger, *Chem. Phys. Lett.* **22**, 257 (1973).

(86) Y. M. Bosworth and R. J. H. Clark, *Chem. Phys. Lett.* **28**, 611 (1974).

(87) O. S. Mortensen, *Chem. Phys. Lett.* **3**, 4 (1969).

(88) O. S. Mortensen, *Chem. Phys. Lett.* **5**, 515 (1970).

(89) R. J. Gillespie and M. J. Morton, *J. Mol. Spectrosc.* **30**, 178 (1969).

(90) W. Holzer and Y. Le Duff, *Adv. Raman Spectrosc.* **1**, 109 (1973); Y. Udagawa, M. Iijima and M. Ito, *J. Raman Spectrosc.* **2**, 313 (1974).

(91) D. G. Rea, *J. Mol. Spectrosc.* **4**, 499 (1960).

(92) J. C. Duinker and C. J. Ballhausen, *Theor. Chim. Acta* **12**, 325 (1968).

(93) L. W. Johnson and S. P. McGlynn, *Chem. Phys. Lett.* **7**, 618 (1970).

(94) C. L. Angell, F. A. Cotton, B. A. Frenz and T. R. Webb, *Chem. Commun.* 399 (1973).

(95) J. S. Fillipo and H. J. Sniadoch, *Inorg. Chem.* **12**, 2326 (1973).

(96) F. A. Cotton and C. B. Harris, *Inorg. Chem.* **6**, 924 (1967).

(97) C. D. Cowman and H. B. Gray, *J. Amer. Chem. Soc.* **95**, 8177 (1973).

(98) W. Holzer, W. F. Murphy and H. J. Bernstein, *J. Mol. Spectrosc.* **32**, 13 (1969).

(99) R. J. H. Clark and M. L. Franks, *Chem. Phys. Lett.* **34**, 69 (1975).

(100) A. Wieckowski, *Phys. Stat. Sol.* **42**, 125 (1970).

(101) M. B. Robin and P. Day, *Adv. Inorg. Chem. Radiochem.* **10**, 247 (1967).

(102) N. S. Hush, *Prog. Inorg. Chem.* **8**, 391 (1967).

(103) R. J. H. Clark and W. R. Trumble, *Chem. Commun.* 318 (1975).

(104) H. Hamaguchi, I. Harada and T. Shimanouchi, *J. Raman Spectrosc.* **2**, 517 (1974).

(105) V. Miskowski, W. H. Woodruff, J. P. Griffin, K. G. Werner and T. G. Spiro, *J. Chem. Phys.* submitted.

Chapter 5

THE METAL ISOTOPE EFFECT ON MOLECULAR VIBRATIONS

N. Mohan and A. Müller

Institute of Chemistry, University of Dortmund, 46 Dortmund, W. Germany

K. Nakamoto

Department of Chemistry, Marquette University, Milwaukee, Wisconsin 53233, U.S.A.

1   INTRODUCTION

Isotope pairs such as (H/D) and ($^{16}O/^{18}O$) have been used routinely by many spectroscopists. However, the use of isotopic pairs of heavy metals such as ($^{58}Ni/^{62}Ni$) and ($^{104}Pd/^{110}Pd$) in vibrational spectroscopy was initiated only recently. In 1969, Nakamoto and his coworkers published the first report on the use of these isotopes in assigning metal–ligand vibrations of phosphine complexes.[1] The delay in their use was probably due to two reasons: (1) it was thought that the magnitude of isotope shifts due to pairs of heavy elements might be too small to be of practical value, and (2) pure metal isotopes were too expensive to use routinely in the laboratory. They have shown, however, that the magnitudes of metal isotope shifts are generally of the order of 2–10 cm^{-1} for a stretching mode and 0–2 cm^{-1} for a bending mode, and the experimental error in measuring the frequency could be as small as ± 0.2 cm^{-1} if proper caution were taken. They have also shown that this 'metal isotope technique' is financially feasible if the compounds are prepared on a milligram scale. Normally, the vibrational spectrum of a compound can be obtained even with 10–20 mg of a sample. Therefore, the cost of preparing it on this scale is not overwhelming (Appendix I, p. 228). Since 1969, Nakamoto and his coworkers as well as others have applied the metal isotope technique to many coordination compounds. In 1972, a short review concerning the metal isotope effect was written by Nakamoto.[2]

Müller *et al.*[3,4] have initiated extensive work on theoretical and experimental aspects of the metal isotope effect on small molecules. They have demonstrated the utility of metal isotope frequency shifts in deriving reliable values of force constants. The same group has also studied the influence of small mass changes due to isotopic substitution on various molecular properties such as isotopic frequency shifts, mean amplitudes of thermal vibrations and Coriolis coupling constants.

2   INFLUENCE OF SMALL MASS CHANGES ON MOLECULAR CONSTANTS

First order perturbation theory has found widespread applications in the study of molecular properties such as isotopic frequency shifts, changes in normal coordinates due to isotopic substitution, Jacobians related to force and compliance constants, and kinematic perturbation etc.[5-12] Here, 'first order perturbation' implies that one considers only the first order terms in the general expanded development of any equation. Hence, suitable approximations are indispensable in this procedure, and the merits of such an attempt have to be assessed in each individual case.[13] In the following, we demonstrate the effect of small mass changes due to isotopic substitution on some molecular constants by using first order perturbation theory.

2.1   Isotopic Frequency Shifts

Considering the standard equation

$$\tilde{\mathbf{L}}_0 \mathbf{F} \mathbf{L}_0 = \mathbf{\Lambda}_0 \tag{1}$$

where $\mathbf{L}_0$ is the eigenvector matrix of $\mathbf{G}_0 \mathbf{F}$, $\mathbf{F}$ is the potential energy matrix and $\mathbf{\Lambda}_0$ is a diagonal matrix containing the frequency parameters $(\lambda_0)_i [= 4\pi^2 c^2 (\nu_0)_i^2]$. The subscript zero indicates the parent molecule. Since, within the Born-Oppenheimer approximation, $\mathbf{F}$ is the same for isotopic molecules, the isotopic frequency shifts $(\Delta \lambda_i)$ should reflect the changes in $\mathbf{L}_0$ only. $\mathbf{L}_0$ in this case is affected only by the mass changes, i.e. $\Delta \mathbf{G}$, due to isotopic substitution. Equation (1) can be rewritten in another equivalent form as

$$\mathbf{L}_0^{-1} \mathbf{G}_0 \mathbf{F} \mathbf{L}_0 = \mathbf{\Lambda}_0 \tag{2}$$

Differentiating eqn. (2) and replacing the differentials by finite increments gives

$$\Delta \mathbf{L}_0^{-1} \mathbf{G}_0 \mathbf{F} \mathbf{L}_0 + \mathbf{L}_0^{-1} \Delta \mathbf{G}_0 \mathbf{F} \mathbf{L}_0 + \mathbf{L}_0^{-1} \mathbf{G}_0 \mathbf{F} \Delta \mathbf{L}_0 = \Delta \mathbf{\Lambda}_0 \tag{3}$$

where $\Delta \mathbf{L}_0^{-1}$ is the increment in $\mathbf{L}_0^{-1}$ by isotopic substitution. Under the assumption that $\mathbf{L}_0$ does not change much due to small mass changes (heavy atom substitution), eqn. (3) can be approximated as

$$\mathbf{L}_0^{-1} \Delta \mathbf{G}_0 \mathbf{F} \mathbf{L}_0 \approx \Delta \mathbf{\Lambda}_0 \tag{4}$$

Substituting eqn. (1) in eqn. (4) gives

$$\mathbf{L}_0^{-1} \Delta \mathbf{G}_0 \tilde{\mathbf{L}}_0^{-1} \approx \Delta \mathbf{\Lambda}_0 \mathbf{\Lambda}_0^{-1} \tag{5}$$

Since $\Delta\Lambda_0$ is a diagonal matrix, it follows that one neglects the off-diagonal elements of $L_0^{-1}\Delta G_0 \tilde{L}_0^{-1}$ in eqn. (5). Equation (5) has been found to be useful in calculating the isotopic shifts. It has already been used to show that the central atom substitution is effective in fixing the precise values of force constants.[14] Some numerical results for $\Delta\lambda$ obtained by using eqn. (5) are presented in Table 1.

2.2 Coriolis Coupling Constants

First order perturbation theory has been successfully used to interpret the influence of mass on the Coriolis coupling constants.[3] This approach is of course valid only for heavy atom isotopic substitution, e.g. $^{14}N/^{15}N$, $^{16}O/^{18}O$, $^{10}B/^{11}B$, $^{96}Ru/^{102}Ru$, $^{116}Sn/^{124}Sn$ etc. The results are presented below.

The ζ matrix containing the Coriolis coupling constant is related to the F matrix through the equation[15]

$$F(G_0 - C_0)(\tilde{L}_0^{-1}) = (\tilde{L}_0^{-1})\Lambda_0(E - \zeta_0) \tag{6}$$

where C_0, like G_0, is determined from the masses of the atoms and the geometrical parameters of the molecule only, and E is the unit matrix.

Equation (6) can be written in an equivalent form as

$$(\tilde{L}_0)F(G_0 - C_0)(\tilde{L}_0^{-1}) = \Lambda_0(E - \zeta_0) \tag{7}$$

Let $(G_0 - C_0) = K_0$. Then eqn. (7) becomes

$$(\tilde{L}_0)FK_0(\tilde{L}_0^{-1}) = \Lambda_0(E - \zeta_0) \tag{8}$$

First order perturbation theory in this case leads to the result (under the assumption that $\Delta L_0 = 0$)

$$(\tilde{L}_0)F\Delta K_0(\tilde{L}_0^{-1}) = -\Lambda_0\Delta\zeta_0 + \Delta\Lambda_0(E - \zeta_0) \tag{9}$$

Substitution of the familiar equation

$$F = \tilde{L}_0^{-1}\Lambda_0 L_0^{-1} \tag{10}$$

into eqn. (9) leads to

$$\Delta\Lambda_0(\Lambda_0^{-1}) = [L_0^{-1}\Delta K_0(\tilde{L}_0^{-1}) + \Delta\zeta_0](E - \zeta_0)^{-1} \tag{11}$$

Using eqn. (5), the above equation can be written as

$$L_0^{-1}\Delta G_0\tilde{L}_0^{-1} = [L_0^{-1}\Delta K_0\tilde{L}_0^{-1} + \Delta\zeta_0](E - \zeta_0)^{-1} \tag{12}$$

Equation (12) leads to some interesting results. Since the sign of $(\zeta_0)_{ij}(i \neq j)$ is not determined by experimental results, one has multiple solutions for $\Delta\zeta_0$ as derived from eqn. (12). For $XY_3(D_{3h})$, $XY_3(C_{3v})$, $XY_4(T_d)$, $XY_6(O_h)$ molecules etc., K_0 contains only the mass of the Y atom for rotation about the symmetry axis, i.e., for first order Coriolis coupling constants. Hence, ΔK_0 is a

TABLE 1
Calculation of isotope shifts (in cm^{-1}) using first order perturbation theory

	$\Delta\nu_1$ (calc.)	$\Delta\nu_1$ (obs.)	$\Delta\nu_2$ (calc.)	$\Delta\nu_2$ (obs.)	Refs. for measured values and force field
$BF_3(e')$	53.5[a]	53.5	1.8[a]	1.77 ± 0.02	d
($^{10}B/^{11}B$)	53.4[b]		1.8[b]		
	54.4[c]		1.6[c]		
$NF_3(a_1)$	24.5	22.98 ± 0.05[e]	2.2	2.32 ± 0.05	f
($^{14}N/^{15}N$)					
$NF_3(e)$	22.1	22.0 ± 0.7	0.6	0.60 ± 0.03	f
($^{14}N/^{15}N$)					
$SiCl_4(t_2)^g$	5.7	5.7	5.0	5.3[h]	i
($^{35}Cl/^{37}Cl$)					
$SF_6(t_{1u})$	17.1	17.4	3.1	3.3	j
($^{32}S/^{34}S$)					
$MoO_4^{2-}(t_2)^g$	7.0	7.0	2.6	2.6[h]	k
($^{92}Mo/^{100}Mo$)					
$MoS_4^{2-}(t_2)^g$	7.0	7.0	1.6	1.7[h]	l
($^{92}Mo/^{100}Mo$)					
$MoO_2S_2^{2-}(b_1)^g$	7.9	8.0	2.3	2.0[k]	m
($^{92}Mo/^{100}Mo$)					
$MoO_2S_2^{2-}(b_2)^g$	7.0	7.0	1.6	1.5[h]	m
($^{92}Mo/^{100}Mo$)					
$RuO_4(t_2)^g$					
($^{16}O/^{18}O$)					
(harmonic frequencies)	43.5	43.6 ± 0.4	15.5	15.6 ± 0.5	n
$RuO_4(t_2)^g$					
($^{16}O/^{18}O$)					
(anharmonic frequencies)	42.1	42.2	15.5	15.1	n
$RuO_4(t_2)$	5.3	5.2$_5$ ± 0.1[o]	2.2	2.1	n
($^{96}Ru/^{102}Ru$)					
$Sn^{35}Cl_4(t_2)^g$	3.6	3.6 ± 0.2	1.2	1.2[h]	p
($^{116}Sn/^{124}Sn$)					
$OsO_4(t_2)$					
($^{16}O/^{18}O$)					
(harmonic frequencies)	50.5	50.3	16.5	16.7	q
$OsO_4(t_2)^g$					
($^{16}O/^{18}O$)					
(anharmonic frequencies)	48.8	48.7 ± 0.2	16.1	16.3 ± 0.2	q
$Ge^{35}Cl_4(t_2)$	7.3	7.2 ± 0.2	—	—	r
($^{70}Ge/^{76}Ge$)					
$Ti^{35}Cl_4(t_2)$	10.7	10.8 ± 0.2	—	—	r
($^{46}Ti/^{50}Ti$)					

[a] Force field from ^{10}B, ζ_{22} (e').
[b] Force field from ^{11}B, ζ_{22} (e').
[c] Force field from $\Delta\nu_2$ (e').
[d] See S. G. W. Ginn, D. Johansen and J. Overend, *J. Mol. Spectrosc.* **36**, 448 (1970).

null matrix in such cases if only the central atom is isotopically substituted. Equation (12) in such cases is reduced to

$$\Delta\Lambda_0\Lambda_0^{-1} = \Delta\zeta_0(E-\zeta_0)^{-1} \tag{13}$$

Hence

$$(\Delta\zeta_0)_{11} = [(\Delta\lambda_0)_1/(\lambda_0)_1][1-(\zeta_0)_{11}] \tag{14}$$

Since $(\zeta_0)_{ij}$ is always $\leqslant 1$, it follows that the sign of $(\Delta\zeta_0)_{ii}$ is determined purely by that of $[(\Delta\lambda_0)_i/(\lambda_0)_i]$. For substitution of the central atom by a heavier isotope, $(\Delta\lambda_0)_1$ is negative and then $(\Delta\zeta_0)_{11}$ also becomes negative. It is generally known that the ζ_0 values of symmetric and spherical rotors pertaining to the stretching vibrations are smaller for isotopic molecules with heavier central atoms than for parent molecules. This can be explained quantitatively by using eqn. (14). For Y atom substitution, however, both K_0 and Λ_0 vary and these effects cannot easily be generalized.

Few experimental determinations of the Coriolis coupling constants for isotopically different molecules have so far been reported. Only recently, Königer and Müller[16] determined the Coriolis coupling constant $\zeta_{33}(t_2)$ for $^{116}SnCl_4/^{124}SnCl_4$ [$\zeta_{33}(^{116}SnCl_4) = 0.26\pm0.08$ and $\zeta_{33}(^{124}SnCl_4) = 0.24\pm 0.08$]. These results are in very good agreement with eqn. (14).

2.3 Mean Square Amplitudes of Vibration

The variation of mean amplitudes of vibration with isotopic substitution can also be studied using first order perturbation theory.[17] Since both the bonded $(l_0^2)_{a-b}$ and the non-bonded $(l_0^2)_{a...b}$ mean square amplitudes of vibration are linear combinations of the symmetrized Σ matrix elements,[15] it is sufficient to carry out the study using this formalism. The mean square amplitude matrix Σ_0 is related to the force constant matrix F through the relation

$$\Sigma_0 F L_0 = L_0\Lambda_0\delta_0 \tag{15}$$

where δ_0 is a diagonal matrix related to the temperature and frequency.[15]

---

e Subject to Fermi resonance correction (see Ref. f below).
f A. Allen, J. L. Duncan, J. H. Holloway and D. C. McKean, *J. Mol. Spectrosc.* **31**, 368 (1969).
g Force field from $\Delta\nu_1(t_2)$, (b_1) and (b_2) respectively.
h Estimated using the product rule.
i See Ref. 29.
j See Ref. 30.
k See Ref. 39.
l See Ref. 38.
m See Ref. 74.
n See Ref. 26.
o The calculated harmonic frequency shift is $\Delta\nu_1(t_2) = 5.3_6\pm0.1$ cm^{-1}, see Ref. 26.
p See Ref. 16.
q See Ref. 27.
r See Ref. 41.

For an isotopically substituted molecule (only small mass changes involved), we have $\Delta L_0 = 0$. Then eqn. (15) is written

$$[\Sigma_0 + \Delta\Sigma_0]FL_0 = L_0[\Lambda_0 + \Delta\Lambda_0][\delta_0 + \Delta\delta_0] \tag{16}$$

Simplifying eqn. (16) with the help of eqn. (15) and neglecting the term $[(\Delta\Lambda_0)(\Delta\delta_0)]$, we have

$$\Delta\Sigma_0 FL_0 = L_0[\Lambda_0(\Delta\delta_0) + \delta_0(\Delta\Lambda_0)] \tag{17}$$

Using eqn. (10), we obtain

$$\Delta\Sigma_0(\tilde{L}_0^{-1})\Lambda_0 = L_0[\Lambda_0(\Delta\delta_0) + \delta_0(\Delta\Lambda_0)] \tag{18}$$

or

$$\Delta\Sigma_0 = L_0[(\Delta\delta_0) + \delta_0(\Delta\Lambda_0)\Lambda_0^{-1}]\tilde{L}_0 \tag{19}$$

Many interesting facts follow from eqn. (19). Since the diagonal $(\Delta\Sigma_0)_{ii}$ elements are functions of $(L_0)_{ij}^2$ only, it follows that the sign of $(\Delta\Sigma_0)_{ii}$ is determined only by the terms inside the bracket. However, the two terms $(\Delta\delta_0)_i$ and $[(\delta_0)_i(\Delta\lambda_0)_i/(\lambda_0)_i]$ have opposite signs. For substitution by a heavier atom $(\Delta\lambda_0)_i$ is negative (for symmetric isotopic substitution) and $(\Delta\delta_0)_i$ is positive. The combination of these two effects determines the sign of $(\Delta\Sigma)_{ii}$. Since the mean square amplitude $(l_0^2)_{a-b}$ is a linear combination of Σ_{ii} only, it follows that $(l_0^2)_{a-b}$ decreases with isotopic substitution if the term $[(\Delta\delta_0) + \delta_0(\Delta\Lambda_0)\Lambda^{-1}]$ is negative. For many $n = 2$ cases, it has already been shown[18] that the L-matrix approximation method[19] ($L_{12} = 0$; $\lambda_1 > \lambda_2$, and L_{11} and L_{22} are positive) yields reasonably accurate values of the bonded and non-bonded mean square amplitudes of vibration even if the mass coupling ($G_{12}/|G|$) is large.

It can be shown from eqn. (19) that the L-matrix approximation method yields

$$(\Delta\Sigma_0)_{12}/(\Delta\Sigma_0)_{11} = (G_0)_{12}/(G_0)_{11} \tag{20}$$

Since $(G_0)_{11}$ is always positive, the sign of $[(\Delta\Sigma_0)_{12}/(\Delta\Sigma_0)_{11}]$ is determined by that of $(G_0)_{12}$.

The numerical results indicate[17] that, for heavy and very heavy atom substitution, the effect of mass on the mean square amplitudes can be studied reasonably accurately using first order perturbation theory. In the case of H/D substitution, however, eqn. (19) is found to be inadequate to explain the influence of mass changes on Σ values. Examples are found in H_2O/D_2O, NH_3/ND_3, PH_3/PD_3 and CH_4/CD_4, etc. This might be easily traced to the fact that, while eqn. (19) neglects the term $[(\Delta\Lambda_0)(\Delta\delta_0)]$, this term is important in dealing with light atom substitution. By taking this into consideration, eqn. (19) is modified as

$$\Delta\Sigma_0 = L_0[(\Delta\delta_0) + \delta_0(\Delta\Lambda_0)\Lambda_0^{-1} + (\Delta\Lambda_0)\Lambda_0^{-1}(\Delta\delta_0)]\tilde{L}_0 \tag{21}$$

This equation can account for the variation of $(\Delta\Sigma_0)$ with the H/D substitution. In all cases studied, it is found that $(\Delta\Sigma)_{ii}$ [hence $(\Delta l_0)_{a-b}^2$] is negative for substitution by an atom of larger mass.

2.4 Summary

First order perturbation theory coupled with the approximation $\Delta L_0 = 0$ is found to be adequate to explain the mass influence on the isotopic frequency shifts, Coriolis coupling constants and mean amplitudes of vibration for isotopic substitutions of heavy and very heavy atoms. Second order perturbation theory is needed only when one deals with Σ for H/D or light atom substitution. Since the validity of any approximation, e.g. $\Delta L_0 = 0$, can be tested only numerically, one should be careful in using such approximations in any general case. While the mass effects on the isotope shifts and the Coriolis coupling constants can be measured experimentally with conventional instruments, our theoretical treatment indicates that its effect on the mean amplitudes of vibration is too small to be detected by electron diffraction studies with the current accuracy.

3 NORMAL COORDINATE ANALYSIS

The assignment of the frequencies observed in the i.r. and Raman spectra to individual valence type vibrations and the calculation of the relative amplitudes of symmetry or valence coordinates in any normal mode constitute normal coordinate analysis. Rigorous calculations of force constants in the General Valence Force Field (GVFF) can only be made if additional data such as isotopic frequency shifts, Coriolis coupling constants, centrifugal distortion constants and mean amplitudes of vibration etc. are available. The general theory underlying the calculation of exact force constants using such additional data has been described elsewhere.[20-22] In the following, the utility of isotopic shifts in fixing the precise values of the force constants will be discussed for coordination compounds.

3.1 Theory and Calculation of Force Constants from Isotopic Shifts

The accuracy of force constant determinations using isotopic shifts depends on (1) the accuracy of the isotopic shift measurements and (2) the sensitivity of the force constants to the isotopic shifts. In the last few years, special attention has been devoted to error assessment in evaluating the force constants obtained by using such data.[13,14,23] Since the force field is the same for a pair of isotopic molecules (Born-Oppenheimer approximation), it follows that the frequencies of the different isotopic molecules give different relations among the same force constants. However, not all frequencies are independent of each other since there exist the well-known product and sum rules[5,24] among the frequencies. The problem of determining the $n(n+1)/2$ symmetry force constants is further complicated by the fact that the dependence of the force constants on the isotopic shifts (or frequencies of different isotopic molecules) is not linear. Hence, more than $n(n+1)/2$ independent relations are generally required to define a force field without any assumptions or ambiguities.

In favourable cases, where the amount of isotopic shift data is sufficient (or more than enough) to determine the GVFF constants, the following procedure is adopted.[13,20-22] The isotopic frequencies and an initial set of force constants are fed into a computer. The differences $[\nu^2(\text{obs}) - \nu^2(\text{calc})]$ are computed for each isotopic molecule and the force constants are adjusted until the above factor becomes negligible for all frequencies involved. This constitutes the well known 'least-squares procedure', the details of which can be found elsewhere.[22,25] By such a procedure, the Jacobians which define the variation of the force constants with respect to the isotopic shifts $\Delta\lambda_i$ (or more precisely $\Delta\lambda_i/\lambda_i$) are needed. For small frequency shifts, these Jacobians can easily be constructed by using first order perturbation theory,[5,7] and are given below:

$$\frac{\partial(\Delta\lambda_a/\lambda_a)}{\partial F_{ij}} = 2 \sum_{a \neq b} \Delta_{ab} L_{ia} L_{jb}/(\lambda_a - \lambda_b) \tag{22}$$

where L_{ia}, L_{jb} etc. are the elements of the transformation matrix

$$\mathbf{S} = \mathbf{LQ} \tag{23}$$

with $\mathbf{S}$ referring to the internal symmetry coordinates, $\mathbf{Q}$ to the normal coordinates, and Δ_{ab} is defined as

$$\Delta_{ab} = [\mathbf{L}^{-1}\Delta\mathbf{G}\tilde{\mathbf{L}}^{-1}]_{ab} \tag{24}$$

For $n = 2$ cases, two methods, known as the 'force constant refinement procedure' and the 'force constant display method (graphical)', are widely used to obtain convergence of the force constants. The details of these procedures are well described in the paper of Chalmers and McKean.[14] The following discussion is divided into three cases (H/D substitution, heavy atom substitution and very heavy atom substitution) since there are differences in the sensitivity of the force constants to these data.

3.1.1 H–D substitution

Very large frequency shifts are associated with H/D or other light atom substitution, e.g. $^6\text{Li}/^7\text{Li}$. The calculation of force constants using large isotopic shifts encounters the following difficulties. Since the difference between the harmonic (ω_i) and the observed (ν_i) frequencies is large for higher frequencies, the force fields determined by using ω and ν separately are widely divergent. Even in cases where the harmonic frequencies have been estimated (mostly empirically), it is generally found difficult to estimate the corresponding uncertainties, so that no 'true error limits' could be given for the values of the harmonic force constants. Secondly, it is found that for light atom substitution, e.g. H/D, the force constants (especially F_{12} in $n = 2$ cases) are very sensitive to $\Delta\omega$, i.e. [$\omega(\text{isotopic}) - \omega(\text{normal})$], and hence the isotopic shifts become virtually useless in fixing the precise values of the force constants. This was first pointed out by Duncan and Mills[21] from their studies of the force field for group IV and V hydrides. This result was put on a more general footing by Chalmers and McKean[14] who pointed out that large frequency shifts (50 cm^{-1} or more) in general are ineffective in the calculation of the exact force constants. To our knowledge no exact force constants with small error limits have been determined by using light atom isotopic substitution alone.

3.1.2 Heavy atom substitution

Small frequency shifts are associated with heavy atom substitution, e.g. $^{10}B/^{11}B$, $^{14}N/^{15}N$, $^{16}O/^{18}O$, $^{35}Cl/^{37}Cl$ etc.

The heavy atom substitution technique has been used successfully in recent years for the evaluation of force constants for compounds like $RuO_4(^{16}O/^{18}O)$[26], $OsO_4(^{16}O/^{18}O)$[27], $XeO_4(^{16}O/^{18}O)$[28], $SiCl_4(^{35}Cl/^{37}Cl)$[29], $SF_6(^{32}S/^{34}S)$[30], CH_3F, CH_3Cl, CH_3Br and $CH_3I(^{12}C/^{13}C$ shifts along with other data)[31], $CHCl_3(^{12}C/^{13}C$ shifts together with additional data)[32], $SiHF_3(^{29}Si/^{30}Si$ splitting with other pieces of data)[33], ONF, ONCl and $ONBr(^{16}O/^{17}O$, $^{16}O/^{18}O$, $^{14}N/^{15}N$ splittings with other data)[34-36] and $CH_2O(^{12}C/^{13}C$ shifts with other data).[37]

3.1.3 Very heavy atom substitution

The substitution of very heavy atoms (metals) such as Ti, (Ge), Ru, Sn, etc. causes only small shifts in the vibrational frequencies. Theoretical interpretation of small isotopic shifts can be given using first order perturbation theory (see p. 174). Using this theory, it was shown that small frequency shifts (associated with heavy and very heavy atom substitution) measured accurately (± 0.2 cm^{-1}) lead to reliable values of the force constants. For this reason, metal isotope substitution is very important in determining accurate values of force constants.

The metal isotope substitution technique has recently been applied with success for the evaluation of force constants for a number of metal inorganic compounds such as $[MoO_4]^{2-}(^{92}Mo/^{100}Mo)$,[38,39] $[MoS_4]^{2-}(^{92}Mo/^{100}Mo)$,[38] $[CrO_4]^{2-}(^{50}Cr/^{53}Cr)$,[39] $RuO_4(^{96}Ru/^{102}Ru)$,[26] $SnCl_4(^{116}Sn/^{124}Sn)$,[16] $GeCl_4(^{70}Ge/^{76}Ge)$,[40,41] $TiCl_4(^{46}Ti/^{50}Ti)$,[41] etc. To illustrate the effectiveness of the metal isotope shifts in fixing the precise values of the force constants, some numerical results are presented in Tables 2–6.

3.2 Calculation of Exact Force Constants for Coordination Compounds with Low Frequency Vibrations

In this section, we deal with the utility of small shifts (2–10 cm^{-1}) determined with a rather large uncertainty, e.g. ± 1 cm^{-1}, in fixing the accurate values of the force constants. We felt it necessary to discuss the normal coordinate analysis of coordination compounds using isotopic shifts as additional data since, under certain circumstances, even a relatively crude estimation of the isotopic shift can be effective in determining accurate values of the force constants.[42,43]

TABLE 2
Error limits (mdyn $Å^{-1}$) in the values of the force constants[a] (t_2 species) due to an uncertainty of 0.2 cm^{-1} in $\Delta\nu_1(t_2)$ for various isotopic species of RuO_4 and OsO_4

	$\Delta m_x(x=Ru, Os)$	$\Delta\nu_1(t_2)$	$|\Delta F_{11}(t_2)|$	$|\Delta F_{12}(t_2)|$	$|\Delta F_{22}(t_2)|$
$^{96}RuO_4/^{104}RuO_4$	8	6.88	0.04	0.06	0.00
$^{96}RuO_4/^{102}RuO_4$	6	5.26	0.06	0.07	0.00
$^{98}RuO_4/^{102}RuO_4$	4	3.43	0.09	0.08	0.01
$^{98}RuO_4/^{100}RuO_4$	2	1.75	0.18	0.22	0.01
$^{98}RuO_4/^{99}RuO_4$	1	0.88	0.39	0.43	0.04
$^{184}OsO_4/^{192}OsO_4$	8	2.12	0.09	0.19	0.01
$^{184}OsO_4/^{190}OsO_4$	6	1.60	0.12	0.25	0.01
$^{186}OsO_4/^{190}OsO_4$	4	1.05	0.19	0.38	0.02
$^{186}OsO_4/^{188}OsO_4$	2	0.53	0.45	0.77	0.09
$^{186}OsO_4/^{187}OsO_4$	1	0.27	1.24	1.67	0.40

[a] The isotopic frequency shifts were determined using the values of the force constants given in Refs. 26 and 27 respectively. The values of the force constants are RuO_4: $F_{11}(t_2) = 6.82$, $F_{12}(t_2) = 0.07$, $F_{22}(t_2) = 0.41$ and OsO_4: $F_{11}(t_2) = 8.11$, $F_{12}(t_2) = 0.10$, $F_{22}(t_2) = 0.47$ (all the above data are in units of mdyn $Å^{-1}$).

Using the Jacobians[7] obtained in eqn. (22) ($n = 2$ case), we have

$$\frac{\partial(\Delta\lambda_1/\lambda_1)}{\partial F_{11}} = 2\Delta_{12}L_{11}L_{12}/(\lambda_1-\lambda_2) \tag{25}$$

$$\frac{\partial(\Delta\lambda_1/\lambda_1)}{\partial F_{12}} = 2\Delta_{12}(L_{11}L_{22}+L_{12}L_{21})/(\lambda_1-\lambda_2) \tag{26}$$

$$\frac{\partial(\Delta\lambda_1/\lambda_1)}{\partial F_{22}} = 2\Delta_{12}L_{21}L_{22}/(\lambda_1-\lambda_2) \tag{27}$$

where

$$\Delta_{12} = (-L_{21}L_{22}\Delta G_{11}+L_{11}L_{22}\Delta G_{12}+L_{12}L_{21}\Delta G_{12}-L_{11}L_{12}\Delta G_{22})/|G| \tag{28}$$

Similar expressions with a negative sign are valid for $\partial(\Delta\lambda_2/\lambda_2)/\partial F_{ij}$. These equations show that the force constants are effectively controlled by the shift (or $\Delta\lambda_i/\lambda_i$) when Δ_{12} is a maximum. A close examination of the expression for Δ_{12} would reveal that for any given sign combination of L_{12} and L_{21} (L_{11} and L_{22} can, without loss of generality, be assumed to have the same sign), two terms inside the bracket in eqn. (28) have a common sign, which is opposite to that of the other two terms. In other words, Δ_{12} is always very small. However, the magnitude of the Jacobian is controlled by the term $(\lambda_1-\lambda_2)$ also. For low frequency vibrations which are close together this factor $(\lambda_1-\lambda_2)$ is very small, with the result that the Jacobians assume large values in this case.

In general, transition metal complexes exhibit their metal–ligand stretching vibrations below 400 cm^{-1}. It is found that, in these cases, even small shifts ($\sim$3 cm^{-1}) determined with a large uncertainty ($\pm$1 cm^{-1}) are useful in fixing accurate values of the force constants. The force constants of [PdCl$_6$]$^{2-}$ (O_h, ^{104}Pd/^{110}Pd), [SnCl$_6$]$^{2-}$ (O_h, ^{116}Sn/^{124}Sn), [Cr(NH$_3$)$_6$]$^{3+}$ (O_h, ^{50}Cr/^{53}Cr), [Ni(NH$_3$)$_6$]$^{2+}$ (O_h, ^{58}Ni/^{62}Ni), [Cu(NH$_3$)$_4$]$^{2+}$ (D_{4h}, ^{63}Cu/^{65}Cu), [Pd(NH$_3$)$_4$]$^{2+}$ (D_{4h}, ^{104}Pd/^{110}Pd) and [Zn(NH$_3$)$_4$]$^{2+}$ (T_d, ^{64}Zn/^{68}Zn) bear testimony to the above conclusion.[42,43] Tables 3 and 4 show the results obtained for these compounds.

The effect of the difference in mass, i.e. ΔG, and the frequency difference, i.e. ($\lambda_1 - \lambda_2$), on the force constants, determined by using isotopic shifts as constraints, can be seen from the approximate relation given below. Considering the expression for the stretching force constant

$$F_{11} = \frac{L_{22}{}^2\lambda_1 + L_{21}{}^2\lambda_2}{|G|} \tag{29}$$

we might assume that any error in F_{11} is caused only by the corresponding ones in $L_{22}{}^2$ and $L_{21}{}^2$, i.e. we ignore the errors in the absolute values of ν_1 and ν_2, but assume cumulative errors in ν_1 and ν_2 (isotope), so that the error in $\Delta\nu_i$ includes that in ν_i (parent). As was shown in Ref. 4, this is justified for metal isotope substitution. Let F_{11} become ($F_{11} + \delta F_{11}$) as $L_{22}{}^2$ and $L_{21}{}^2$ become ($L_{22}{}^2 + K_1$) and ($L_{21}{}^2 + K_2$) respectively. Then we have

$$F_{11} + \delta F_{11} = \frac{L_{22}{}^2\lambda_1 + L_{21}{}^2\lambda_2}{|G|} + \frac{K_1\lambda_1 + K_2\lambda_2}{|G|} \tag{30}$$

Because of the normalization condition, we have

$$(L_{22}{}^2 + K_1) + (L_{21}{}^2 + K_2) = L_{22}{}^2 + L_{21}{}^2 = G_{22} \tag{31}$$

Thus,

$$K_1 = -K_2 \tag{32}$$

Substituting this in eqn. (9), we have

$$\delta F_{11} = \frac{K_1(\lambda_1 - \lambda_2)}{|G|} \tag{33}$$

From first order perturbation theory[5] we have for $n = 2$ cases

$$\frac{\Delta\lambda_1}{\lambda_1} = (L_{22}{}^2\Delta G_{11} - 2L_{12}L_{22}\Delta G_{12} + L_{12}{}^2\Delta G_{22})/|G| \tag{34}$$

For XY$_n$ type molecules with small mass coupling, i.e., $m_x \gg m_y$, the solution for ($\Delta\lambda_1/\lambda_1$) corresponding to $L_{12} \approx 0$ is found to be quite reasonable.[6] This constraint leads to the result

$$\frac{\Delta\lambda_1}{\lambda_1} \approx \frac{L_{22}{}^2\Delta G_{11}}{|G|} \tag{35}$$

TABLE 3
Force constants of $SnCl_4$ and RuO_4

	mdyn Å^{-1}	
	set I	set II[c]
$SnCl_4$		
$F_{11}(a_1)$	2.80 ±0.04[a]	
$F_{22}(e)$	0.07$_2$±0.006$_6$[a]	
$F_{33}(t_2)$	2.57 ±0.05[b]	2.50 ±0.07
$F_{44}(t_2)$	0.10$_7$±0.004[b]	0.10$_7$±0.00$_2$
$F_{34}(t_2)$	0.10 ±0.05	0.06 ±0.06
RuO_4[d]		
$F_{11}(a_1)$	7.50 ±0.08	
$F_{22}(e)$	0.33$_5$±0.01$_3$	
$F_{33}(t_2)$	6.82 ±0.05	
$F_{44}(t_2)$	0.41$_0$±0.01$_5$	
$F_{34}(t_2)$	0.07 ±0.04	

[a] See Ref. 58.

[b] This set corresponds to the very accurately measured ^{116}Sn/^{124}Sn isotopic frequency shift = 3.6±0.2 cm^{-1}; see Ref. 16.

[c] Calculated from the ^{35}Cl/^{37}Cl isotopic shift = 8.0±0.2 cm^{-1} in ^{116}SnCl$_4$; see Ref. 16.

[d] These values are the averages of those determined from different additional data like ^{96}Ru/^{102}Ru isotopic frequency shift, ^{16}O/^{18}O isotopic shift, Coriolis coupling constants etc.; see Ref. 26.

TABLE 4
Force constants of $[CrO_4]^{2-}$, $[MoO_4]^{2-}$ and $[MoS_4]^{2-}$

		mdyn Å^{-1}
$CrO_4]^{2-}$	$F_{33}(t_2)$	5.23±0.17[a]
	$F_{44}(t_2)$	0.43±0.02[a]
	$F_{34}(t_2)$	0.08±0.12[a]
$[MoO_4]^{2-}$	$F_{33}(t_2)$	5.28±0.12[b] (5.14±0.21)[c]
	$F_{44}(t_2)$	0.36±0.01[b] (0.38±0.03)[c]
	$F_{34}(t_2)$	−0.08±0.13[b] (−0.22±0.21)[c]
$[MoS_4]^{2-}$	$F_{33}(t_2)$	2.84±0.12[d]
	$F_{44}(t_2)$	0.20±0.01[d]
	$F_{34}(t_2)$	−0.01±0.07[d]

[a] Derived from the isotopic frequency shift $\Delta\nu_3$(^{50}Cr/^{53}Cr) = 8.5±0.6 cm^{-1} in Cs_2CrO_4. For details see Ref. 39.

[b] Estimated from $\Delta\nu_3$(^{92}Mo/^{100}Mo) = 7.3±0.6 cm^{-1} in Cs_2MoO_4. For details see Ref. 39.

[c] This set of force constants corresponds to $\Delta\nu_3$(^{92}Mo/^{100}Mo) = 8.0±1.0 cm^{-1} in Na_2MoO_4 reported in Ref. 38.

[d] Determined from $\Delta\nu_3$(^{92}Mo/^{100}Mo) = 7.0±0.5 cm^{-1} in K_2MoS_4; see Ref. 38.

Hence the error associated with $L_{22}{}^2$ due to an error in $(\Delta\lambda_1/\lambda_1)$ is found to be

$$\delta\left(\frac{\Delta\lambda_1}{\lambda_1}\right) \approx \frac{K_1\Delta G_{11}}{|G|} \tag{36}$$

Substituting this result in eqn. (11), we have

$$\delta F_{11} = \frac{\delta(\Delta\lambda_1/\lambda_1)(\lambda_1-\lambda_2)}{\Delta G_{11}} \tag{37}$$

Hence,

$$\frac{\delta F_{11}}{\delta(\Delta\lambda_1/\lambda_1)} = \frac{\lambda_1-\lambda_2}{\Delta G_{11}} \tag{38}$$

Equation (38) shows that large errors in $(\Delta\lambda_1/\lambda_1)$ have relatively small influence on the value of F_{11} if $(\lambda_1-\lambda_2)$ is small and also if ΔG_{11} is large.

For coordination compounds with a heavy central atom (XY_n type), eqn. (38) is valid. In this case, accurate values of F_{11} can be obtained even with large errors in $\Delta\lambda_1/\lambda_1$ if ΔG_{11} is large and if the factor $(\lambda_1-\lambda_2)$ is small. This conclusion is valid mainly for coordination compounds with a small value of the stretching force constant (~ 1 mdyn Å^{-1}), because for such compounds $(\lambda_1-\lambda_2)$ is very small.

3.3   Pseudo-exact Force Constants

Sets of force constants obtained from experimental data (isotopic shifts and Coriolis coupling constants) in conjunction with certain plausible approximations are called 'pseudo-exact force constants' (for definition, see Müller et al.[42-44]). The 'high–low frequency separation method' (HLFS) and the 'point mass model' (PMM) constitute good examples of such a procedure, and the force constants thus obtained might be termed as 'pseudo-exact' force constants. In principle, the order of the secular equation is reduced in this procedure to any desired extent so that the force constants involved in the reduced block can be calculated without any approximation. This implies that the vibrations involved in the factored-out block do not have much influence on those involved in the remaining block (90% or more pure).

3.3.1   Theory of high–low frequency separation

The theory of the high–low frequency separation (HLFS) method developed by Crawford and Edsall[45] and Wilson[46] is well described in the book of Wilson et al.[5] Hence, only a brief outline is given below.

The separation of the high frequency vibration in a secular determinant of order n is accomplished by dropping the corresponding rows and columns in G^{-1} and F, and then by solving the reduced secular determinant. This leads to modified expressions for the G matrix elements of the reduced block (G^0) given by

$$G_{tt''}{}^0 = G_{tt'} - \sum_{ss'} G_{ts}X_{ss'}G_{s't'} \tag{39}$$

where s covers the symmetry coordinates S_s to be held rigid while $X_{ss'}$ elements satisfy the relation

$$\sum_{s'} X_{ss'} G_{s's''} = \delta_{ss''} \qquad (40)$$

where $\delta_{ss''}$ is the Kronecker delta.

The factoring of the low frequency vibrations on the other hand is achieved simply by dropping the corresponding rows and columns in $\mathbf{G}$ and $\mathbf{F}$, and then by solving the reduced secular determinant to obtain the corresponding force constants. Hence, no change in the $\mathbf{G}$ matrix elements corresponding to the reduced block is necessary in this case.

The similarity between this and other approximate methods (L-matrix method,[19] PED method,[47] etc.) in calculating force constants was pointed out by Müller et al.[48] who stressed the special suitability of this method in reducing the order of secular determinants from three to two. The pseudo-exact force constants of the reduced 2×2 block after factoring out either the highest or the lowest frequency vibration can be calculated without approximation by using experimental data (isotopic shifts or Coriolis coupling constants). Müller et al.[49,50] have applied this method to a number of molecules which contain 3 or 4 vibrations of a single symmetry species.

3.3.2 The point mass model

This model is specially suited for molecules and ions containing hydrogen. In XH_nZ_m type molecules the XH_n group can, under certain circumstances, be assumed to be a point with an aggregate mass of $(M_X + nM_H)$. This assumption is valid only when the frequencies of the XZ_m group vibrations are far from those of the XH_n group vibrations. The pseudo-exact force constants of the reduced molecular model can be determined in most cases using available experimental data. The force constants of tetrammine and hexammine complexes of Cr(III)(O_h), Cu(II)(D_{4h}), Pd(II)(D_{4h}), Ni(II)(O_h) and Zn(II)(T_d) have been obtained by assuming the NH_3 group as a point mass and by using the metal isotope shifts as constraints on the force field[42] (see Tables 5 and 6). This method cannot be applied to molecules of the type $E(CH_3)_n$ (E = light element belonging to the main group), as in these cases strong vibrational coupling between $\rho(CH_3)$ and $\nu(E-C)$ is possible.

3.4 Summary

The discussion presented above covers the utility of metal isotope substitution in the calculation of force constants. In dealing with the evaluation of force constants using metal isotope shifts, it is important to note the limitations of the metal isotope technique in normal coordinate analyses.

The accuracy of the force constants estimated using isotopic shifts depends (1) on the accuracy of the shifts themselves and (2) on the sensitivity of the force

TABLE 5
Pseudo-exact force constants for $[MX_6]^{2+}$ type ions[a] from metal isotope shifts (M = Pd, Sn, Cr, Ni; X = Cl, NH$_3$)

	Nature of isotope substitution	Shift of $\nu_3(t_{1u})$ (in cm^{-1})	Force constants in mdyn Å^{-1} F_{11}	F_{22}	F_{12}
$[PdCl_6]^{2-}$	^{104}Pd/^{110}Pd	3.5 ± 0.5[b]	1.80 ± 0.11	$0.18 \pm 0.00_3$	0.14 ± 0.07
$[SnCl_6]^{2-}$	^{116}Sn/^{124}Sn	3.2 ± 0.8[b]	1.55 ± 0.10	0.18 ± 0.02	0.21 ± 0.09
$[Cr(NH_3)_6]^{3+\ d}$	^{50}Cr/^{53}Cr	4.0 ± 1.0[c]	1.59 ± 0.09	0.21 ± 0.02	0.26 ± 0.04
$[Ni(NH_3)_6]^{2+\ d}$	^{58}Ni/^{62}Ni	2.2 ± 0.6[c]	0.87 ± 0.02	0.15 ± 0.01	0.18 ± 0.02

[a] The force constants correspond to the t_{1u} species.
[b] The isotopic frequency shift, observed in K$_2$PdCl$_6$ and K$_2$SnCl$_6$; see Ref. 43.
[c] This shift, observed in [Cr(NH$_3$)$_6$](NO$_3$)$_3$ and [Ni(NH$_3$)$_6$]Cl$_2$, is taken from Ref. 42.
[d] Refer to the skeletal vibrations.

constants to the isotopic shifts. While the accuracy of the shifts depends on the experimental conditions, the sensitivity varies from molecule to molecule. While gas phase and matrix spectra can provide very accurate values for the isotopic shifts (this is no longer true in the case of gas phase spectra, when the bands are overlapped by 'hot bands'), and in some cases pure metal isotopes, such as ^{116}Sn, ^{124}Sn, ^{70}Ge, ^{74}Ge, might be needed for the purpose (pp. 188–195), the spectra of complexes usually contain rather broad bands in solution or as solids. Hence, the isotopic shifts of complexes in general carry large uncertainties ($\sim \pm 0.5$–1.0 cm^{-1}). In such cases, the spectra of species doped in host-lattices (see p. 199) might be measured. In such cases, the choice of relatively less

TABLE 6
Pseudo-exact force constants for $[M(NH_3)_4]^{2+}$ skeletons[a] (M = Cu, Pd, Zn)

	Nature of isotope substitution	Shift[c] (in cm^{-1})	Force constants in mdyn Å^{-1} F_{11}	F_{22}	F_{12}
$[Cu(NH_3)_4]^{2+}$	^{63}Cu/^{65}Cu	2.0 ± 1.0[b]	1.29 ± 0.22	0.25 ± 0.06	-0.17 ± 0.18
	^{14}N/^{15}N	7.7 ± 0.8[d]	(1.14 ± 0.23)[d]	(0.25 ± 0.07)[d]	(-0.04 ± 0.10)[d]
$[Pd(NH_3)_4]^{2+}$	^{104}Pd/^{110}Pd	3.0 ± 1.0[b]	1.93 ± 0.14	0.34 ± 0.04	$-0.17_5 \pm 0.17$
$[Zn(NH_3)_4]^{2+}$	^{64}Zn/^{68}Zn	2.0 ± 0.6[b]	$1.38 \pm 0.03_5$	0.10 ± 0.02	0.19 ± 0.07

[a] The assumed symmetry is square planar (D_{4h}) for [Cu(NH$_3$)$_4$]$^{2+}$ and [Pd(NH$_3$)$_4$]$^{2+}$ and tetrahedral (T_d) for [Zn(NH$_3$)$_4$]$^{2+}$. The force constants correspond to the (2×2) block (e_u or t_2) of the skeletal vibrations.
[b] The experimental results are from Ref. 42.
[c] Corresponds to the ν(M—N) vibration in e_u or t_2.
[d] Data from Müller et al. (Ref. 80).

polarizing host-lattices is important, so that the isotopic shifts correspond as nearly as possible to those of the free ion.

Regarding the sensitivity of the force constants to the isotopic shifts, it might be noted (see p. 185) that the error limits in the force constants (for a given value of the isotopic shift) are directly proportional to the term $(\lambda_i - \lambda_j)$ and inversely proportional to ΔG_{ij}. In other words, least error in the force constants, e.g. F_{ii} corresponding to stretching, is expected with an isotopic substitution producing the largest possible change in G, e.g. in G_{11} for F_{11}, when the stretching frequency corresponding to the molecule is rather low ($\sim 400 \text{ cm}^{-1}$). The validity of the above conclusion has been thoroughly verified.[4]

4   GAS PHASE SPECTRA

In the gas phase, small molecules containing light atoms exhibit rotation–vibration spectra which provide precise information about the molecular structure, moments of inertia and Coriolis coupling constants, etc. Since the vapour pressures of most inorganic and coordination compounds are relatively low, their gas phase spectra cannot be measured unless they are heated to high temperatures. Furthermore, the rotation–vibration fine structures are seldom observed, because of the usually large moments of inertia. Even so, it is still possible to obtain such information from the observation of the rotation–vibration band contour. For more detailed theories of gas phase rotation–vibration spectra, the reader should consult recent articles on this subject.[51–53] In this section, we review a few studies of gas phase spectra of inorganic and coordination compounds, with special reference to the observation of metal isotope frequency shifts.

The spectra of RuO_4 with ^{16}O and ^{18}O in the gas and liquid phase were reported by McDowell et al.[26] They have estimated the harmonic frequency shifts. Figure 1, curve A, illustrates the band contour of the $\nu_3(t_2)$ fundamental in the gas phase.[54] According to McDowell et al., this complex contour can be explained satisfactorily on the assumption that each Ru isotope Q-branch is separated by about 0.9 cm^{-1} per mass unit with each being accompanied by a series of hot bands on the low frequency side, separated by $1.2–1.3 \text{ cm}^{-1}$ $(\nu_3 + n\nu_4 - n\nu_4)$. They[26] report a shift of $5.2_5 \pm 0.1 \text{ cm}^{-1}$ due to the $^{96}Ru/^{102}Ru$ substitution. Using the anharmonicity constants derived by using Dennison's rule[55] and the data obtained from some combination and overtone bands, McDowell et al.[26] estimated the harmonic frequency shift of ν_3 by the above substitution to be $5.36 \pm 0.1 \text{ cm}^{-1}$. This was the first observation of metal isotope peaks of any inorganic or coordination compound in the gaseous phase. McDowell et al. have also demonstrated that the observed Ru isotopic shift provides the most effective constraint in calculating the force constants. This conclusion is in agreement with that of Müller et al.,[3,4] who emphasized the importance of very small metal isotope shifts arising from central atom substitution in estimating a reliable set of force constants (if metal atom substitution causes sufficiently large change in the inverse kinetic energy matrix G; see pp. 174–185). McDowell *et al.* could not resolve the isotope peaks in $\nu_4(t_2)$ which are expected to be separated only by *ca.* 0.35 cm^{-1} per mass unit.

Königer *et al.*[54] have measured the i.r. spectra of pure $^{96}RuO_4$ and $^{104}RuO_4$. As is seen in Fig. 1, curve B, a well defined QR structure with a hot band progression $(\nu_3+n\nu_4-n\nu_4)$ was observed for these isotopically pure compounds. The spectrum of RuO_4 shown in Fig. 1, curve A, is a superposition of

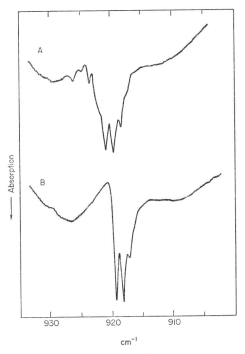

Fig. 1. The $\nu_3(t_2)$ bands of (A) RuO_4 and (B) $^{104}RuO_4$ in the gaseous phase at 301 K.

seven spectra such as shown in Fig. 1, curve B, since naturally abundant Ru consists of seven isotopes: ^{96}Ru, ^{98}Ru, ^{99}Ru, ^{100}Ru, ^{101}Ru, ^{102}Ru and ^{104}Ru. It is, therefore, rather difficult to interpret the spectrum of RuO_4. McDowell and Asprey[28] studied the i.r. and Raman spectra of $Xe^{16}O_4$ and $Xe^{18}O_4$ in the gaseous phase. Similar to RuO_4, XeO_4 exhibits five peaks in the Q-branch region of $\nu_3(t_2)$ which are due to the presence of the ^{129}Xe, ^{131}Xe, ^{132}Xe, ^{134}Xe and ^{136}Xe isotopes in the naturally occuring XeO_4. These bands are again overlaid by hot bands. The shift of ν_3 due to the ^{129}Xe–^{132}Xe substitution was estimated to be 1.56 ± 0.06 cm^{-1}. The corresponding harmonic frequency shift was calculated to be 1.64 ± 0.07 cm^{-1}.

The gas phase spectrum of $SnCl_4$ has been reported by several authors.[56–58] Figure 2, curve A, illustrates the gas phase spectrum of the $\nu_3(t_2)$ band of natural $SnCl_4$. It is seen that the P- and R-branches are observed only as shoulders and not as separate maxima. This is partly due to the fact that $SnCl_4$ consists of fifty isotopic molecules (a combination of ten Sn isotopes with two Cl isotopes gives fifty isotopic $SnCl_4$ molecules), many of which are abundant

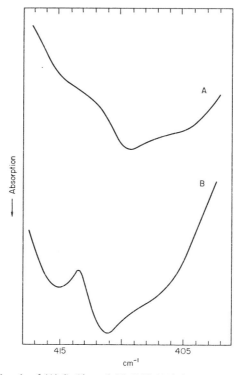

Fig. 2. The $\nu_3(t_2)$ bands of (A) $SnCl_4$ and (B) $^{116}Sn^{35}Cl_4$ in the gaseous phase at 301 K.

enough to influence the shape of the band contour. In order to improve the spectrum, Müller et al.[58] previously measured the ν_3 band of $^{116}SnCl_4$ and $^{124}SnCl_4$ (each still consists of five isotopic molecules) in the gaseous phase. However, their spectra were almost identical to that of $SnCl_4$; no clear maxima were observed for the P- and R-branches. Recently, Königer and Müller[16] obtained the gas phase as well as the matrix isolation spectra of isotopically pure $^{116}Sn^{35}Cl_4$ (also $^{116}SnCl_4$ in Ar matrix) and $^{116}Sn^{37}Cl_4$ (also $^{124}SnCl_4$ in Ar matrix). Figure 2, curve B, shows the gas phase spectrum of $\nu_3(t_2)$ corresponding to $^{116}Sn^{35}Cl_4$. It is seen that the R-branch is well separated from the Q-branch. This is not true, however, for the P-branch. The failure to observe the P-branch maximum might be due to the overlap of hot bands. Similar observations were made previously by McDowell and his coworkers for RuO_4,[26] XeO_4[28] and WF_6.[59] Since the $Q-R$ separations are not so strongly perturbed by hot bands, they were used[16] to calculate the Coriolis coupling constants. It was noted that the Q-branch frequencies of $^{116}Sn^{35}Cl_4$ and $^{116}SnCl_4$ are identical within the error limit (0.4 cm^{-1}). Thus, it may be sufficient to record the spectra of $^{116}SnCl_4$ and $^{124}SnCl_4$ to discuss the Sn isotope shift. In sharp contrast to the above case, the $P-Q-R$ structure of the $\nu_4(t_2)$ band is well resolved for an isotopically pure compound such as $^{116}Sn^{35}Cl_4$. Similar investigations have been extended[40,41] to $GeCl_4$ and $TiCl_4$.

5 MATRIX ISOLATION SPECTRA

The technique of matrix isolation was originally developed by Pimentel and his coworkers[60] in order to study the infrared spectra of unstable and stable species. In this method, sample vapour and inert gas molecules such as Ar and N_2 are deposited simultaneously on an alkali halide window (such as CsBr and CsI) at a ratio of 1:500 or greater. Since the sample molecules trapped in a rare gas matrix are completely isolated from each other, the matrix isolation spectrum is basically similar to the gas phase spectrum. However, the lack of rotational fine structure and 'hot bands' in the matrix isolation spectrum makes it simpler than the gas phase spectrum. Also, the sharpness of the bands observed at low temperatures (10–20 K) gives the best chance of resolving closely located peaks of individual isotopic molecules mixed in natural abundance. Application of this method is not limited to volatile compounds. Linevsky[61] developed a technique to produce vapours of non-volatile compounds (such as metal halides) in a Knudsen cell at high temperatures and co-condense them with inert gas molecules on a cold window. This method has been used by many other investigators to study the structure and spectra of simple inorganic compounds. For details of experimental techniques and for a compilation of references on matrix isolation spectroscopy, the reader should consult a recent book edited by Hallam.[62]

As stated in other chapters, the metal isotope technique is very useful in assigning metal–ligand vibrations and in calculating the metal–ligand force constants. This technique is rather difficult to apply to gaseous and liquid compounds because it is difficult to prepare these compounds on a milligram scale. For example, it is technically difficult to synthesize $Ni(CO)_4$ with a pure Ni isotope on a milligram scale. However, the matrix isolation spectrum of $Ni(CO)_4$ in natural abundance can be obtained easily. Figure 3 illustrates the infrared spectrum of $Ni(CO)_4$ in an Ar matrix obtained by Cormier et al.[63] The band near 430 cm^{-1} clearly exhibits at least four peaks due to the presence of ^{58}Ni, ^{60}Ni, ^{62}Ni and ^{64}Ni isotopes. This result suggests that the vibration involves a substantial motion of the Ni atom and should be assigned as the Ni—C stretching. On the other hand, the band near 460 cm^{-1} does not show any indication

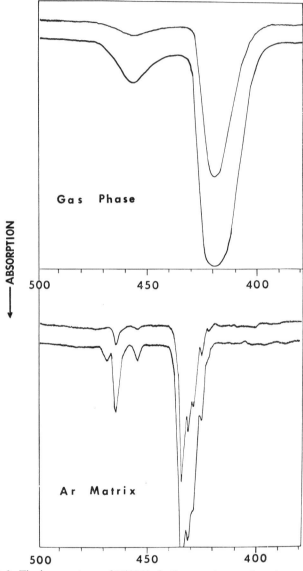

Fig. 3. The i.r. spectrum of Ni(CO)$_4$ in the gas phase and an Ar matrix.

of isotopic splitting under the same condition. Thus, it may be assigned to the Ni—C≡O bending mode in which the C and O atoms are displaced in the direction perpendicular to the Ni—C≡O bond. Needless to say, metal isotope frequencies obtained from the matrix isolation spectrum provide vital information on the potential energy constants.

For a tetrahedral MCl_4 molecule, where M is isotopically pure and Cl is in natural abundance (^{35}Cl, 75.5% and ^{37}Cl, 24.5%), we have a mixture of five isotopic species: $M^{35}Cl_4$, $M^{35}Cl_3{}^{37}Cl$, $M^{35}Cl_2{}^{37}Cl_2$, $M^{35}Cl^{37}Cl_3$ and $M^{37}Cl_4$. If M is isotopically mixed, however, the spectrum becomes very complicated even in inert gas matrices. For example, Sn consists of ten stable isotopes, none of which is predominant: ^{112}Sn (0.96%), ^{114}Sn (0.66%), ^{115}Sn (0.35%), ^{116}Sn (14.30%), ^{117}Sn (7.61%), ^{118}Sn (24.03%), ^{119}Sn (8.58%), ^{120}Sn (32.85%), ^{122}Sn (4.72%) and ^{124}Sn (5.94%). Since each Sn isotope forms five isotopically different tetrachlorides, natural $SnCl_4$ consists of fifty isotopic molecules. Thus, it is almost impossible to resolve individual isotopic peaks even in an inert gas matrix at low temperatures. This difficulty can be overcome if we combine the matrix isolation technique with the metal isotope technique. Thus, Königer and Müller[16] prepared isotopically pure $^{116}SnCl_4$ and $^{124}SnCl_4$ on a milligram scale and measured their spectra in an Ar matrix.

Figure 4 illustrates the infrared spectrum of $^{116}SnCl_4$ in an Ar matrix and Table 7 lists the observed frequencies and band assignments. Each isotopic molecule exhibits one to three peaks in the 415–400 cm^{-1} region, depending upon its symmetry, and the whole spectrum is interpreted as an overlap of bands due to these isotopic molecules. The vertical lines in Fig. 4 indicate the predicted frequency and intensity of the five peaks resulting from such overlap. As

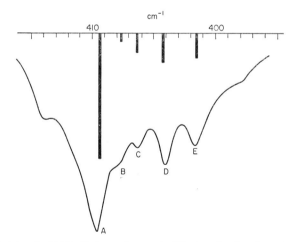

Fig. 4. The $\nu_3(t_2)$ band of $^{116}SnCl_4$ in an Ar matrix. The length of each vertical line is proportional to the theoretical intensity predicted from the percentage abundance of each isotopic molecule.

expected, the matrix isolation spectrum of $^{116}Sn^{35}Cl_4$ (prepared from the reaction of pure ^{116}Sn with $Ag^{35}Cl$) shows only a single band at 409.8 cm^{-1}. The splitting pattern of $^{124}SnCl_4$ in an Ar matrix was similar to that of $^{116}SnCl_4$.

Germanium consists of five isotopes, none of which is predominant (as in the case of Sn): ^{70}Ge (20.52%), ^{72}Ge (27.43%), ^{73}Ge (7.76%), ^{74}Ge (36.54%) and ^{76}Ge (7.76%), and therefore natural $GeCl_4$ consists of twenty-five isotopic molecules. Thus, it is still not simple to interpret the spectrum of $GeCl_4$ without additional data. Königer et al.[40,41] prepared isotopically pure $^{74}GeCl_4$ and $Ge^{35}Cl_4$ (enriched with ^{74}Ge) and measured their i.r. spectra in Ar matrices.

TABLE 7
Observed (adjusted to the gas phase)^a and calculated^b frequencies of matrix isolation spectrum of $^{116}SnCl_4$ in the 415–400 cm^{-1} region (cm^{-1})

	$^{116}Sn^{35}Cl_4$	$^{116}Sn^{35}Cl_3^{37}Cl$	$^{116}Sn^{35}Cl_2^{37}Cl_2$	$^{116}Sn^{35}Cl^{37}Cl_3$	$^{116}Sn^{37}Cl_4$
% abundance	32.5%	42.2%	20.5%	4.4%	0.4%
	411.3(t_2)	411.3(e)	411.3(b_1)	409.2(a_1)	
	(411.3)	(411.3)	(411.3)	(409.6)	
			407.8(a_1)		
			(407.7)		
		405.7(a_1)	403.3(a_2)	403.3(e)	403.3(t_2)
		(405.6)	(403.1)	(403.1)	(403.1)

^{a} $\nu_3(t_2)$ of $^{116}Sn^{35}Cl_4$ given here is from the gas phase spectrum; the other frequencies have been deduced from the isotopic shifts observed in an Ar matrix. For details, see Ref. 16.

^{b} The calculated values (given in brackets) are the ones derived from the force field determined from the frequencies of $^{116}Sn^{35}Cl_4$ and $^{124}Sn^{35}Cl_4$.

Figure 5, curve A, shows the i.r. spectrum (ν_3) of the former. As expected, the whole spectral pattern is similar to that of $^{116}SnCl_4$ (Fig. 4). However, the isotopic peaks are more clearly separated in $^{74}GeCl_4$ than in $^{116}SnCl_4$ because the isotopic shift per unit mass difference (in a.u.) is larger in the former than in the latter.

Figure 5, curve B, shows the spectrum of $Ge^{35}Cl_4$ enriched with ^{76}Ge. In this case, one expects five peaks due to five Ge isotopes with the intensity of each peak proportional to its percentage natural abundance. This was proved to be the case. The i.r. spectrum of $GeCl_4$ can now be interpreted by superimposing the spectra of isotopic species such as $^{74}GeCl_4$ and $Ge^{35}Cl_4$.

Related work has been carried out[41] for $TiCl_4$ using Ti and Cl isotopes. The accurately measured isotopic shifts have been used to determine the exact force constants within narrow error limits for the first time. The examples given above demonstrate that the combination of matrix isolation and isotopic substitution techniques developed by Müller and coworkers[16,40,41] is very useful in the determination of accurate force constants.

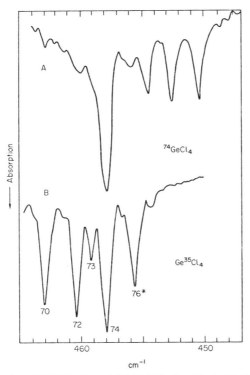

Fig. 5. The $\nu_3(t_2)$ bands of (A) $^{74}GeCl_4$ and (B) $Ge^{35}Cl_4$ (enriched with ^{76}Ge) in Ar matrices.

6   SOLID STATE SPECTRA

The spectra observed from the crystalline state exhibit lattice vibrations in addition to internal vibrations. In general, the lattice modes appear below 400 cm^{-1} where some low frequency internal modes also appear. Thus, the distinction of these two types of vibration is very important in interpreting the spectra of solid compounds. It should be pointed out also that these two types of vibration may couple with each other if they have close frequencies in the same symmetry species. Thus, it is not simple and straightforward to assign the solid state spectra on an empirical basis. This situation has resulted in conflicting and controversial assignments as will be shown later. In order to aid the interpretation of the solid state spectra, Tarte and Preudhomme,[64] for the first time, used the method of metal isotope substitution.

As an example, consider the i.r. spectra of Ni_2GeO_4. This compound crystallizes in the spinel structure in which the Ge atom is coordinated tetrahedrally and the Ni atom is coordinated octahedrally by oxygen atoms. Theoretically, four fundamentals are expected to be i.r. active for this structure, and the observed shifts of these bands due to $^{70}Ge/^{76}Ge$ and $^{58}Ni/^{62}Ni$ substitutions are listed in Table 8.

TABLE 8
Observed infrared frequencies and band assignments of Ni_2GeO_4 (cm^{-1})

	^{70}Ge	^{76}Ge	$\Delta\nu$	^{58}Ni	^{62}Ni	$\Delta\nu$	Assignment from isomorphic replacement	Assignment from isotopic substitution
ν_1	694	688	6	692	692	0	GeO$_4$ str.	GeO$_4$ str.
ν_2	456	451	5	455	453	2	Essentially but not purely GeO$_4$ bending	Essentially GeO$_4$ bend, but with some non-negligible contribution from Ni—O bonds
ν_3	336	334	2	337	330.5	6.5	Vibration of NiO$_6$ octahedra	Mixed vibration, predominantly Ni—O, but some contribution from Ge—O
ν_4	201	199.5	1.5	201	196.5	4.5		

Previously, the occurrence of vibrational interaction between the GeO$_4$ and NiO$_6$ groups was deduced by the isomorphous substitution (replacement of the metal by a different metal without changing the crystal structure). However, the isotopic substitution can provide more detailed information about the nature of vibrational coupling. A similar study has been made on CaMoO$_4$ by using the ^{40}Ca/^{44}Ca substitution. It was found that the band near 230 cm^{-1} which was previously assigned to a rotatory lattice mode must be assigned to a translatory lattice mode because it gives a large isotopic shift (13 cm^{-1}). The metal isotope method provides a definitive tool in distinguishing between these two types of lattice modes since the translatory mode should be mass-sensitive whereas the rotatory mode is not. Tarte and Preudhomme[64] also studied the vibrational spectra of Zn$_2$GeO$_4$, α-CaSiO$_3$ and Ca$_3$Cr$_2$Ge$_3$O$_{12}$ using the method of metal isotope substitution (isotopically substituted metals are underlined).

They have also studied simple oxides such as GeO$_2$ and Cr$_2$O$_3$ (see Table 9). In the case of GeO$_2$, nearly all the bands are shifted by 5–10 cm^{-1} whereas a

TABLE 9
Infrared isotopic shifts of GeO$_2$ and Cr$_2$O$_3$ (cm^{-1})

^{70}GeO$_2$	^{76}GeO$_2$	$\Delta\nu$	^{50}Cr$_2$O$_3$	^{54}Cr$_2$O$_3$	$\Delta\nu$
962	960	2	643	643	0
885	877	8	583	583	0
592	584	8	444.5	444.5	0
559	551	8	414.5	415	−0.5
523	516	7	312	302	10
345	339	6			
265.2	262	3 ?			
258	251	7			
215	212	3			

significant shift was observed only for one band in Cr_2O_3. This unexpected result was explained by assuming that the high frequency vibrations (700–400 cm^{-1}) of metal oxides of this type are practically independent of the mass of the metal atom. In fact, Rh_2O_3 and Al_2O_3 exhibit bands of similar frequencies to those of Cr_2O_3. On the other hand, the low frequency band is strongly mass sensitive.

Tarte and Liegeois-Duyckarts[65] carried out detailed studies on $CaMoO_4$ and $CaWO_4$ (scheelite-type compound) by using the Ca and Mo isotopes. As is seen in Table 9, previous band assignments made by four groups of workers are contradictory. Group theoretical analysis predicts five internal modes [$\nu_1(a_u)$stretch, $\nu_4(e_u)$stretch, $\nu_2(a_u)$bend, $\nu_3(a_u)$bend, $\nu_5(e_u)$bend] and three lattice modes [rotation (e_u), translation (a_u) and translation (e_u)] to be infrared active. We eliminate two stretching modes from our discussion since these assignments are not controversial. Since the symmetry species of other modes can be determined from previous single crystal work, we are concerned only with the assignments of six vibrations below 450 cm^{-1}. First, two bands at 232 and 199 cm^{-1} must be assigned to the translatory modes because they are highly sensitive to the Ca and Mo isotope substitution. On the other hand, the band at 153 cm^{-1} must be assigned to a rotatory lattice mode since it is not mass sensitive at all. The remaining three bands must be due to internal bending modes which should not be sensitive to the $^{40}Ca/^{44}Ca$ substitution if they were not coupled with lattice modes. Table 10 shows, however, that the bands at 329 and 284 cm^{-1} are slightly shifted by this substitution. This result indicates the presence of vibrational coupling between these internal modes and translatory lattice modes. The distinction between the ν_2 and ν_4 internal modes (both a_u species) has been made by comparing the spectrum of $CaMoO_4$ with that of $CaWO_4$. Liegeois-Duyckarts and Tarte[66] have obtained the Raman spectra of $CaMoO_4$ and its Ca and Mo

TABLE 10
Infrared frequencies and band assignments of Ca[MoO₄] (cm⁻¹)

Previous workers				Tarte *et al.*				Assignment
a	b	c	d	^{40}Ca	$\Delta\nu$	^{92}Mo	$\Delta\nu$	
245	324	425	404	431	0	432	1	$\nu_2(a_u)$
425	436	245	282	284	2	281	1	Essentially $\nu_3(a_u)$, but small contribution from a translation
318	436	318	332	329	2	327	3	Essentially $\nu_5(e_u)$, but small contribution from a translation
208	272	208	204	153	0	153	0	Rotation (e_u)
150	218	150	234	237	13	232	7	Translation (e_u)
196	150	196	150	200	7	199	3	Translation (a_u)

a: S. P. S. Porto and J. F. Scott, *Phys. Rev.* **157**, 716 (1967).
b: R. K. Khanna and E. R. Lippincott, *Spectrochim. Acta* **24A**, 905 (1968).
c: J. F. Scott, *J. Chem. Phys.* **48**, 874 (1968).
d: R. G. Brown, J. Denning, A. Hallett and S. D. Ross, *Spectrochim. Acta* **26A**, 963(1970).

isotope compounds also, and made definitive assignments of all the bands observed.

$Na_2[MoO_4]$ and $Ag_2[MoO_4]$ crystallize in the spinel structure. Since the four i.r. active modes (two internal and two lattice) all belong to the t_{1u} species, vibrational coupling may occur between internal and lattice modes although the bonding between the Na^+ ion and the MoO_4^{2-} ion is rather ionic. According to Preudhomme and Tarte,[67] $Na_2[^{92}MoO_4]$ exhibits four i.r. bands at 840, 319, 231 and 178.5 cm^{-1} which are shifted by 8, 2, 0.5 and 2.5 cm^{-1}, respectively, by the $^{92}Mo/^{100}Mo$ substitution. The former two are the stretching and bending modes of the MoO_4^{2-} ion, whereas the latter two are translatory lattice modes (rotatory modes are inactive in this case). The 231 cm^{-1} band is due to the translatory motion of the Na^+ ion while the 178.5 cm^{-1} band is due to a simultaneous displacement of the Na^+ ion and of the Mo atom. If the product rule is applied to the t_{1u} species of the 'isolated tetrahedral group', such as MoO_4^{2-} in $Na_2[MoO_4]$, we obtain

$$\frac{\nu_3 \nu_4}{\nu_3^* \nu_4^*} = \frac{m_x^* (m_x + 64)}{m_x (m_x^* + 64)}$$

(m_x^* represents the heavy isotope). The product rule is satisfactorily verified for the molybdate; the internal and lattice modes are well isolated in this case. This is no longer true for the germanates; vibrational coupling is not negligible because of the increased valency of the cation and of the decreased valency of the central metal in the complex anion.

Tetrahedral oxo-anions of the MO_4^{n-} type are expected to show four Raman active fundamentals; $\nu_1(a_1$, stretch), $\nu_2(e$, bend), $\nu_3(t_2$, stretch) and $\nu_4(t_2$, bend). Although the assignments of ν_1 and ν_3 are straightforward, those of ν_2 and ν_4 have been controversial in some cases. Müller et al.[68] have shown that ν_2 and ν_4 should be assigned to the bands of higher and lower intensity, respectively, in aqueous solution (and sometimes in the crystalline state). Some anions such as VO_4^{3-} and MoO_4^{2-} exhibit only one bending band due to accidental overlap of ν_2 and ν_4.[68]

Preudhomme and Tarte[69,70] also studied the infrared spectra of normal (II-III) spinels of the type $M(II)M(III)O_4$ where M(II) is Zn, Mg, Co or Ni and M(III) is Al, Cr or Rh. These compounds exhibit four bands as predicted by the theory. They have proposed the following assignments based on their systematic and detailed investigations involving metal isotope shifts, solid, solution and metal substitution: ν_1 and ν_2 depend on the chemical nature of the octahedral $M(III)O_6$ unit and should be assigned to the internal vibrations of this group. ν_3 and ν_4 involve the simultaneous vibration of both $M(II)O_4$ (tetrahedral) and $M(III)O_6$ (octahedral) groups. The metal isotope data presented in Table 11 support these assignments.

Finally, Tarte and Preudhomme[71] carried out a detailed infrared study on lithium spinels of the type $LiX(III) Y_4(III)O_8$ (X = Ga, Fe, In; Y = Cr, Rh) which exhibit six bands below 700 cm^{-1}. For example, $LiGaCr_4O_8$ exhibits

TABLE 11
Observed frequencies and isotope shifts of various normal II–III
spinels (cm^{-1})

	ν_1	ν_2	ν_3	ν_4
ZnAl$_2$O$_4$	690	575	518	227
Zn<u>Cr</u>$_2$O$_4$	635(1)^a	537(1)	375(1)	188(4)
Mg<u>Cr</u>$_2$O$_4$	650(1)	527(1)	435(7.5)	254(2.5)
<u>Mg</u>Cr$_2$O$_4$	646(0)	523(−1)	431(6)	253.5(3.5)
MgRh$_2$O$_4$	605	540	353	242

^{a} The numbers in the brackets indicate isotopically substituted shifts (cm^{-1}) such as ^{50}Cr/^{54}Cr and ^{24}Mg/^{26}Mg. The isotope elements are underlined.

bands at 652(0), 612(1), 565(1), 515(0), 470(25) and 220(0) cm^{-1} (the numbers in the brackets indicate the ^{6}Li/^{7}Li isotopic shifts). Four bands between 700 and 500 cm^{-1} have been assigned to complex motions of the whole spinel lattice involving both GaO$_4$ and CrO$_6$ groups, whereas the bands at 470 and 220 cm^{-1} have been assigned to the Li and GaO$_4$ group translatory lattice modes, respectively. These assignments are markedly different from previous ones by DeAngelis et al.[72]

In the case of solid state spectra, one important fact is to be noted. One can measure the spectrum of either the pure solid or of the solid doped in a host lattice. It is found[39,73] that in general the i.r. peaks of ions doped in host lattices are rather sharp in comparison to those of the normal solids. One of the reasons for this effect is that site group splittings of degenerate vibrations of ions doped in host lattices become prominent.

Thus, for example, Müller et al.[39] could measure the $\nu_3(t_2)$ fundamental frequencies of [^{50}CrO$_4$]$^{2-}$, [^{53}CrO$_4$]$^{2-}$, [^{92}MoO$_4$]$^{2-}$ and [^{100}MoO$_4$]$^{2-}$ (all T_d point group) doped in the corresponding alkali sulphates (K$_2$SO$_4$, Rb$_2$SO$_4$ and Cs$_2$SO$_4$) to ± 0.2 cm^{-1}. The frequencies are listed in Table 12; see also Fig. 6.

The main difficulty in such studies is that the frequencies and the isotopic shifts associated with the anion might depend to a great extent on the cation chosen, e.g. for [MoO$_4$]$^{2-}$; see Ref. 39. Hence, the measurements may not provide the true values of the shifts corresponding to the free anion. The best way of minimizing the perturbation of the potential field on the free anion, due to the interaction with the crystal field around the site of the anion, appears to lie in the choice of host lattices with the minimum polarizing effect (non-polarizing cations). In this case, the frequencies and the shifts associated with the anion doped in host lattices correspond as nearly as possible to those of the free ion. The isotopic shifts of the relatively less polarizable anions in comparison to those of the more polarizable ones (e.g. [CrO$_4$]$^{2-}$ in comparison to [MoO$_4$]$^{2-}$) can, on the other hand, be relatively independent of the host lattice chosen.

TABLE 12
Observed frequencies[a] of $[MoO_4]^{2-}$ and $[CrO_4]^{2-}$ in host lattices (cm^{-1})

		ν_3	ν_4
$Na_2[MoO_4]$	doped in	884.7[b] (6.4)[c]	—
	K_2SO_4	866.7[b] (7.0)[c]	—
		863.3[b] (5.6)[c]	—
	doped in	873.0[b] (6.4)[c]	—
	Rb_2SO_4	858.3[b] (7.0)[c]	—
		855.7[b] (7.0)[c]	—
	doped in	858.0[b] (7.7)[c]	—
	Cs_2SO_4	844.7[b] (7.3)[c]	—
		841.7[b] (7.0)[c]	—
$Cs_2[CrO_4]$	doped in	931.3[b] (8.1)[c]	—
	K_2SO_4	911.3[b] (8.3)[c]	—
		905.7[b] (8.1)[c]	—
	doped in	918.8[b] (8.8)[c]	—
	Rb_2SO_4	903.0[b] (8.2)[c]	—
		898.2[b] (8.1)[c]	—
	doped in	905.1[b] (8.8)[c]	389.8[b,d] (2.2)[c,d]
	Cs_2SO_4	892.1[b] (8.4)[c]	382.4[b,d] (2.2)[c,d]
		887.1[b] (8.2)[c]	380.1[b,d] (2.3)[c,d]

[a] The individual frequencies are accurate to ± 0.2 cm^{-1} and the data are from Ref. 39.

[b] Corresponds to the lighter isotope (^{92}Mo or ^{50}Cr).

[c] The number in brackets indicates the isotopic shift ($^{92}Mo/^{100}Mo$ or $^{50}Cr/^{53}Cr$).

[d] Observed for a Nujol mull; see Ref. 39.

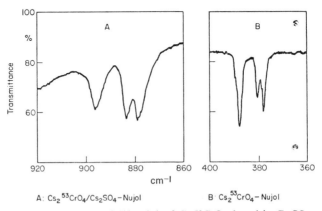

Fig. 6. The Nujol-mull spectrum of (A) $\nu_3(t_2)$ of $Cs_2{}^{53}CrO_4$ doped in Cs_2SO_4 and (B) $\nu_4(t_2)$ of $Cs_2{}^{53}CrO_4$.

7 SPECTRA OF SIMPLE ANIONS

Metal isotope data have been obtained for simple inorganic anions mainly to aid in the refinement of force constants. Müller et al.[38] obtained the infrared and Raman spectra of $Na_2[MoO_4]$, $K_2[MoS_4]$ and $Cs_2[MoS_4]$ with ^{92}Mo and ^{100}Mo isotopes. Later, they[39] reported the i.r. spectra of several alkali sulphates doped with $Cs_2[^{50}CrO_4]$, $Cs_2[^{53}CrO_4]$, $Na_2[^{92}MoO_4]$ and $Na_2[^{100}MoO_4]$ (for details, see p. 200). As the alkali sulphates crystallize in the β-K_2SO_4 type with space group D_{2h}^{16}-$Pnma$, the metal–oxygen stretching and bending vibrations split into three components. All three components were observed as extremely sharp bands so that the metal isotope shifts could be measured accurately. It should be noted (see Table 12) that the magnitude of the isotopic shift is almost the same for the three components. Therefore, the average value of these three shifts was used for the force constant calculation.[39]

The Raman spectrum of $(NH_4)_2MoO_2S_2$ with ^{92}Mo and ^{100}Mo has also been measured.[74] The observed frequencies and isotopic shifts for the stretching modes are: $v_s(MoO) = 819(4)$ cm^{-1}; $v_{as}(MoO) = 801(8)$ cm^{-1}; $v_{as}(MoS) = 506(7)$ cm^{-1}; $v_s(MoS) = 473(3)$ cm^{-1}. It was shown that the frequencies calculated by using the L-matrix approximation method are in good agreement with these values. A similar study has also been made for the $[MoOS_3]^{2-}$ ion.[49] The observed frequencies and isotopic shifts for the stretching modes are: $v(MoO) = 853.5(5)$; $v_{as}(MoS) = 479.3(7)$; $v_s(MoS) = 464.8(1.3)$ (all in cm^{-1}).

The vibrational spectra of $Cs_2{}^{50}Cr_2O_7$ and its ^{53}Cr analog have been measured and approximate normal coordinate analysis has been carried out by using the metal isotope shift data.[75] The observed fundamental frequencies and shifts are

a_1 species (in cm^{-1}):
952.5(7.5); 907(3); 539(6.2); 376.2(2.4); 216(0)

b_2 species (in cm^{-1}):
938.5(8.2); 885.5(1.7); 784(4.0); 389.5(1.5); 369.2(2.0); 216(0).

The Cr—O—Cr bending mode was predicted to be at ca. 80 cm^{-1} with a theoretical isotopic frequency shift of 0.3 cm^{-1}.

8 SPECTRA OF COORDINATION COMPOUNDS

Vibrational spectra of coordination compounds consist of bands due to ligand vibrations, metal–ligand vibrations and lattice vibrations (if the spectra are related to the crystalline state). The metal–ligand vibrations are the most important since they provide direct information about the structure of the complex and the nature of the metal–ligand bonds. These vibrations appear in the low frequency region (generally below 500 cm^{-1}) where ligand as well as lattice vibrations appear. Thus, the interpretation of the low frequency spectra of coordination compounds is not straightforward. The conventional methods used to assign the metal–ligand vibrations have been one or a combination of the following methods:

(1) A comparison of spectra of a free ligand and its metal complex; only the latter exhibits metal–ligand vibrations. This method often fails to give a clearcut assignment since some ligand vibrations which are absent in free ligand are activated by complex formation and may appear in the same region as the metal–ligand vibrations.

(2) The metal–ligand vibrations are metal sensitive and shifted by changing the metal atom or its oxidation state. In order to apply this method, it is necessary to prepare a pair or series of compounds which have exactly the same structure with only the metal atom different. This is technically inconvenient.

(3) Metal–ligand vibrations appear in the same frequency region if the metal is the same and the ligands are similar. For example, the Zn—N stretching frequencies of Zn(II) α-picoline complexes are similar to those of Zn(II) pyridine complexes. This method is applicable only when the metal–ligand frequency is known for one parent compound.

(4) The metal–ligand vibration shows an isotopic shift if the ligand is isotopically substituted. For example, the Ni—N stretching mode of $[Ni(NH_3)_6]Cl_2$ at 334 cm^{-1} is shifted to 318 cm^{-1} upon deuteration of the NH_3 ligand. Similar shifts are observed if the atom directly bonded to the metal, i.e. the α-atom, is isotopically substituted. In this method, however, metal–ligand vibrations as well as ligand vibrations involving the motion of the α-atom are shifted by isotopic substitution. Thus this method does not provide a clearcut result if these two types of vibration appear in the same region.

(5) The frequency of a metal–ligand vibration may be predicted if the metal–ligand and ligand force constants are known *a priori*. At present, this method is not practical since only a few force constants are known with certainty.

It is obvious that none of these methods is satisfactory. Furthermore, these methods encounter more and more difficulties as the structure of the complex, and hence the spectrum, becomes more complicated. On the other hand, the metal isotope method[1] provides definitive band assignments of the metal–ligand vibrations. In this method, two metal complexes in which only the metal atoms are isotopically substituted are prepared, and their spectra are compared in the low frequency region. Only the vibrations involving the motion of the metal atom are shifted by metal isotope substitution. If the metal–ligand vibrations are coupled with ligand or lattice vibrations, the latter also become metal isotope sensitive. If the degree of coupling is appreciable, band assignments can only be made through rigorous normal coordinate analysis.

8.1 Ammine Complexes

The vibrational spectra of metal ammine complexes have already been studied extensively by many investigators. The band assignments have been well established except for the case of $[Co(NH_3)_6]^{3+}$ which is still controversial. The metal isotope technique has been applied to the six compounds listed in Table 13 mainly to refine the force constants.[42] As expected, all the NH_3

TABLE 13
Observed frequencies, isotopic shifts[a] and band assignments for ammine complexes (cm^{-1})[b]

	$\delta_a(NH_3)$	$\delta_s(NH_3)$	$\rho_r(NH_3)$	$\nu(MN)$	$\delta(NMN)$ or $\gamma(NMN)$	Isotope substitution
$[^{58}Ni(NH_3)_6]Cl_2$	$1,605\pm3$ (0)	$1,186\pm2$ (0)	684 ± 1 (2 ± 1)	335.2 ± 0.3 (2.2 ± 0.6)	$217\ \pm1$ (3 ± 2)	$^{58}Ni/^{62}Ni$
$[^{50}Cr(NH_3)_6](NO_3)_3$	$1,627\pm5$ (0)	$1,290\pm10$ (0)	770 ± 2 (0)	471.0 ± 0.5 (4.0 ± 1)	$270\ \pm0.5$ (3.0 ± 1)	$^{50}Cr/^{53}Cr$
$[Co(NH_3)_6]Cl_3$	$1,620\pm5$	$1,327\pm3$	830 ± 2	$498\ \pm0.5$ $477\ \pm0.5$ $449\ \pm0.5$	$331\ \pm1$	
$Cu(NH_3)_4]SO_4\cdot H_2O$	$1,670\pm5$ (0) $1,640\pm5$ (0)	$1,280\pm2$ (0)	735 ± 2 (0)	$426\ \pm0.5$ (2.0 ± 1)	$256\ \pm0.5$ $(1\ \pm1)$ 226.5 ± 0.5 (0.5 ± 1)	$^{63}Cu/^{65}Cu$
$Pd(NH_3)_4]Cl_2\cdot H_2O$	$1,630\pm5$ (0)	$1,279\pm2$ (0)	849 ± 2 (0) 802 ± 2 (0)	495.0 ± 0.5 (3.0 ± 1)	291.0 ± 0.5 (0.5 ± 1) 237.5 ± 0.5 (1.5 ± 1)	$^{104}Pd/^{110}Pd$
$[Zn(NH_3)_4]I_2$	$1,596(0)$	$1,253(0)$ $1,239(0)$	$685(0)$	$432(0.5)$ $412(2.0)$	$156(1.0)$	$^{64}Zn/^{68}Zn$

[a] The number in the brackets indicates the isotopic shift.
[b] See Ref. 42.

vibrations are insensitive to metal isotope substitutions. This indicates the presence of very little vibrational coupling between NH_3 vibrations and MN_x skeletal vibrations. Only the NH_3 rocking vibration of the Ni complex shows some metal isotope sensitivity indicating the presence of vibrational coupling with the NiN_6 skeletal modes.

All the M—N stretching and NMN bending modes are metal isotope sensitive. The magnitude of the shift ranges from 2 to 4 cm^{-1} for the stretching and from 0.5 to 3 cm^{-1} for the bending mode. It is somewhat surprising to see relatively large shifts for bending modes.

Table 14 lists the complete assignments of all the skeletal modes made by Schmidt and Müller.[42] Although no isotope data are available for the Co(III) complex, three weak i.r. bands near 470 cm^{-1} were assigned to the $\nu_3(t_{1u})$

TABLE 14
Observed skeletal frequencies of octahedral ammine complexes (cm^{-1})^a

	$[Cr(NH_3)_6]^{3+}$	$[Ni(NH_3)_6]^{2+}$	$[Co(NH_3)_6]^{3+}$
$\nu_1(a_{1g})$, $\nu_s(MN)$	465 (R)	370 (R)	494 (R)
$\nu_2(e_g)$, $\nu(MN)$	412 (R)	(251)^{calc}	442 (R)
$\nu_3(t_{1u})$, $\nu_{as}(MN)$	468.5 (i.r.)	334 (i.r.)	477 (i.r.)
$\nu_4(t_{1u})$, $\delta(NMN)$	268 (i.r.)	215 (i.r.)	331 (i.r.)
$\nu_5(t_{2g})$, $\delta(NMN)$	270 (R)	235 (R)	322 (R)
$\nu_6(t_{2u})$, $\delta(NMN)$	206 (R)^b	(158)^{calc}	(240)^{calc}

^{a} See Ref. 42.

^{b} This is in aqueous solution. Though this mode is active neither in i.r. nor in Raman, it can become active due to the Jahn–Teller effect. The value listed here was deduced from a Raman spectrum, superimposed partly by a resonance phosphorescence spectrum. For details see T. V. Long, II and D. J. B. Penrose, *J. Am. Chem Soc.* **93**, 632 (1971).

Co—N stretching mode (which splits into 3 peaks (498, 477 and 449 cm^{-1}) due to the crystal field effect).[76] However, Swaddle *et al.*[77] have assigned the weak infrared bands at 495 and 473 cm^{-1} to the Raman active $\nu_1(a_{1g})$ mode.

The Raman spectrum of a tetrahedral $[Zn(NH_3)_4]^{2+}$ ion with ^{64}Zn and ^{68}Zn isotopes has been measured by Takemoto and Nakamoto.[78] The ^{64}Zn complex exhibits two Zn—N stretching modes at 432.0 and 412.0 cm^{-1}. By ^{64}Zn/^{68}Zn substitution the former gives almost no shift whereas the latter gives a shift of 2.0 cm^{-1}. Thus, it was possible to assign these vibrations to the a_1 and t_2 species, respectively. This assignment is also supported by the fact that the former band is strong and polarized. Later, Nakamoto *et al.*[79] carried out a complete normal coordinate analysis on the $[Zn(NH_3)_4]^{2+}$ ion by considering all isotope data including the H/D substitution.

8.2 Cyano Complexes

Thus far, the only cyano complex studied by the metal isotope technique is $K_2[Zn(CN)_4]$.[80] The $[Zn(CN)_4]^{2-}$ ion (T_d) is expected to show four infrared active t_2 modes. Three of them were located at 2153.6$\pm$0.3 (ν_6, CN stretching), 356.4$\pm$0.3 (ν_7, ZnCN bending+Zn—N stretching), and 315.4$\pm$0.3 cm^{-1} (ν_8, Zn—N stretching+ZnCN bending) for the ^{64}Zn complex. Due to strong coupling between ν_7 and ν_8, these fundamentals exhibit shifts of 2.1$\pm$0.6 and 1.4$\pm$0.6 cm^{-1}, respectively, by the ^{64}Zn–^{68}Zn substitution. This result clarifies ambiguities which could not be solved previously.[81]

8.3 Complexes of Thioanions

The infrared spectra of the CS_3^{2-} complexes with Ni and Zn isotopes have been measured, and a normal coordinate analysis has been carried out on the D_{2h} model of the planar $[Ni(CS_3)_2]^{2-}$ ion:[82]

Table 15 gives the observed the calculated frequencies, isotope shifts and band assignments. Two Ni—S stretching bands were clearly located at about 385 and 366 cm^{-1}, with isotope shifts of 7.2 and 5.0 cm^{-1}, respectively, by the ^{58}Ni/^{62}Ni substitution. The Zn—S stretching bands of the $[Zn(CS_3)_2]^{2-}$ ion (D_{2d} symmetry) were assigned at 260 and 205 cm^{-1}.

TABLE 15
Comparison of observed and calculated frequencies and band assignments for the $[Ni(CS_3)_2]^{2-}$ ion (cm^{-1})^a

| | | Observed | | Calculated | | Predominant |
		$\tilde{v}(^{58}Ni)$	$\Delta\tilde{v}^b$	$\tilde{v}(^{58}Ni)$	$\Delta\tilde{v}^b$	mode
a_{1u}*	v_1	—	—	25	0.0	inter-ring torsion
b_{1u}	v_2	485.0	0.0	480	0.0	$\pi(CS_3)$
	v_3	—	—	60	1.0	$\pi(NiCS_3)$
	v_4	—	—	14	0.0	$\pi(NS_4)$
b_{2u}	v_5	857.5	0.0	856.1	0.0	$v(C—S)$
	v_6	365.9	5.0	366.0	4.8	$v(Ni—S)$
	v_7	297.1	1.5	297.1	1.8	$\delta(S_1CS_3)$
	v_8	55.5	0.5	55.0	0.4	$\delta(S_1NiS)$
b_{3u}	v_9	1,010.0	0.0	1,009.0	0.0	$v(C\!=\!S)$
	v_{10}	507.0	0.5	504.8	0.1	$v(C—S)$
	v_{11}	384.7	7.2	384.3	7.1	$v(Ni—S)$
	v_{12}	185.7	2.0	186.4	1.4	$\delta(S_1NiS_2)$

^{a} PPh$_4$$^+$ salt.
^{b} $\Delta\tilde{v} = \tilde{v}(^{58}Ni) - \tilde{v}(^{62}Ni)$.
* Not i.r. active.

The infrared spectra of WS_4^{2-} and MoS_4^{2-} complexes with Ni and Zn isotopes have been measured, and normal coordinate analysis has been carried out on the $[Ni(WS_4)_2]^{2-}$ ion of D_{2h} symmetry.[83]

M = W or Mo

The same authors[84] extended their calculations to the $[Ni(MoS_4)_2]^{2-}$ ion for which both Ni and Mo isotope data were obtained. Table 16 summarizes the observed frequencies of the stretching fundamentals and their isotopic shifts.

Phosphine sulfides SPR_3(R = alkyl or phenyl) are known to form planar

TABLE 16
Observed frequencies and isotope shifts^a for $[M(M'S_4)_2]$ type complexes (cm^{-1})

		Terminal ν(M=S)	Bridging ν(M—S)	ν(Ni—S)	Reference
$[Ni(WS_4)_2]^{2-}$	$^{58}Ni/^{62}Ni$	496(0.5)	449 (0.5)	328 (4.5)	83
		490 (0)	449 (0.5)	320 (4.0)	
$[Ni(^{92}MoS_4)_2]^{2-}$	$^{58}Ni/^{62}Ni$	494 (0)	455.5(0.5)	331.5(4.7)	84
			442.5(0.0)	323.8(4.7)	
$[^{58}Ni(MoS_4)_2]^{2-}$	$^{92}Mo/^{100}Mo$	494 (6.0)	455.5(6.0)	331.5(0)	84
			442.5(1.7)	323.8(0.3)	
$[^{64}Zn(MoS_4)_2]^{2-}$	$^{92}Mo/^{NA}Mo^b$	516(6.0)	456 (1)	—	85
		499(2.5)	434.5(0)	—	

^{a} Numbers in brackets indicate isotopic shifts.
^{b} NA = natural abundance.

three coordinate complexes of the type $[Cu(SPR_3)_3]Y$ (Y = BF_4 or ClO_4). The Cu—S stretching fundamentals of these complexes have been assigned to bands in the region of 315 ~ 295 cm^{-1} based on isotopic shifts due to the $^{63}Cu/^{65}Cu$ substitution.[86]

8.4   Complexes of Phosphines and Arsines

The vibrational spectra of metal–phosphine complexes have been studied by many investigators. The metal–phosphorus stretching frequencies assigned by previous investigators cover a wide frequency range from 460 to 90 cm^{-1}. This scattering of frequencies has been attributed to the differences (1) in the nature of the metals (oxidation state, electronic structure, etc.); (2) in the phosphine ligands PR_3 in which R is CH_3, C_2H_5 or C_6H_5, etc., and (3) in the structure of the complex (stereochemistry, coordination number, etc.). All metal–phosphorus stretching frequencies reported previously, however, were assigned empirically. Thus, the first series of compounds to be chosen for the metal isotope study was one of phosphine complexes.[87]

The infrared spectrum of *trans*-planar $Ni(PEt_3)_2X_2$ (X = Cl or Br) is expected to show one Ni—P and one Ni—X stretching band, both of which should be shifted by the ^{58}Ni-^{62}Ni substitution. Figure 7 gives the actual tracing of the spectra of the chloro and bromo complexes with ^{58}Ni and ^{62}Ni isotopes, and Table 17 lists the observed band frequencies, isotopic shifts and band assignments. The bands above 420 cm^{-1} are not listed here since they are due to

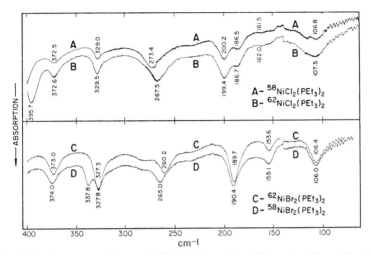

Fig. 7. The i.r. spectra of *trans*-58NiX$_2$(PEt$_3$)$_2$ and its 62Ni analog (X = Cl or Br).

almost pure ligand vibrations which are not sensitive to metal isotope substitution. It is seen that each compound exhibits two bands which give large isotope shifts relative to others. The band at 403.3 cm^{-1} of ^{58}Ni(PEt$_3$)$_2$Cl$_2$ is due to the Ni—Cl stretching mode, since this band is strong and not present in the bromo complex. Then another band at 273.4 cm^{-1} must be assigned to the Ni—P stretching mode. This band remains at 265.0 cm^{-1} in the bromo complex.

TABLE 17
Observed frequencies and band assignments of triethylphosphine complexes (cm^{-1})

^{58}Ni(PEt$_3$)$_2$Cl$_2$		^{58}Ni(PEt$_3$)$_2$Br$_2$		^{104}Pd(PEt$_3$)$_2$Cl$_2$		Assignment
$\tilde{\nu}$	$\Delta\tilde{\nu}^a$	$\tilde{\nu}$	$\Delta\tilde{\nu}^a$	$\tilde{\nu}$	$\Delta\tilde{\nu}^b$	
416.7	0.0	413.6	1.2	413.0	−0.3	δ(CCP)
403.3	6.7	337.8	10.5^c	358.3	3.4	ν_a(M—X)
372.5	−0.1	374.0	1.1	375.7	−0.3	δ(CCP)
329.0	−0.5	327.8	0.5	330.8	0.0	δ(CCP)
273.4	5.9	265.0	4.7	234.5	2.5	ν_a(M—P)
— ^{d}		— ^{d}		272.0	0.0	δ(CCP)
200.2	0.8	190.4	0.7	183.8	1.8	δ(CPC)
186.5	−0.2	155.1	1.5	168.0	−0.5	δ(MX)
161.5	−0.5	— ^{d}		152.0	1.0	δ(MP)
106.8	−0.7	106.0	−0.4	105.5	0.5	?

^{a} $\Delta\tilde{\nu} = \tilde{\nu}(^{58}Ni) - \tilde{\nu}(^{62}Ni)$.

^{b} $\Delta\tilde{\nu} = \tilde{\nu}(^{104}Pd) - \tilde{\nu}(^{110}Pd)$.

^{c} This value is only approximate since two bands are overlapped.

^{d} Hidden bands.

Table 17 also includes the results obtained from *trans*-Pd(PEt$_3$)$_2$Cl$_2$ by using the ^{104}Pd/^{110}Pd pair. The Pd—Cl and Pd—P stretching modes were assigned at 358.3 and 234.5 cm^{-1}, respectively, for the ^{104}Pd complex.

It is known that Ni(PPh$_3$)$_2$Cl$_2$ is tetrahedral whereas Pd(PPh$_3$)$_2$Cl$_2$ is *trans*-planar. Thus, the former should exhibit two Ni—Cl and two Ni—P stretching, whereas the latter should exhibit one Pd—Cl and one Pd—P stretching band in the infrared. It was found that four bands at 341.2, 305.0, 189.6 and 164.0 cm^{-1} of the ^{58}Ni complex are relatively sensitive to the metal isotope substitution. The former two bands are assigned to the Ni—Cl stretching modes since they are relatively strong and absent in the bromo analog. Then the latter two bands must be assigned to the Ni—P stretching modes. In *trans*-^{104}Pd(PPh$_3$)$_2$Cl$_2$, two bands at 360.3 and 191.2 cm^{-1} are metal isotope sensitive. The former is strong and in the region of terminal Pd—Cl stretching frequencies. Therefore, it is safely assigned to the Pd—Cl stretching mode. Then, the latter must be assigned to the Pd—P stretching mode.

Complexes of the type Ni(PPh$_2$R)$_2$Br$_2$ (R = alkyl) exist in two isomeric forms: tetrahedral (green) and *trans*-planar (brown). The distinction between these two forms can be made easily by infrared spectroscopy since the numbers of infrared active Ni—Br and Ni—P stretching modes are different in each case. Wang *et al.*[88] studied the infrared spectra of a series of compounds of this type, where R is C$_2$H$_5$, *n*-C$_3$H$_7$, *i*-C$_3$H$_7$ or *i*-C$_4$H$_9$, and confirmed that the Ni—Br and Ni—P stretching bands are at *ca.* 330 and 260 cm^{-1}, respectively, for the planar form and at about 270 to 230 and 200 to 160 cm^{-1}, respectively, for the tetrahedral form. The presence or absence of the 330 cm^{-1} band is particularly useful in distinguishing these two forms.

According to X-ray analysis,[89] the green form of Ni(PPh$_2$Bz)$_2$Br$_2$ (Bz = benzyl) is a mixture of the planar and tetrahedral molecules in a 1:2 ratio. Ferraro *et al.*[90] have studied the effect of high pressure on the infrared spectra of this compound. It was found that all the bands characteristic of the tetrahedral form disappear as the pressure is increased to about 20,000 atm. This result indicates that the tetrahedral molecule can be converted to the planar form under high pressure. This conversion is completely reversible; the original green form is recovered as the pressure is reduced.

In planar NiX$_2$L$_2$ (X = halogen; L = phosphine) type complexes, the Ni—X stretching frequencies are very useful in distinguishing the *cis* and *trans* configurations. For example, the Ni—Cl stretching bands of the *trans* complexes are at about 400 cm^{-1} whereas those of the *cis* complexes are at about 320 cm^{-1}. The Ni—P stretching bands of the *trans* complexes are at about 260 cm^{-1}. The Ni—P stretching frequencies of the *cis* complexes are expected to be higher than those of the *trans* complexes, since the chlorine atom (ligand of relatively weak *trans* effect) is *trans* to the phosphine ligand in the *cis* complex. Udovich *et al.*[91] confirmed this expectation by studying the infrared spectra of a series of complexes of the type Ni(DPE)X$_2$, where DPE is 1,2-bis-(diphenylphosphine)ethane and X is Cl, Br or I, by using the metal isotope technique. Square-planar complexes of the type $[M(dias)_2]X_2$, where $M = $ Ni(II), Co(II), Pd(II) or Pt(II) and $X = $ Cl, Br or ClO_4, exhibit two M—As stretching bands at about 317 and 303 cm^{-1} for Ni(II), 315 and 298 cm^{-1} for Co(II), 271 and 247 cm^{-1} for Pd(II) and 235 and 216 cm^{-1} for Pt(II). Since all these bands are very weak, it was extremely difficult to assign them on an empirical basis. *trans* Octahedral complexes of the type $[MX_2(dias)_2]Y_n$ ($M = $ Ni(III), Ni(IV), Fe(II), Fe(III) or Fe(IV), $X = $ Cl or Br, $Y = $ Cl, Br, ClO_4 or BF_4 and $n = 0$, 1 or 2) have also been studied. Table 18 summarizes the M—As and M—X stretching frequencies

TABLE 18
Metal–ligand stretching frequencies and electronic structure of $[MX_2(dias)_2]^{n+}$ type complexes (cm^{-1})

	d^4 Fe(IV)	d^5 Fe(III)	d^6 Ni(IV)	d^7 Ni(III)
ν(M—Cl)	390	384	421	240
ν(M—As)	$\{$320 $\{$—	$\{$324 $\{$—	$\{$— $\{$296	$\{$316 $\{$—
ν(M—Br)		308	312	183
ν(M—As)		$\{$324 $\{$—	$\{$341 $\{$295	$\{$313 $\{$—

determined by the metal isotope technique.[92] It is seen that the M—X stretching frequencies for these complexes are relatively constant for those with the d^4, d^5 and d^6 configurations but drop sharply for that with the d^7 configuration, whereas the M—As stretching frequencies remain fairly constant throughout the series. A similar sharp drop of the M—N stretching frequencies in going from complexes with the d^6 to those with the d^7 configuration was noted previously for a series of tris-bipy complexes (see p. 214). This result was interpreted[92] on the basis that the accommodation of one electron in the antibonding orbital greatly reduces the strength of the metal–ligand bond. According to Manoharan and Rogers,[93] the ground state of the Ni(III) complex is described as

$$^2A_g = [b_{3g}(yz)]^2[b_{2g}(xz)]^2[a_g(x^2-y^2)]^2[a_g(z^2)]^1$$

In the case of the Ni(IV) complex, the electron may be removed from the antibonding $a_g(z^2)$ orbital. Then, the Ni—Cl bond may become considerably stronger whereas the Ni—As bond does not change appreciably in going from Ni(III) to Ni(IV).

8.5 Complexes of Pyridine and Related Ligands

The infrared and Raman spectra of $Zn(py)_2X_2$ (py = pyridine; X = halogen) have been studied by Saito et al.[94] Complete assignments have been made based on the ^{64}Zn/^{68}Zn and py/py(d$_5$) shifts and approximate normal coordinate calculations. Figure 8 shows the infrared and Raman spectra of the chloro complexes with ^{64}Zn and ^{68}Zn, and Table 19 lists the observed and calculated frequencies, isotopic shifts and band assignments. As is seen in Fig. 8, two Zn—Cl and two Zn—N stretching bands are shifted by the metal isotope substitution, as predicted from its C_{2v} symmetry. In the bromo and iodo complexes, only three stretching bands were observed due to overlapping of the bands.

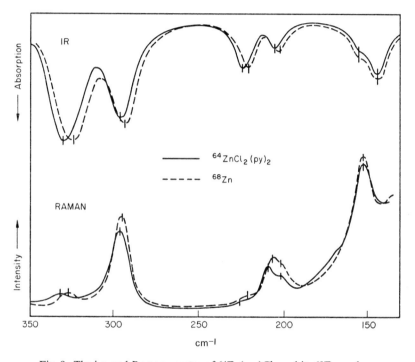

Fig. 8. The i.r. and Raman spectra of ^{64}Zn(py$_2$)Cl$_2$ and its ^{68}Zn analog.

Saito et al.[95] also studied the infrared spectra of trans Ni(py)$_4$X$_2$ (X = Cl, Br or I) by using the ^{58}Ni/^{62}Ni, py/py(d$_5$) and ^{35}Cl/^{37}Cl substitutions. In the case of the chloro complex, two bands at 249.0 and 238.8 cm^{-1} are sensitive to the Ni and py substitutions. Thus, these bands have been assigned to the Ni—N stretching modes. The 207.0 cm^{-1} band is assigned to the Ni—Cl stretching mode since it is sensitive to the Ni as well as Cl isotope substitution. However, it is also slightly sensitive to the py/py(d$_5$) substitution indicating the presence of coupling with the Ni—N stretching mode. Table 20 lists the observed frequencies, isotopic shifts and band assignments for all three compounds studied.

The infrared spectra of SnX$_4$(py)$_2$ (X = Cl, Br or I) and related compounds

TABLE 19
Infrared frequencies, isotopic shifts and assignments of tetrahedral $Zn(py)_2X_2$ type complexes (cm^{-1})

$Zn(py)_2Cl_2$				$Zn(py)_2Br_2$				$Zn(py)_2I_2$				Assignment
^{64}Zn	$\Delta\tilde{\nu}^a$	$\Delta\tilde{\nu}^b$	$\Delta\tilde{\nu}^c$	^{64}Zn	$\Delta\tilde{\nu}^a$	$\Delta\tilde{\nu}^b$	$\Delta\tilde{\nu}^c$	^{64}Zn	$\Delta\tilde{\nu}^a$	$\Delta\tilde{\nu}^b$	$\Delta\tilde{\nu}^c$	
329.2(vs)	4.8	3.8	0.8	257.0(vs)	5.3	5.4	1.7	227.4(vs)	4.4	5.9	1.9	$\nu(Zn—X), b_2$
296.5(s)	2.4	1.7	0.2	223.0(vs)	3.7	4.4	1.3	—^{a}	—	4.5	—	$\nu(Zn—X)+\nu(Zn—N), a_1{}^f$
222.4(m)	3.6	4.4	3.8	—^{a}	—	4.7	—	217.0(vs)	2.5	4.6	3.1	$\nu(Zn—N), b_1$
203.9(w)	2.4	2.2	4.1	184.5(w)	0.7	0.2	3.2	168.8 (m)	0.7	0.4	7.2	$\nu(Zn—N)+\nu(Zn—X), a_1{}^f$
154(sh)	—^{e}	0.7	≈7	152.5(w)	0.0	0.4	9.1	161.0(sh)	—	0.1	6.0	$\delta(NZnN), a_1$
143.3(m)	0.3	1.0	4.7	131.6(m)	0.0	0.6	6.8	126.0(w)	0.2	0.2	6.1	$\delta(XZnN), b_1, b_2$
		0.1				0.3			0.4			
109.9(m)	0.0	0.3	0.7	77.0(m)	0.0	0.2	0.5	66.5(m)	0.0	0.1	0.3	$\delta(XZnX), a_1$

^{a} $\tilde{\nu}(^{64}Zn)-\tilde{\nu}(^{68}Zn)$—observed.
^{b} $\tilde{\nu}(^{64}Zn)-\tilde{\nu}(^{68}Zn)$—calculated. No shift is predicted for the a_2 mode.
^{c} $\tilde{\nu}(py)-\tilde{\nu}(py\text{-}d_5)$—observed.
^{a} Hidden or overlapped by other bands.
^{e} Isotopic shifts could not be determined because of poor shape of band maxima.
^{f} This coupling does not occur in the chloro complex.

have been studied by using the ^{116}Sn and ^{124}Sn isotopes.[96] In the case of SnCl$_4$(py)$_2$, two bands at 323 and 227.5 cm^{-1} give large isotopic shifts relative to others. These have been assigned to the Sn—Cl and Sn—N stretching modes, respectively. This result definitely supports the *trans* structure of this complex. The *trans* structure was also proposed for other compounds based on similar results.

TABLE 20
Observed frequencies and band assignments for Ni(py)$_4$X$_2$ (cm^{-1})

$\tilde{\nu}(^{58}$Ni)	$\Delta\tilde{\nu}^a$	$\Delta\tilde{\nu}^b$	$\Delta\tilde{\nu}^c$	$\tilde{\nu}(^{58}$Ni)	$\Delta\tilde{\nu}^a$	$\Delta\tilde{\nu}^b$	$\tilde{\nu}(^{58}$Ni)	$\Delta\tilde{\nu}^a$	$\Delta\tilde{\nu}^b$	Assignment
	Ni(py)$_4$Cl$_2$				Ni(py)$_4$Br$_2$			Ni(py)$_4$I$_2$		
249.0(s)	4.2 ~7		0.7	238.0(vs)	5.0	5.5	229.0(vs)	3.0	5.3	ν(Ni—N), e_u
238.8(vs)	5.4	4.5	0.2				241.0(vs)	4.5	8.0	ν(Ni—N),a_{2u}
207.0(s)	1.8	2.5	2.3	140.5(s)	~1	3.0	104.5(vs)^f	>1	2.0	ν(Ni—X)
— ^{d}	—	—	—	200.5(w)	0.9	7.5	189.3(vw)	—^{e}	9.5	δ(NNiN)
194(sh)	—^{e} ~6		—^{e}	185(vw)	—^{e} ~5		172.5(vw)	—^{e}	~10	δ(NNiN)
176.1(m)	0.0	8.7	0.7	151(sh)	—^{e} ~7		140.3(vw)	—^{e}	~12	δ(XNiN)
154.8(s)	0.1	2.5	2.0	120.2(m)	~0	3.4	104.5(vs)^f	>1	2.0	δ(XNiX)

^{a} $\tilde{\nu}(^{58}$Ni$)-\tilde{\nu}(^{62}$Ni$)$.
^{b} $\tilde{\nu}$(py)$-\tilde{\nu}$(py$-$d$_5$).
^{c} $\tilde{\nu}$(Cl)$-\tilde{\nu}(^{37}$Cl$)$.
^{d} Hidden by the neighbouring band.
^{e} Isotope shifts could not be determined because of poor shape of band maxima.
^{f} Overlapped band.

Lever and Ramaswamy[97] studied the infrared spectra of ML$_2$X$_2$ (M = Cu(II) or Co(II); L = 2-picoline or ethylpyridine; X = Cl, Br or I) and NiL$_4$X$_2$ (L = 3-picoline). The M—X and M—L stretching bands have been assigned by using Cu and Ni isotopes. For all the compounds studied the M—N stretching bands were assigned in the region from 300 to 220 cm^{-1}.

The infrared spectra of sixteen imidazole complexes with Ni(II), Cu(II) and Zn(II) have been studied by using the metal isotope technique.[98] It was noted that the M—N stretching frequencies of the imidazole complexes are always higher by 10 to 30 cm^{-1} than those of the corresponding pyridine complexes.

Ferraro et al.[99] studied the far-infrared spectra of ZnX$_2$·L (X = Cl or Br) where L is

2, 2′ - Dithiodipyridine   4, 4′ - Dithiodipyridine

by using $^{64}Zn-^{68}Zn$ isotopes. Two Zn—Cl stretching bands are observed for both ligands at 290 to 340 cm^{-1}, indicating the tetrahedral environment of the Zn atom. The Zn—N stretching bands were located near 210 cm^{-1}. However, the sulphur atoms are not coordinated to the metal atom. Polymeric structures involving the bridging L ligand and tetrahedral ZnX_2N_2 skeletons were proposed.

The infrared spectra of metal complexes with 8-hydroxyquinoline(Q) have been studied by Ohkaku and Nakamoto.[100] The M—O and M—N stretching fundamentals have been assigned to bands in the regions 210 to 332 and 190 to 300 cm^{-1} for the MQ_2 type (M = $^{58,62}Ni$, $^{63,65}Cu$ or $^{64,68}Zn$) and the $MQ_2 2H_2O$ type (M = $^{54,57}Fe$, $^{58,62}Ni$ or $^{64,68}Zn$) complexes. The structures of these complexes have been elucidated from the number of metal–ligand stretching bands observed.

8.6   Complexes of Ethylenediamine and Related Ligands

Canham and Lever[101] studied the far-infrared spectra of *trans*-octahedral $Cu(N—N)_2X_2$ and planar $Cu(N—N)X_2$[(N—N)=ethylenediamine, *sym* or *asym* N,N'-dimethyl or N,N'-diethylethylenediamine; X = Cl, Br or NO_3] by using the $^{63}Cu/^{65}Cu$ substitution method. Table 21 lists the observed frequencies and band assignments for some ethylenediamine complexes.

TABLE 21
Observed M—N stretching frequencies of metal complexes with ethylenediamine and related ligands (cm^{-1})

Structure	Compound	ν(M—N)
Monomeric	$Cu(en)_2Cl_2$	403.5, 326
Dimeric-octahedral	$Ni(en)_2Cl_2$	355.5, 308
		302 , 276
	$Ni(en)_2Br_2$	357 , 310
		295(sh), 275
Trans-octahedral	$Ni(en)_2(NCS)_2$	345 , 305
	$Ni(en)_2(NO_3)_2$	380, 292.5
Monomeric-planar	$[Ni(en)_2](AgI_2)_2$	554 , 525
	$[Cu(en)_2](AgI_2)_2$	414.5

Lever and Mantovani[102] also carried out an extensive study on the far-infrared spectra of the $M(L—L)X_2$ type complexes where L—L is the same as before, M is Ni(II) or Co(II) and X is Cl, Br, I, NO_3, SCN, ClO_4 or AgI_2. Since the structure of $Ni(en)_2X_2$ (X = Cl, Br or probably I) is *cis* dimeric, their far-infrared spectra are markedly different from that of monomeric $Cu(en)_2X_2$. As is seen in Table 21, the former exhibit four Ni—N stretching bands due to their low symmetry and no terminal Ni—X bands above 200 cm^{-1}. On the other hand, Ni(en)$_2$X$_2$ (X = NCS or probably NO$_3$) show simpler spectra because they are monomeric *trans*-octahedral(D_{2h}). Only two Ni—N stretching bands are identified for these complexes by the Ni isotope substitution. The M—N stretching frequencies of square-planar [M(en)$_2$]$^{2+}$ type complexes are much higher than those of octahedral complexes since the M—N bonds of the former are stronger than those of the latter. These authors presented detailed discussion on the variation of M—N stretching frequencies in terms of electronic and geometric factors.

8.7 Complexes of 2,2′-Bipyridine and 1,10-Phenanthroline

Because of the complicated structures of the ligands, the M—N stretching bands of metal complexes of 2,2′-bipyridine (bipy) and 1,10-phenanthroline (phen) were difficult to assign on an empirical basis. This difficulty was first overcome by Hutchinson et al.,[103] who studied the far-infrared spectra of [M(bipy)$_3$]$^{2+}$ (M = Fe(II), Ni(II) or Zn(II)) and [M(phen)$_3$]$^{2+}$ (M = Ni(II) or Zn(II)) by using the metal isotope technique. It was found that all these complexes exhibit two M—N stretching bands in the following regions: 360 to 375, Fe(II); 240 to 300, Ni(II); 175 to 240 cm^{-1}, Zn(II).

Saito et al.[104] extended this work to bipy and phen complexes with many other transition metals in various oxidation states. The final results tabulated in Table 22 for tris-bipy complexes reveal several interesting relationships between the electronic structure of the metal ion and the M—N stretching frequency. (1) In terms of a simple MO theory, Cr(III), Cr(II), Cr(I), Cr(0), V(II), V(0), Ti(0), Ti(−I), Fe(III), Fe(II) and Co(III) have filled or partly filled t_{2g} (bonding) orbitals and empty e_g (antibonding) orbitals. The M—N stretching frequencies of these metals (group A) are in the 300 to 390 cm^{-1} region. (2) On the other hand, Co(II), Co(I), Co(0), Mn(II), Mn(0), Mn(−I), Ni(II), Cu(II) and Zn(II) have filled or partly filled e_g orbitals. The M—N stretching frequencies of these metals (group B) are in the 180 to 290 cm^{-1} region. (3) Thus, no marked changes in frequencies are seen in the Cr(III) ∼ Cr(0) and Co(II) ∼ Co(0) series, although a dramatic decrease in frequencies is observed in going from Co(III) to Co(II). (4) The fact that the M—N stretching frequencies do not change appreciably in the Cr(III) ∼ Cr(0) or Co(II) ∼ Co(0) series indicates that the M—N bond strength remains approximately the same in these series. This result suggests that as the oxidation state is lowered, increasing numbers of electrons of the metal would reside in essentially ligand orbitals which do not affect the metal–ligand bond strength.

The far-infrared spectra of SnX$_4$(bipy) and SnX$_4$(phen) (X = Cl, Br or I) have been studied by using Sn isotopes.[105] These complexes exhibit the Sn—N stretching bands in the region between 210 and 150 cm^{-1}. The number of Sn—X stretching bands observed were four to two depending upon the halogen.

TABLE 22
M—N stretching frequency and electronic structure in the
$[M(bipy)_3]^{n+}$ type compounds $(cm^{-1})^a$

	-1	0	I	II	III
d^3				V 374 / 335 (3.67) $(t_{2g})^3$	Cr 385 / 349 (3.78) $(t_{2g})^3$
d^4		Ti 374 / 339 (0) $(t_{2g})^4$-ls		Cr 351 / 343 (2.9) $(t_{2g})^4$-ls	
d^5	Ti 365 / 322 (1.74) $(t_{2g})^5$-ls	V 371 / 343 (1.68) $(t_{2g})^5$-ls	Cr 371 / 343 (2.0) $(t_{2g})^5$-ls	Mn 224 / 191 (5.95) $(t_{2g})^3(e_g)^2$-hs	Fe 384 / 367 (?)
d^6		Cr 382 / 308 (0) $(t_{2g})^6$		Fe 386 / 376 (0) $(t_{2g})^6$	Co 378^b / 370 (0) $(t_{2g})^6$
d^7		Mn 258 / 227 (4.10) $(t_{2g})^5(e_g)^2$		Co 266 / 228 (4.85) $(t_{2g})^5(e_g)^2$	
d^8	Mn 235 / 184 (3.71) $(t_{2g})^6(e_g)^2$		Co 244 / 194 (3.3) $(t_{2g})^6(e_g)^2$	Ni 282 / 258 (3.10) $(t_{2g})^6(e_g)^2$	
d^9		Co 280 / 257 (2.23) $(t_{2g})^6(e_g)^3$		Cu 291 / 268 (?) $(t_{2g})^6(e_g)^3$	
d^{10}				Zn 230 / 184 (0) $(t_{2g})^6(e_g)^4$	

[a] The numbers at upper right of each group indicate the M—N stretching frequencies (cm^{-1}). The number in parentheses gives the observed magnetic moment in Bohr magnetons. ls = low spin; hs = high spin.
[b] Values for $[Co(phen)_3](ClO_4)_3$.

Takemoto[106] studied the far-infrared spectra of the Ni(II) and Ni(IV) complexes of 2,6-diacetylpyridine dioxime.

$$H_3C \underset{\underset{HO}{\overset{\|}{N_2}}}{\overset{}{C}} \underset{N_1}{\overset{}{}} \underset{\underset{N_2}{\overset{\|}{}}}{\overset{}{C}} CH_3 \qquad (DAPDH)$$

In $[Ni(DAPDH)_2](ClO_4)$, the $Ni—N_2$ stretching bands are at about 230 cm^{-1} and the $Ni—N_1$ stretching band is at about 276 cm^{-1}. In $[(Ni(DAPD)_2]$ where the Ni is in the $+IV$ state, the $Ni—N_2$ stretching bands are located at 509.8 and 472.0 cm^{-1} and the $Ni—N_1$ stretching band is at 394.8 cm^{-1}. These high frequency shifts in going from $Ni(II)(d^8)$ to $Ni(IV)$ (d^6, diamagnetic) are in complete agreement with the results of previous investigations on tris-bipy complexes.[104]

Hutchinson and Sunderland[107] determined the M—N stretching frequencies of tris-(2,7-dimethyl-1,8-naphthyridine) complexes by using the Ni and Zn isotopes.

$$H_3C \underset{M}{\overset{N \quad N}{}} CH_3$$

It was found that the M—N stretching frequencies of the former complexes are lower than those of the corresponding tris-bipy complexes by 16 to 24%. This lowering was attributed to the weakening of the M—N bond due to the strain in the four membered ring of the former complex. Hutchinson et al.[108] also assigned the M—N stretching bands of $[M(1,8\text{-naphthyridine})_4]$ (M = Ni, Cu or Zn) at 252, 269 and 221 cm^{-1}, respectively.

Wilde and Srinivasan[109] studied the infrared and Raman spectra of $M(bipy)X_2$ and $M(phen)X_2$ where M = Mn(II), Ni(II), Cu(II) or Zn(II) and X = Cl or Br by using the metal isotope technique. The M—N stretching frequencies of the $M(phen)Cl_2$ complexes are similar to those of the corresponding tris-(phen) complexes.

Takemoto and Hutchinson[110] used the metal isotope technique to study the magnetic spin crossover of $[Fe(phen)_2(NCS)_2]$. In the high spin state (298 K), the complex exhibits two metal isotope sensitive bands at 252 and 220 cm^{-1} which are assigned to the Fe—NCS and Fe—N(phen) stretching modes, respectively. By cooling the complex to ~ 100 K (low spin state) the complex exhibits two Fe—N stretching bands at 379 and 371 cm^{-1} in agreement with previous investigations.[104] In addition, it exhibits two metal isotope sensitive bands at 532.6 and 528.5 cm^{-1} which are assigned to the Fe—NCS stretching modes. These large shifts to higher frequencies in going from the high to low spin state have been attributed to the strengthening of the Fe—N and Fe—NCS bonds, this conclusion being supported by X-ray studies. Takemoto and

TABLE 23
Skeletal stretching frequencies of Fe(phen)$_2$X$_2$ and Fe(bipy)$_2$X$_2$ type complexes in high and low spin states (cm^{-1})

		Fe(phen)$_2$(NCS)$_2$	Fe(phen)$_2$(NCSe)$_2$	Fe(bipy)$_2$(NCS)$_2$
High spin	ν(Fe—NCS (or Se))	252.0	228.0	253.0
	ν(Fe—N)	220.0	218	(hidden)
Low spin	ν(Fe—NCS (or Se))	$\begin{cases} 532.6 \\ 528.5 \end{cases}$	$\begin{cases} 530.5 \\ 527.0 \end{cases}$	$\begin{cases} 498.3 \\ 492.0 \end{cases}$
	ν(Fe—N)	$\begin{cases} 379.0 \\ 371.0 \end{cases}$	$\begin{cases} 366.0 \\ 360.0 \end{cases}$	$\begin{cases} 393.0 \\ 374.7 \end{cases}$

Hutchinson[111] extended their work to other Fe(phen)X$_2$ and Fe(bipy)X$_2$ type complexes. Table 23 summarizes their results on skeletal stretching frequencies. Takemoto et al.[112] also assigned skeletal frequencies of high spin complexes such as Fe(phen)$_2$X$_2$ (X = Cl, Br, N$_3$, NCO or OAc) and of low spin complexes such as Fe(phen)$_2$X$_2$ (X = CNO or CN) by using the metal isotope technique. For the high spin complexes, the Fe—N stretching bands are at ~ 225 cm^{-1} and the Fe—X stretching bands at 180 to 325 cm^{-1}. For the low spin complexes, however, the corresponding bands are at much higher frequencies; Fe—N stretch, 350 to 395 cm^{-1} and Fe—X stretch, 520 to 540 cm^{-1}.

8.8 Complexes of Amino Acids

The infrared spectra of amino acid complexes have been studied extensively by many investigators. However, the assignments of the M—N and M—O stretching bands has been a subject of controversy in the past ten years. Rayner Canham and Lever[113] carried out the ^{63}Cu/^{65}Cu substitution on cis-Cu(gly)$_2$ · H$_2$O, and assigned the bands at 379.5 ($\Delta\nu$ = 1.5) and 334.0 ($\Delta\nu$ = 2.5) cm^{-1} to the Cu—N and Cu—O stretching modes, respectively.

Kincaid and Nakamoto[114] carried out extensive i.r. and Raman studies on trans-Ni(gly)$_2$·2H$_2$O and trans-Cu(gly)$_2$·2H$_2$O by using frequency shift data due to the ^{58}Ni/^{62}Ni; ^{63}Cu/^{65}Cu, ^{14}N/^{15}N and H/D substitutions. Table 24

TABLE 24
Metal–ligand vibrations of glycino complexes (cm^{-1})

trans-Ni(gly)$_2$·2H$_2$O		trans-Cu(gly)$_2$·2H$_2$O		
IR	R	IR	R	Assignment
442.0 (2.1)^a	441 (6.1)	483 (0.8)	462	ν(M—N)
341.5 (3.5)	282 (3.8)	377.5 (1.7)	—	ring deformation
289.0 (5.0)	227 (—)	337 (4.0)	230	ν(M—O)

^{a} The number in parentheses indicates the metal isotope shift.

summarizes their results. Their assignments are in good agreement with previous assignments by Condrate and Nakamoto[115] and are also supported by normal coordinate analysis.

8.9   Complexes of β-Diketones

The infrared spectra of acetylacetonate (acac) complexes have been studied extensively by many investigators, the M—O stretching bands being assigned mainly on the basis of normal coordinate analyses. Recently, the direct experimental evidence to support these assignments was provided by using the metal isotope technique.[116] Table 25 lists the M—O stretching frequencies. Each

TABLE 25
Metal isotope sensitive bands of acac complexes (cm^{-1})

	$Fe(acac)_3$	$Cr(acac)_3$	$Cu(acac)_2$	$Pd(acac)_2$
	$^{54}Fe/^{57}Fe$	$^{50}Cr/^{53}Cr$	$^{63}Cu/^{65}Cu$	$^{104}Pd/^{110}Pd$
A	436.0 (1.5)^a	463.4 (3.0)	455 (1.0)	677.8 (1.3)
				466.8 (1.5)
B	300.5 (5.0)	358.4 (3.9)	290.5 (3.5)	297 (3.6)
				266 (2.9)

^{a} Numbers in parentheses indicate isotopic shifts.

complex exhibits metal isotope sensitive bands in two frequency regions. The higher frequency bands (A) show small isotope shifts due to vibrational coupling with the C—CH$_3$ bending mode. The lower frequency bands (B), on the other hand, are very sensitive to the metal isotope substitution and are almost pure metal–oxygen stretching modes. The Pd(II) complex is unusual in that it exhibits four metal isotope sensitive bands. This indicates the presence of extensive vibrational coupling between metal–ligand and ligand vibrations in this complex.

Recently, Nakamura and Nakamoto[117] studied the infrared spectra of M[PtCl(acac)$_2$]$_n$ [M = VO(II), Co(II), Ni(II), Cu(II), Zn(II), Pd(II) for $n = 2$ and M = Fe(III) for $n = 3$] by using the metal isotope technique.

The M$\cdots$O stretching bands of ring B were located at the following frequencies (cm^{-1}):

$$VO(II) > Pd(II) > Cu(II) > Ni(II) > Co(II) > Zn(II)$$

VO(II)	Pd(II)	Cu(II)	Ni(II)	Co(II)	Zn(II)
355	309	283	279	270	233
	297	269	266	264	

In general, the C=O and C—C stretching frequencies of ring B are higher and lower, respectively, than those of M(acac)$_2$ (ring A) for the same metal. This result combined with the trend observed for the M—O stretching frequencies suggests that ring B has strong keto character.

8.10   Complexes of S-containing Ligands

The far-infrared spectrum of bis-(N,N-dimethyl-2-mercaptoethylamine)nickel exhibits two bands at 397 and 371 cm^{-1} which give large metal isotope shifts relative to other bands.[118] These are reasonably assigned to the Ni—S and Ni—N stretching modes, respectively, in agreement with its *trans*-planar coordination.

Ni(MMEA)$_2$ Ni(MEA)$_2$

On the other hand, the spectrum of bis-(mercaptoethylamine)nickel is more complicated than that of Ni(MMEA)$_2$. It exhibits at least seven bands in the 290 to 560 cm^{-1} region which give isotopic shifts of more than 1.7 cm^{-1} by the ^{58}Ni/^{62}Ni substitution. Based on their intensities and the magnitude of shifts due to deuteration of the NH$_2$ hydrogens, two bands at 377 and 336 cm^{-1} were assigned to the Ni—N stretching modes. Two Ni—S stretching bands were assigned at 377 cm^{-1} (overlapped by the Ni—N stretching band) and 330 cm^{-1} from the comparison of its spectrum with that of Ni(MPA)$_2$ (MPA: 3-mercapto-propylamine). These results suggest the *cis* structure for Ni(MEA)$_2$ and Ni(MPA)$_2$. The structure of Pd(MEA)$_2$ is also *cis*-planar since its spectrum is similar to that of the Ni complexes.

Jayasooriya and Powell[119] have found that the 377 cm^{-1} band of Ni(MEA)$_2$ with Ni in natural abundance splits into two peaks due to the presence of ^{58}Ni (67.8%) and ^{60}Ni (26.2%) when the spectrum is measured at liquid nitrogen temperature under high resolution. The observed separation between the peaks of 3 cm^{-1} is about half of that observed for the corresponding band of ^{58}Ni and

[62]Ni substituted complexes (5–9 cm^{-1}). The relative intensity ratio of these peaks is about 7:3 with the more intense component at the higher frequency. These results suggest that in favourable cases the metal isotope splitting can be observed for complexes containing metals in natural abundance if the spectra are measured at low temperatures.

Schläpfer et al.[120] also studied the infrared spectra of octahedral Ni(II) halide complexes containing nitrogen and sulphur donors. The complexes studied are of the trans-Ni(L—L')$_2$X$_2$ type where L—L' is 2,5-dithiahexane(dth), 2-(ethylthio)ethylamine (ete.), 2-(methylthio)ethylamine(mte) or asym-N,N'-dimethylethylenediamine (adimen), and X is Cl, Br, I or NO$_3$. The Ni—N, Ni—S and Ni—Cl stretching frequencies occur at 360–400, 210–260 and 210–260 cm^{-1}, respectively, based on observed shifts due to metal isotope ([58]Ni/[62]Ni) substitution, chlorine isotope ([35]Cl/[37]Cl) substitution, ligand deuteration (NH$_2$/ND$_2$ and SCH$_3$/SCD$_3$) and ligand substitution (Cl/Br/I/NO$_3$). Table 26 summarizes the results.

TABLE 26
Skeletal stretching frequencies of octa-hedral Ni chloride complexes containing N and S donors (cm^{-1})

	ν(Ni—N)	ν(Ni—S)	ν(Ni—Cl)
Ni(dth)$_2$Cl$_2$	—	$\begin{cases} 232 \\ 209.6 \end{cases}$	266.7
Ni(ete)$_2$Cl$_2$	405.8	263.5	212.2
Ni(adimen)$_2$Cl$_2$	$\begin{cases} 378.1 \\ 360.0 \end{cases}$	—	260.7

Dithiolato complexes[121] of the type [Ni(S$_2$C$_2$R$_2$)$_2$]^x where R is H, C$_6$H$_5$, CF$_3$ or CN and $x = 0$, 1− or 2− have been studied by using the metal isotope technique. Normal coordinate analyses have been made on these complexes to calculate the force constants. The effect of changing the charge x of the complex and the substituent R on the stretching force constants have been discussed in terms of the Hückel MO theory.

8.11 Metalloporphyrins

Bürger et al.[122] carried out the [64]Zn/[68]Zn substitution for Zn(OEP) (OEP = octaethylporphin) and found that the 203 cm^{-1} band of [64]Zn(OEP) shows the largest isotopic shift among the bands below 500 cm^{-1}. In the case of Ni(OEP), they found that the 360 cm^{-1} band shows the largest shift by the [58]Ni/[62]Ni substitution.

Ogoshi et al.[123] carried out normal coordinate analysis on the planar D_{4h} model of Zn–porphin by using the [64]Zn–[68]Zn isotopic shift data. As is seen in

Table 27, the 202.8 cm^{-1} band of Zn–porphin gives the largest shift and is assigned to a Zn–N stretching coupled with a CCN bending mode. This is in good agreement with Bürger et al. In the case of Ni–porphin strong coupling exists between two bands at 360 and 290 cm^{-1}. Thus, the bands near 360 cm^{-1} of Ni–porphin could be more sensitive to the metal isotope substitution than the band at 290 cm^{-1}.

TABLE 27
Observed frequencies and band assignments for metal porphin complexes (cm^{-1})

Zn-porphin	Cu-porphin	Ni-porphin	Assignment
385.0 (w) (0.3)^a	392 (m)	420 (s)	δ(CCN)
348.0 (m) (0.5)^a	346 (s)	366 (m)	δ(CCN)+ν(CC, CN)
		356 (s)	+ν(M—N)
202.8 (w) (3.5)^a	246 (w)	290 (m)	ν(M—N)+δ(CCN)
182.0 (m) (0)^a	234 (m)	282 (m)	π (ring)
167.0 (w) (0)^a	223 (w)	243 (m)	π (ring)

^{a} Isotopic shifts due to the ^{64}Zn/^{68}Zn substitution.

Ogoshi et al.[124] also studied the vibrational spectra of Fe(OEP)X, where X = F, Cl, Br, I, N$_3$ or NCS, by using ^{54}Fe and ^{56}Fe isotopes. Most of these high spin Fe(III) complexes exhibit two Fe—N stretching bands in the region from 250 to 280 cm^{-1}, suggesting a square-pyramidal structure for the FeN$_4$X core. On the other hand, low-spin octahedral [Fe(OEP)L$_2$]ClO$_4$ (L = γ-picoline, imidazole or benzimidazole) complexes exhibit one Fe—N stretching band in the 294–319 cm^{-1} region. This result favours the $trans$-octahedral structure in which the FeN$_4$ core is coplanar with the porphin core, thus strengthening the Fe—N bonds through π-electron conjugation.

The spectra of tetraphenylporphin(TPP) complexes are much more complicated than those of OEP complexes. Recently, Kincaid and Nakamoto[125] located two (sometimes three) bands which shift markedly on changing the metal atom and give relatively large shifts when metal isotope substitution is performed:

Pd(II)   Ni(II)   Co(II)   Cu(II)   Ag(II)   Zn(II)

302 < 322(2.5) > 313 > 268(1.5) > 245 ≈ 245(0.5)

218 < 255(0.7) > 252 > 219(1.0) > 200 ≈ 203(3.0)

207(0.5) > 188 ≈ 188(1.0)

(The numbers in parentheses indicate metal isotope shifts due to ^{58}Ni/^{62}Ni, ^{63}Cu/^{65}Cu or ^{64}Zn/^{68}Zn substitution.)

It is seen that the relative magnitude of isotopic shifts between the two bands is reversed in going from Zn(II) to Ni(II). The same trend was noted previously for porphin and OEP complexes.

8.12   Complexes of Olefins and Metal Sandwich Compounds

The metal isotope technique has not been applied to complexes of olefins mainly because olefin complexes with transition metals such as Cr, Fe and Ni etc. are not common. Shobatake and Nakamoto[126] studied the infrared spectra of $[^{104}Pd(\pi C_3H_5)Cl_2]$ and its ^{110}Pd analogue, and assigned the bands at about 400 and 370 cm^{-1} to Pd–allyl stretching fundamentals. The Pd—Cl (bridging) and Pd—Br (bridging) stretching bands were located at about 240 to 260 and 170 to 190 cm^{-1}, respectively.

Nakamoto et al.[127] obtained the infrared spectra of $^{54}Fe(C_5H_5)_2$ and its ^{57}Fe analog. It has been found that three bands at 790.5, 497.0 and 482.0 cm^{-1} give shifts of 6.9, 6.9 and 8.0 cm^{-1}, respectively, by the $^{54}Fe/^{57}Fe$ substitution. This result confirms previous assignments[128] that the bands are due to $\nu_3 + \nu_1$, ν_5 and ν_3, respectively, where ν_1 is symmetric metal ring stretch (Raman active), ν_3 is asymmetric metal ring stretch and ν_5 asymmetric ring tilt.

9   CONCLUSIONS

We have reviewed the applications of the metal isotope effect to theoretical and inorganic vibrational spectroscopy and shown that metal isotope data are extremely useful and almost indispensable in refining force and other molecular constants and in making empirical assignments of metal–ligand vibrations. It is anticipated that the use of metal isotope data will become more routine in the near future.

It should be noted, however, that this method has its own limitations. First of all, pure and stable metal isotopes are available only for certain elements. Thus, the complexes of Co and Mn for which stable isotopes are not available cannot be studied by this method. Second, this method may not be applicable to very heavy metals such as Pt and Os since observed shifts might be too small for these isotopes. Finally, some compounds are difficult to prepare on a milligram scale. This is particularly true if a compound is gaseous or liquid at room temperature. This shortcoming can partly be remedied by measuring the spectrum of a compound containing a metal atom in natural abundance in an inert gas matrix (see p. 191). Even this method cannot provide isotopic frequencies of $SnCl_4$ since it is a mixture of fifty isotopic molecules. Individual isotope peaks can only be observed by measuring the spectrum of a partly isotopically pure compound such as $^{116}SnCl_4$ in an inert gas matrix. Thus, a combination of the metal isotope technique with the matrix isolation spectroscopy is necessary in this case.

ACKNOWLEDGMENT

The authors express their sincere thanks to the Deutsche Forschungs-gemeinschaft, the Fonds der Chemischen Industrie and NATO (Scientific Affairs

Division) for their financial support. One of the authors (K.N.) would like to thank the Alexander von Humboldt Stiftung for making it possible for him to stay in Germany for six months, during which time this manuscript was written.

REFERENCES

(1) K. Nakamoto, K. Shobatake and B. Hutchinson, *Chem. Commun.* 1451 (1969).

(2) K. Nakamoto, *Angew. Chem.* **84**, 755 (1972); *ibid. Int. Ed. Engl.* **11**, 666 (1972).

(3) A. Müller, K. H. Schmidt and N. Mohan, *J. Chem. Phys.* **57**, 1752 (1972).

(4) A. Müller, N. Mohan and F. Königer, *J. Mol. Struct.* (in press).

(5) E. B. Wilson, Jr., J. C. Decius and P. C. Cross, *Molecular Vibrations*, McGraw-Hill, New York, 1955, p. 188.

(6) I. M. Mills, *J. Mol. Spectrosc.* **5**, 334 (1960); Erratum: *ibid.* **17**, 164 (1965).

(7) M. Tsuboi, *J. Mol. Spectrosc.* **19**, 4 (1966); T. Shimanouchi, I. Nakagawa, J. Hiraishi and M. Ishi, *J. Mol. Spectrosc.* **19**, 78 (1966).

(8) T. Miyazawa, *J. Mol. Spectrosc.* **13**, 321 (1964); T. Miyazawa and J. Overend, *Bull. Chem. Soc. Japan* **39**, 1410 (1966).

(9) G. Strey, *J. Mol. Spectrosc.* **17**, 265 (1965); *ibid.* **19**, 229 (1966).

(10) Ch. V. S. Ramachandra Rao, *J. Mol. Spectrosc.* **41**, 105 (1972).

(11) B. Jordanov and B. Nikalowa, *J. Mol. Struct.* **13**, 21 (1972).

(12) J. K. G. Watson, *J. Mol. Spectrosc.* **48**, 479 (1973).

(13) A. Alix, N. Mohan, A. Müller and S. N. Rai, *Z. Naturforsch.* **28a**, 1408 (1973).

(14) A. A. Chalmers and D. C. McKean, *Spectrochim. Acta* **22A**, 251 (1966); D. C. McKean, *Spectrochim. Acta* **22A**, 267 (1966).

(15) S. J. Cyvin, *Molecular Vibrations and Mean Square Amplitudes*, Elsevier, Amsterdam. 1968, p. 85.

(16) F. Königer and A. Müller, *J. Mol. Spectrosc.* **56**, 200 (1975).

(17) A. Müller and N. Mohan, *J. Chem. Phys.* **58**, 2994 (1973).

(18) S. J. Cyvin, B. N. Cyvin and A. Müller, *J. Mol. Struct.* **4**, 341 (1969).

(19) A. Müller, *Z. Phys. Chem.* (*Leipzig*) **238**, 116 (1968); C. J. Peacock and A. Müller, *J. Mol. Spectrosc.* **26**, 454 (1968).

(20) I. M. Mills, *Spectrochim. Acta* **16**, 35 (1960); J. Aldous and I. M. Mills, *Spectrochim. Acta* **18**, 1073 (1962); **19**, 1567 (1963).

(21) J. L. Duncan and I. M. Mills, *Spectrochim. Acta* **20**, 523 (1964).

(22) T. Shimanouchi, in *Physical Chemistry*, Vol. 4 (H. Eyring, D. Henderson and W. Jost, Eds.), Academic Press, New York, 1970, p. 233.

(23) L. H. Jones, *Inorganic Vibrational Spectroscopy*, Vol. 1, Marcel Dekker Inc., New York, 1971.

(24) N. T. Kyong and D. C. Bystrov, *Optics and Spectrosc.* **35**, 613 (1973).

(25) D. Steele, *Theory of Vibrational Spectroscopy*, W. B. Saunders, London, Philadelphia, 1971, p. 87.

(26) R. S. McDowell, L. B. Asprey and L. C. Hoskins, *J. Chem. Phys.* **56**, 5172 (1972).

(27) R. S. McDowell and M. Goldblatt, *Inorg. Chem.* **10**, 625 (1971).

(28) R. S. McDowell and L. B. Asprey, *J. Chem. Phys.* **57**, 3062 (1972).

(29) N. Mohan and A. Müller, *J. Mol. Spectrosc.* **42**, 203 (1972).

(30) S. N. Thakur, *J. Mol. Struct.* **7**, 315 (1971).

(31) J. L. Duncan, A. Allan and D. C. McKean, *Mol. Phys.* **18**, 289 (1970); J. L. Duncan, D. C. McKean and G. K. Speirs, *Mol. Phys.* **24**, 553 (1972).

(32) K. H. Schmidt and A. Müller, *J. Mol. Spectrosc.* **50**, 115 (1974).

(33) H. Bürger, S. Biedermann and A. Ruoff, *Spectrochim. Acta* **30A**, 1655 (1974).
(34) L. H. Jones, L. B. Asprey and R. R. Ryan, *J. Chem. Phys.* **47**, 3371 (1967).
(35) L. H. Jones, R. R. Ryan and L. B. Asprey, *J. Chem. Phys.* **49**, 581 (1968).
(36) J. Laane, L. H. Jones, R. R. Ryan and L. B. Asprey, *J. Mol. Spectrosc.* **30**, 485 (1969).
(37) H. J. Becher and A. Adrian, *J. Mol. Struct.* **7**, 323 (1971).
(38) A. Müller, N. Weinstock, N. Mohan, C. W. Schläpfer and K. Nakamoto, *Z. Naturforsch.* **27a**, 542 (1972); *Appl. Spectrosc.* **27**, 257 (1973).
(39) A. Müller, F. Königer and N. Weinstock, *Spectrochim. Acta* **30A**, 641 (1974).
(40) F. Königer, A. Müller and K. Nakamoto, *Z. Naturforsch.* (in press).
(41) F. Königer, R. O. Carter and A. Müller (to be published).
(42) A. Müller, K. H. Schmidt and G. Vandrish, *Spectrochim. Acta* **30A**, 651 (1974); K. H. Schmidt and A. Müller, *J. Mol. Struct.* **22**, 343 (1974).
(43) A. Müller, N. Mohan, F. Königer and M. C. Chakravorti, *Spectrochim. Acta* **31A**, 107 (1975).
(44) A. J. P. Alix, H. H. Eysel, R. Kebabcioglu, N. Mohan, B. Jordanov and A. Müller, *J. Mol. Struct.* **27**, 1 (1975).
(45) B. L. Crawford, Jr and J. T. Edsall, *J. Chem. Phys.* **7**, 223 (1939).
(46) E.B. Wilson, Jr., *J. Chem. Phys.* **7**, 1047 (1939); **9**, 76 (1941).
(47) H. J. Becher and K. Ballein, *Z. Phys. Chem.* (*Frankfurt am Main*) **54**, 802 (1967); M. Pfeiffer, *Z. Phys. Chem.* (*Frankfurt am Main*) **61**, 253 (1968).
(48) A. Müller, N. Mohan, K. H. Schmidt and I. W. Levin, *Chem. Phys. Lett.* **15**, 127 (1972).
(49) K. H. Schmidt and A. Müller, *J. Mol. Struct.* **18**, 135 (1973).
(50) A. Müller and S. N. Rai, *J. Mol. Struct.* **24**, 59 (1975).
(51) W. A. Seth-Paul, *J. Mol. Struct.* **3**, 403 (1969).
(52) F. N. Masri and W. H. Fletcher, *J. Raman Spectrosc.* **1**, 221 (1973).
(53) R. J. H. Clark and D. M. Rippon, *J. Mol. Spectrosc.* **44**, 479 (1972).
(54) F. Königer, E. Diemann and A. Müller (to be published).
(55) D. M. Dennison, *Rev. Mod. Phys.* **12**, 175 (1940).
(56) R. Kebabcioglu, A. Müller, C. J. Peacock and L. Lange, *Z. Naturforsch.* **23a**, 703 (1968).
(57) H. Bürger and A. Ruoff, *Spectrochim. Acta* **24A**, 1863 (1968).
(58) A. Müller, F. Königer, K. Nakamoto and N. Ohkaku, *Spectrochim. Acta* **28A**, 1933 (1972).
(59) R. S. McDowell and L. B. Asprey, *J. Mol. Spectrosc.* **48**, 254 (1973).
(60) M. von Thiel, E. D. Becker and G. C. Pimentel, *J. Chem. Phys.* **27**, 95 (1957).
(61) M. J. Linevsky, *J. Chem. Phys.* **34**, 587 (1961).
(62) H. E. Hallam, Ed., *Vibrational Spectroscopy of Trapped Species*, John Wiley, New York, 1973.
(63) A. D. Cormier, J. D. Brown and K. Nakamoto, *Inorg. Chem.* **12**, 3011 (1973).
(64) P. Tarte and J. Preudhomme, *Spectrochim. Acta* **26A**, 2207 (1970).
(65) P. Tarte and M. L. Duyckarts, *Spectrochim. Acta* **28A**, 2029 (1972).
(66) M. L. Duyckarts and P. Tarte, *Spectrochim. Acta* **28A**, 2037 (1972).
(67) J. Preudhomme and P. Tarte, *Spectrochim. Acta* **28A**, 69 (1972).
(68) N. Weinstock, H. Schulze and A. Müller, *J. Chem. Phys.* **59**, 5063 (1973).
(69) J. Preudhomme and P. Tarte, *Spectrochim. Acta* **27A**, 1817 (1971).
(70) P. Tarte and J. Preudhomme, *Spectrochim. Acta* **29A**, 1301 (1973).
(71) J. Preudhomme, *Thèse*, Liège (Belgium), 1970.
(72) B. A. DeAngelis, V. G. Keramidas and W. B. White, *J. Solid State Chem.* **3**, 358 (1971).
(73) H. J. Becher, F. Friedrich and H. Willner, *Z. Anorg. Allg. Chem.* **395**, 134 (1973).
(74) A. Müller, N. Weinstock, K. H. Schmidt, K. Nakamoto and C. W. Schläpfer, *Spectrochim. Acta* **28A**, 2289 (1972).
(75) R. Mattes, F. Königer and A. Müller, *Z. Naturforsch.* **29b**, 58 (1974).
(76) K. Nakamoto, *Infrared Spectra of Inorganic and Coordination Compounds*, 2nd ed., John Wiley, 1970, p. 152.

(77) T. W. Swaddle, P. J. Craig and P. M. Boorman, *Spectrochim. Acta* **26A**, 1559 (1970).
(78) J. Takemoto and K. Nakamoto, *Chem. Commun.* 1017 (1970).
(79) K. Nakamoto, J. Takemoto and J. L. Chow, *Appl. Spectrosc.* **25**, 352 (1971).
(80) K. H. Schmidt and A. Müller (to be published).
(81) See D. M. Adams and R. E. Christopher, *Inorg. Chem.* **12**, 1609 (1973).
(82) A. Cormier, K. Nakamoto, P. Christophliemk and A. Müller, *Spectrochim. Acta* **30A** 1059 (1974).
(83) A. Cormier, K. Nakamoto, E. Ahlborn and A. Müller, *J. Mol. Struct.* **25**, 43 (1975).
(84) A. Cormier, K. Nakamoto, E. Ahlborn and A. Müller (to be published).
(85) A. Müller, H. H. Heinsen, K. Nakamoto, A. D. Cormier and N. Weinstock, *Spectrochim. Acta* **30A**, 1661 (1974).
(86) J. Kincaid, K. Nakamoto, J. A. Tiethof and D. W. Meek, *Spectrochim. Acta* **30A**, 2091 (1974).
(87) K. Shobatake and K. Nakamoto, *J. Am. Chem. Soc.* **92**, 3332 (1970).
(88) J. T. Wang, C. Udovich, K. Nakamoto, A. Quattrochi and J. R. Ferraro, *Inorg. Chem.* **9**, 2675 (1970).
(89) B. T. Kilbourn and H. M. Powell, *J. Chem. Soc.* A, 1688 (1970).
(90) J. R. Ferraro, K. Nakamoto, J. T. Wang and L. Lauer, *Chem. Commun.* 266 (1973).
(91) C. Udovich, J. Takemoto and K. Nakamoto, *J. Coord. Chem.* **1**, 89 (1971).
(92) K. Konya and K. Nakamoto, *Spectrochim. Acta* **29A**, 1965 (1973); R. J. H. Clark, in *Halogen Chemistry* (V. Gutmann, Ed.), Academic Press, London, 1967, p 85.
(93) P. T. Manoharan and M. T. Rogers, *J. Chem. Phys.* **53**, 1682 (1970).
(94) Y. Saito, M. Cordes and K. Nakamoto, *Spectrochim. Acta* **28A**, 1459 (1972).
(95) Y. Saito, C. W. Schläpfer, M. Cordes and K. Nakamoto, *Appl. Spectrosc.* **27**, 213 (1973).
(96) N. Ohkaku and K. Nakamoto, *Inorg. Chem.* **12**, 3440 (1973).
(97) A. B. P. Lever and B. S. Ramaswamy, *Can. J. Chem.* **51**, 1582 (1973).
(98) B. C. Cornilsen and K. Nakamoto, *J. Inorg. Nucl. Chem.* **36**, 2467 (1974).
(99) J. R. Ferraro, B. B. Murray and N. J. Wieckowicz, *J. Inorg. Nucl. Chem.* **34**, 231 (1972).
(100) N. Ohkaku and K. Nakamoto, *Inorg. Chem.* **10**, 798 (1971).
(101) G. W. Rayner Canham and A. B. P. Lever, *Can. J. Chem.* **50**, 3866 (1972).
(102) A. B. P. Lever and E. Mantovani, *Can. J. Chem.* **51**, 1567 (1973).
(103) B. Hutchinson, J. Takemoto and K. Nakamoto, *J. Am. Chem. Soc.* **92**, 3335 (1970).
(104) Y. Saito, J. Takemoto, B. Hutchinson and K. Nakamoto, *Inorg. Chem.* **11**, 2003 (1972).
(105) N. Ohkaku and K. Nakamoto, *Inorg. Chem.* **12**, 2446 (1972).
(106) J. Takemoto, *Inorg. Chem.* **12**, 949 (1973).
(107) B. Hutchinson and A. Sunderland, *Inorg. Chem.* **11**, 1948 (1972).
(108) B. Hutchinson, A. Sunderland, M. Neal and S. Olbricht, *Spectrochim. Acta* **29A**, 2001 (1973).
(109) R. Wilde, T. K. K. Srinivasan and S. N. Ghosh, *J. Inorg. Nucl. Chem.* **35**, 1017 (1973).
(110) J. Takemoto and B. Hutchinson, *Inorg. Nucl. Chem. Lett.* **8**, 769 (1972).
(111) J. Takemoto and B. Hutchinson, *Inorg. Nucl. Chem.* **12**, 705 (1973).
(112) J. Takemoto, B. Streusand and B. Hutchinson, *Spectrochim. Acta* **30A**, 827 (1974).
(113) G. W. Rayner Canham and A. B. P. Lever, *Spectrosc. Lett.* **6**, 109 (1973).
(114) J. Kincaid and K. Nakamoto, *Spectrochim. Acta* (in press).
(115) R. A. Condrate and K. Nakamoto, *J. Chem. Phys.* **42**, 2590 (1965).
(116) K. Nakamoto, C. Udovich and J. Takemoto, *J. Am. Chem. Soc.* **92**, 3973 (1970).
(117) Y. Nakamura and K. Nakamoto, *Inorg. Chem.* **14**, 63 (1975).
(118) C. W. Schläpfer and K. Nakamoto, *Inorg. Chim. Acta* **6**, 177 (1972).
(119) V. A. Jayasooriya and D. B. Powell, *Spectrosc. Lett.* **6**, 763 (1973).
(120) C. W. Schläpfer, Y. Saito and K. Nakamoto, *Inorg. Chim. Acta* **6**, 284 (1972).
(121) C. W. Schläpfer and K. Nakamoto, *Inorg. Chem.* **14**, 1338 (1975).
(122) H. Bürger, K. Burczyk and J. H. Fuhrhop, *Tetrahedron* **27**, 3257 (1971).

(123) H. Ogoshi, Y. Saito and K. Nakamoto, *J. Chem. Phys.* **57**, 4194 (1972).

(124) H. Ogoshi, E. Watanabe, Z. Yoshida, J. Kincaid and K. Nakamoto, *J. Am. Chem. Soc.* **95**, 2485 (1973).

(125) J. Kincaid and K. Nakamoto, *J. Inorg. Nucl. Chem.* **37**, 85 (1975).

(126) K. Shobatake and K. Nakamoto, *J. Am. Chem. Soc.* **92**, 3339 (1970).

(127) K. Nakamoto, C. Udovich, J. R. Ferraro and A. Quattrochi, *Appl. Spectrosc.* **24**, 606 (1970).

(128) E. R. Lippincott and R. D. Nelson, *Spectrochim. Acta* **10**, 307 (1958).

LIST OF ABBREVIATIONS

i.r. = infrared; R = Raman; s = symmetric; as = asymmetric; str = stretching; bend = bending; r = rocking; vs = very strong; s = strong; m = medium; w = weak; vw = very weak; sh = shoulder; NA = natural abundance.

APPENDIX I

Some stable metal isotopes available at Oak Ridge National Laboratory

Element	Atomic weight	Inventory form	Isotope	% Natural abundance	Purity (%)	1974 Price ($/mg)
Lithium	6.941	Li_2CO_3	^6Li	7.42	99	8.25/g
			^7Li	92.58	99.99	3.00/g
Magnesium	24.305	MgO	^{24}Mg	78.70	> 99	0.35
			^{25}Mg	10.13	> 90	2.50
			^{26}Mg	11.17	> 95	1.45
Potassium	39.102	KCl	^{39}K	93.10	> 99	0.80
			^{40}K	0.012	> 80	250.00
			^{41}K	6.88	> 95	5.00
Calcium	40.08	$CaCO_3$	^{40}Ca	96.97	99–99.9	0.25
			^{42}Ca	0.64	> 90	8.50
			^{43}Ca	0.145	> 80	41.00
			^{44}Ca	2.06	> 98	2.30
			^{48}Ca	0.185	> 95	26.00
Titanium	47.90	TiO_2	^{46}Ti	7.93	80–90	1.40
			^{47}Ti	7.28	75–90	1.35
			^{48}Ti	73.94	> 96.5	0.20
			^{49}Ti	5.51	75–90	2.30
			^{50}Ti	5.34	75–85	2.30
Vanadium	50.9414	V_2O_5	^{50}V	0.24	35–45	120.00
			^{51}V	99.76	> 99.95	0.65
Chromium	51.996	Cr_2O_3	^{50}Cr	4.31	> 95	2.60
			^{52}Cr	83.76	99–99.9	0.55
			^{53}Cr	9.55	> 95	0.90
			^{54}Cr	2.38	70–95	4.40
Iron	55.847	Fe_2O_3	^{54}Fe	5.82	> 90	1.50
			^{56}Fe	91.66	> 99.8	0.10
			^{57}Fe	2.19	> 90	3.30
			^{58}Fe	0.33	> 70	26.00
Nickel	58.71	Ni	^{58}Ni	67.88	98–99	0.15
			^{60}Ni	26.23	86–99.8	0.20
			^{61}Ni	1.19	85–95	4.20
			^{62}Ni	3.66	> 95	1.25
			^{64}Ni	1.08	> 95	5.90
Copper	63.546	CuO	^{63}Cu	69.09	> 98	0.20
			^{65}Cu	30.91	> 90	0.35

Element	Atomic weight	Inventory form	Isotope	% Natural abundance	Purity (%)	1974 Price ($/mg)
Zinc	65.37	ZnO	^{64}Zn	48.89	>99	0.35
			^{66}Zn	27.81	90–99	0.55
			^{67}Zn	4.11	75–95	3.70
			^{68}Zn	18.57	95–99	0.80
			^{70}Zn	0.62	65–75	14.25
Gallium	69.72	Ga_2O_3	^{69}Ga	60.4	>95	0.55
			^{71}Ga	39.6	>90	1.00
Germanium	72.59	GeO_2	^{70}Ge	20.52	80–90	0.30
			^{72}Ge	27.43	>85	0.30
			^{73}Ge	7.76	85–95	1.40
			^{74}Ge	36.54	>90	0.20
			^{76}Ge	7.76	70–85	2.20
Rubidium	85.4678	RbCl	^{85}Rb	72.15	>98	0.55
			^{87}Rb	27.85	>99	1.15
Strontium	87.62	$Sr(NO_3)_2$	^{84}Sr	0.56	60–80	27.00
			^{86}Sr	9.86	90–98	0.50
			^{87}Sr	7.02	>90	0.70
			^{88}Sr	82.56	>98	0.05
Zirconium	91.22	ZrO_2	^{90}Zr	51.46	90–99	0.25
			^{91}Zr	11.23	>85	1.05
			^{92}Zr	17.11	>95	0.95
			^{94}Zr	17.40	90–99	1.05
			^{96}Zr	2.80	75–90	9.00
Molybdenum	95.94	Mo	^{92}Mo	15.84	90–98	0.30
			^{94}Mo	9.04	85–95	0.50
			^{95}Mo	15.72	85–99	0.30
			^{96}Mo	16.53	85–97	0.25
			^{97}Mo	9.46	85–95	0.45
			^{100}Mo	9.63	80–99	0.50
Ruthenium	101.07	Ru	^{96}Ru	5.51	>90	10.00
			^{98}Ru	1.87	55–65	13.50
			^{99}Ru	12.72	>90	3.80
			^{100}Ru	12.62	>90	5.25
			^{101}Ru	17.07	>95	3.00
			^{102}Ru	31.61	>90	1.55
			^{104}Ru	18.58	90–99	2.60
Palladium	106.4	Pd	^{104}Pd	10.97	>85	4.25
			^{105}Pd	22.23	>80	2.25
			^{106}Pd	27.33	>90	1.95
			^{108}Pd	26.71	>95	1.60
			^{110}Pd	11.81	>95	3.65
Silver	107.868	Ag	^{107}Ag	51.35	>98	0.45
			^{109}Ag	48.65	>99	0.60
Cadmium	112.40	CdO	^{110}Cd	12.39	95–98	0.65
			^{111}Cd	12.75	>90	0.65
			^{112}Cd	24.07	>95	0.30
			^{113}Cd	12.26	>90	0.75
			^{114}Cd	28.86	>95	0.25
			^{116}Cd	7.58	>90	1.00

Element	Atomic weight	Inventory form	Isotope	% Natural abundance	Purity (%)	1974 Price ($/mg)
Indium	114.82	In_2O_3	^{113}In	4.28	85–98	12.00
			^{115}In	95.72	> 99.5	0.45
Tin	118.69	SnO_2	^{116}Sn	14.30	> 95	0.30
			^{117}Sn	7.61	> 75	0.50
			^{118}Sn	24.03	> 90	0.15
			^{119}Sn	8.58	> 75	0.35
			^{120}Sn	32.85	> 90	0.25
			^{122}Sn	4.72	85–95	0.65
			^{124}Sn	5.94	> 80	0.55
Antimony	121.75	Sb	^{121}Sb	57.25	> 95	1.55
			^{123}Sb	42.75	> 95	1.55
Tellurium	127.60	Te	^{123}Te	0.87	> 85	6.25
			^{124}Te	4.61	> 95	1.15
			^{125}Te	6.99	> 90	0.70
			^{126}Te	18.71	> 85	0.60
			^{128}Te	31.79	80–97	0.25
			^{130}Te	34.48	90–99	0.25
Barium	137.34	$Ba(NO_3)_2$	^{134}Ba	2.42	> 70	2.40
			^{135}Ba	6.59	> 90	1.90
			^{136}Ba	7.81	> 90	1.40
			^{137}Ba	11.32	80–90	1.05
			^{138}Ba	71.66	> 95	0.15
Lanthanum	138.9055	La_2O_3	^{139}La	99.911	> 99.95	1.00
Cerium	140.12	CeO_2	^{140}Ce	88.48	> 98	0.15
			^{142}Ce	11.07	> 80	1.40
Neodymium	144.24	Nd_2O_3	^{142}Nd	27.11	> 95	0.60
			^{143}Nd	12.17	> 85	0.90
			^{144}Nd	23.85	85–99	0.30
			^{145}Nd	8.30	> 85	1.25
			^{146}Nd	17.22	> 95	0.45
			^{148}Nd	5.73	> 90	1.15
			^{150}Nd	5.62	> 90	1.35
Samarium	150.4	Sm_2O_3	^{144}Sm	3.09	> 95	0.95
			^{148}Sm	11.24	95–99.9	0.40
			^{149}Sm	13.83	> 90	0.35
			^{150}Sm	7.44	85–99.7	0.50
			^{152}Sm	26.72	> 90	0.15
			^{154}Sm	22.71	> 96	0.15
Europium	151.96	Eu_2O_3	^{151}Eu	47.82	90–99	0.45
			^{153}Eu	52.18	90–99	0.50
Gadolinium	157.25	Gd_2O_3	^{154}Gd	2.15	60–75	3.10
			^{155}Gd	14.73	80–95	1.40
			^{156}Gd	20.47	80–97	0.90
			^{157}Gd	15.68	80–95	1.00
			^{158}Gd	24.87	90–99	0.55
			^{160}Gd	21.90	> 99.9	0.70

Element	Atomic weight	Inventory form	Isotope	% Natural abundance	Purity (%)	1974 Price ($/mg)
Dysprosium	162.50	Dy_2O_3	^{160}Dy	2.294	70–80	2.80
			^{161}Dy	18.88	> 90	0.40
			^{162}Dy	25.53	> 90	0.40
			^{163}Dy	24.97	> 90	0.35
			^{164}Dy	28.18	> 95	0.25
Erbium	167.26	Er_2O_3	^{166}Er	33.41	95–99	0.30
			^{167}Er	22.94	85–95	0.60
			^{168}Er	27.07	> 90	0.50
			^{170}Er	14.88	90–97	0.50
Ytterbium	173.04	Yb_2O_3	^{170}Yb	3.03	75–85	2.80
			^{171}Yb	14.31	85–95	0.45
			^{172}Yb	21.82	85–98	0.40
			^{173}Yb	16.13	75–95	0.40
			^{174}Yb	31.84	85–99	0.25
			^{176}Yb	12.73	85–99	0.35
Lutetium	174.97	Lu_2O_3	^{175}Lu	97.41	> 99.8	0.35
Hafnium	178.49	HfO_2	^{177}Hf	18.50	85–95	1.10
			^{178}Hf	27.14	80–96	0.50
			^{179}Hf	13.75	> 85	1.55
			^{180}Hf	35.24	90–99	0.50
Tungsten	183.85	WO_3	^{182}W	26.41	90–95	0.25
			^{183}W	14.4	> 70	0.60
			^{184}W	30.64	80–96	0.30
			^{186}W	28.41	> 96	0.35
Rhenium	186.2	Re	^{185}Re	37.07	> 85	0.65
Osmium	190.2	Os	^{186}Os	1.59	60–65	20.00
			^{189}Os	16.1	> 85	2.25
			^{190}Os	26.4	> 95	1.60
Iridium	192.22	Ir	^{191}Ir	37.3	85–90	30.00
			^{193}Ir	62.7	> 95	30.00
Platinum	195.09	Pt	^{194}Pt	32.9	40–60	0.40
			^{195}Pt	33.8	55–65	2.75
			^{196}Pt	25.3	60–70	2.00
Mercury	200.59	HgO	^{198}Hg	10.02	> 99	11.00
			^{199}Hg	16.84	80–90	6.50
			^{200}Hg	23.13	70–90	4.00
			^{201}Hg	13.22	> 95	12.75
			^{202}Hg	29.80	70–85	1.55
			^{204}Hg	6.85	> 80	10.00
Thallium	204.37	Tl_2O_3	^{203}Tl	29.50	> 90	0.30
			^{205}Tl	70.50	> 95	0.20
Lead	207.2	$Pb(NO_3)_2$	^{204}Pb	1.48	80–99	6.50
			^{206}Pb	23.6	88–99.9	0.60
			^{207}Pb	22.6	90–98	0.40
			^{208}Pb	52.3	99–99.9	0.20

Chapter 6

ABSOLUTE ABSORPTION INTENSITIES AS MEASURED IN THE GASEOUS PHASE

Derek Steele

Department of Chemistry, Royal Holloway College, Egham, Surrey TW20 0EX, England

1 INTRODUCTION

Any absorption band is characterized by three parameters—its frequency, its intensity and its shape. In infrared spectroscopy the frequency has been of prime importance throughout the history of the subject. It is the easiest quantity to measure, especially with spectrometer systems with resolutions similar to the band widths, and it is directly related to the separation between the energy levels of the molecular system. The last ten years have seen a breakthrough in the understanding of band contours.[1-4] The factors governing the intensities of absorption bands have been understood in a general sense since the early 1950s, but for a variety of reasons the ability to understand quantitatively and predict the intensities has eluded investigators in the field for many years. Despite the obvious importance of the field the subject fell into disfavour. It cannot be claimed that any breakthrough has been made which is as dramatic as that for band contours in liquids. Nevertheless the persistent attention of a few small groups of investigators, notably in the United States and Russia, has advanced the subject to the point where the errors and misconceptions of earlier years can be discarded and a firm basis for future studies now exists.

   In this review the aim will be to summarize the development of the subject as I see it today. A comprehensive review of the recent literature has been recently produced,[5] so no attempt will be made to duplicate that effort. The subject matter will be further restricted to gas phase studies, *cf.* earlier reviews by Gribov,[6] Overend[7] and Steele.[8] A comprehensive review on the use of intensities in the analysis of liquids has been published by Wexler.[9]

1.1 The Definition of a Band Intensity

   There are various possible definitions of a band intensity. Provided that the absorption of electromagnetic radiation is insufficient to upset the Boltzmann distribution between the energy levels then Lambert's Law holds rigorously. We can therefore write:

$$I^\nu = I_0^\nu \exp\left(-k^\nu l\right)$$

where I_0^ν and I^ν are the incident and transmitted intensities of radiation of frequency ν passing through a sample of path length l. It follows that for a given sample under specific conditions the absorbance at a given frequency, $\ln\left(I_0^\nu/I^\nu\right)$, is a constant. Provided that the absorbance is proportional to concentration, then Beer's extension to Lambert's Law can be invoked and we can define a molar absorbance, or absorption coefficient, equal to $\ln\left(I_0^\nu/I^\nu\right)/C_n l$ where C_n is the concentration of the species n responsible for the absorption. Of course absorbance is not a truly extensive property. Band widths, and more noticeably line widths, are pressure dependent. Also the distribution of intensity within a band is determined by a number of factors, the most important of which is usually molecular rotation. In fact, the Fourier transform of $\ln\left(I_0^\nu/I^\nu\right)/\nu$ with respect to the frequency shift from the band centre is equal to the expectation value of $\mu(0)\cdot\mu(t)^{(1)}$ where $\mu(t)$ is $\langle\psi'(t)|p|\psi''(t)\rangle$, p being the molecular dipole moment and t the time. Thus

$$\langle\mu(0)\cdot\mu(t)\rangle = \int \{\ln\left(I_0^\nu/I^\nu\right)/\nu\} \exp\left(-i2\pi(\nu-\nu_0)t\right) d\nu \tag{1}$$

It is obvious that a change of environment of the absorbing species will affect the ability of the molecule to rotate and hence alter the intensity distribution within the band, though it will not necessarily alter the moment $\mu(t)$. One might consider that the integral of the absorption coefficient over the band was a reasonable molecular parameter which might be accepted as the definition of band intensity. Indeed, throughout most of the spectroscopic literature, band intensities are expressed in terms of the integrated absorption coefficient. This is designated by the symbol A.

If we examine the expressions which can be derived for transition probabilities of a rovibrator from quantum mechanical arguments it will be seen that these are all proportional to the frequency multiplied by a function of the rotational quantum members. The expression for A, derived say for a linear molecule,[10] is relatively complex. If, however, the integration is over $\ln\left(I_0^\nu/I^\nu\right)/\nu$ then a considerable simplification is attained.[11] In accord with normal practice this modified integral will be designated

$$\Gamma = \int \ln\left(I_0^\nu/I^\nu\right) d\ln\nu \tag{2}$$

Two more convincing arguments can be put forward for the selection of Γ rather than A as the fundamental quantity. It has been shown[11] that in the case of resonance interaction between energy states the sum of the Γ is invariant. No such invariance exists for A. Isotopic sum rules exist for Γ which are exact within the harmonic approximations,[12] but for A the sum rules can only be approximated to the extent that $A_i \sim \Gamma_i\nu_i^0$ where ν_i^0 is the band centre. For an isotropic distribution of harmonic oscillators in which the electric dipole moment can be expressed by

$$p = p_0 + \sum_i (\partial p/\partial Q_i)Q_i \tag{3}$$

Q_i being the ith normal coordinate, then

$$\Gamma_i = \frac{N\pi g_i}{3c\nu_i} \left(\frac{\partial p}{\partial Q_i}\right)^2 \tag{4}$$

g_i is the degeneracy of the transition. The sum rules are readily obtained by use of well known rules relating the Q_i to the internal coordinate set $\{R_k\}$, to the inverse kinetic energy matrices, $\mathbf{G}$, and the force constant matrix, $\mathbf{F}$. The $\mathbf{F}$ rule states that

$$\sum_i \frac{\Gamma_i}{\nu_i} = \frac{N\pi}{3c} \sum_{jk} F_{jk}^{-1} \frac{\partial p}{\partial R_j} \frac{\partial p}{\partial R_k} \tag{5}$$

and the $\mathbf{G}$ rule

$$\sum_i \Gamma_i \nu_i = \frac{N\pi}{3c} \sum_{jk} G_{jk} \frac{\partial p}{\partial R_j} \frac{\partial p}{\partial R_k} \tag{6}$$

The $\mathbf{F}$ sums will be isotopically invariant in the Born–Oppenheimer approximation and the corresponding $\mathbf{G}$ sum changes can be readily computed if the symmetry species over which the summation is made does not include a rotation or if the molecule is non polar. When these conditions are not satisfied it is necessary to correct the $\partial p/\partial R_j$ for contributions due to the vibrational angular momentum.

Additional sum rules for intensities have been given by Decius.[13] These rules depend on the linear dependence of the $\mathbf{G}$ matrix elements on the reciprocal masses of the atoms and bear a strong analogy to the Decius–Wilson–Sverdlov[14,15] sum rules for vibrational frequencies. These intensity rules, as they stem from eqn. (6), are also strictly valid only for the Γ of harmonic oscillators.

In summary it appears that Γ is a more fundamental quantity than A. It has been argued however that the reporting of intensities in terms of A has certain advantages.

(i) Interest is in the derivatives $\partial p/\partial Q$ which are directly proportional to the square root of A. This is true only within the approximation $\Gamma \nu = A$. Comparative values of the derivatives are therefore apparent at a glance. The frequency dependence in the relation between Γ and $\partial p/\partial Q$ destroys this convenience.

(ii) A is directly related to the plot of absorbance against frequency. Since the usual spectral presentation is on a transmittance or absorbance scale it is certainly more convenient to report the A values.

(iii) Since the widths of vibrational bands are usually small compared with the frequency of the band centre it is usually an adequate approximation to take

$$A_i \simeq \Gamma_i \nu_i$$

to deduce a set of dipole parameters in terms of the internal coordinates. This problem has been discussed in detail and applied to the methyl halides.[17] The variance–covariance matrix (designated Σ) of the dipole derivative in terms of the internal valence coordinate R_j is related to the variance–covariance matrices of the L^{-1} (defined by $Q = L^{-1}R$), of the measured intensities and of the harmonic frequencies, ω, by the relations

$$\Sigma\left(\frac{\partial p}{\partial Q}\right) = C\omega\Sigma(\Gamma)\tilde{\omega}\tilde{C} + C\Gamma\Sigma(\omega)\tilde{\Gamma}\tilde{C} \tag{7}$$

where C is a diagonal matrix with entries $3c^2/N\Pi g_i$, g_i being the degeneracy of the vibrational state, and

$$\Sigma\left(\frac{\partial p}{\partial R_j}\right)_{mn} = \sum_{ik}\left\{(L^{-1})_{im}(L^{-1})_{kn}\Delta\left(\frac{\partial p}{\partial Q_i}\right)\Delta\left(\frac{\partial p}{\partial Q_k}\right)\right.$$

$$\left. + \left[\Delta(L^{-1})_{im}\Delta(L^{-1})_{kn}\left(\frac{\partial p}{\partial Q_i}\right)\left(\frac{\partial p}{\partial Q_k}\right)\right]\right\} \tag{8}$$

$\Delta(x)$ represents the uncertainty in x. Equation (8) can be written in matrix form as

$$\Sigma\left(\frac{\partial p}{\partial R_j}\right) = (\tilde{L}^{-1})\Sigma\left(\frac{\partial p}{\partial Q}\right)L^{-1} + Y\Sigma(L^{-1})\tilde{Y} \tag{9}$$

where Y is an $M \times (MN)$ matrix, M being the number of internal (or symmetry) coordinates and N the number of normal coordinates. A tilde ($\sim$) is used to denote a matrix transpose. The size of the matrices involved can render this analysis formidable. An alternative approach to computing the contribution to $\Sigma(\partial p/\partial R_j)$ arising from the uncertainty in the coordinate transformation [that is, the second term in eqn. (9)] was proposed. For small uncorrelated changes in G and F the changes in the elements of L are given by

$$L_{ij} - L_{ij}^{0} = \sum_{kl}\frac{\partial L_{ij}}{\partial F_{kl}}\Delta F_{kl} + \sum_{kl}\frac{\partial L_{ij}}{\partial G_{kl}}\Delta G_{kl} \tag{10}$$

It follows that an estimate to the maximum error in L_{ij} can be calculated by summing the moduli of all the terms in the right-hand side of eqn. (10). The best data available on intensities and force fields of CH_3X and CD_3X ($X = F$, Cl, Br or I) were analysed using the above methods. The results for CH_3F and CD_3F are shown in Table 2 for a given choice of signs for the $\partial p/\partial Q$. The $\partial p/\partial S_j$ are corrected for rotational angular momenta (which affect the e species results only), S_j being a set of symmetrized internal valence coordinates. The use of the more rigorous eqn. (9) leads to a statistical dispersion in $\partial p/\partial S_j$ denoted by $\sigma(p_j)$. Use of eqn. (10) gives $\sigma_{FG}(p_j)$. For CH_3F the results agree very well indeed, but for CD_3F there is a tendency, which could well be anticipated, for eqn. (10) to give a rather high dispersion. Nevertheless even for CD_3F the agreement between the two sets shows that the simpler method is quite satisfactory—especially when the difficulties in estimating the dispersions in the

TABLE 2
Dipole-moment derivatives p_j, their dispersions, $\sigma\{p_j\}$, and their approximate dispersions, $\sigma_{FG}\{p_j\}$, for one $\partial p/\partial Q$ sign choice. The approximate dispersions, $\sigma_{FG}\{p_j\}$, were obtained by using the approximate dispersions in L^{-1}, $\sigma_{FG}\{L^{-1}\}$, as the square root of the diagonal terms of $\Sigma\{L^{-1}\}$ and setting all off-diagonal terms to zero. Contributions to $\sigma\{p_j\}$ due to uncertainties in experimental intensities ($\sigma\{int\}$), harmonic frequencies ($\sigma\{freq\}$), force constants ($\sigma\{f.c.\}$), and molecular geometry ($\sigma\{geom\}$) are listed separately. All units are in debyes per ångstrom[a]

j		p_j	$\sigma\{p_j\}$	$\sigma_{FG}\{p_j\}$	$\sigma\{int\}$	$\sigma\{freq\}$	$\sigma\{f.c.\}$	$\sigma\{geom\}$
CH_3F	1	−0.5668	0.2435	0.2440	0.2435	0.0000	0.0000	0.0000
	2	0.1420	0.1670	0.1781	0.1666	0.0000	0.0000	0.0003
	3	4.3391	0.3879	0.3882	0.3850	0.0016	0.0011	0.0002
	$4_{a,b}$	−0.9073	0.0954	0.0954	0.0952	0.0000	0.0000	0.0001
	$5_{a,b}$	0.0983	0.0218	0.0229	0.0208	0.0000	0.0010	0.0000
	$6_{a,b}$	0.4180	0.0226	0.0188	0.0166	0.0000	0.0060	0.0001
CD_3F	1	−0.4810	0.0665	0.1287	0.0658	0.0001	0.0002	0.0004
	2	0.2416	0.1012	0.2368	0.0949	0.0004	0.0010	0.0049
	3	4.9407	0.3254	0.6403	0.3165	0.0015	0.0066	0.0009
	$4_{a,b}$	−1.0285	0.0555	0.0560	0.0554	0.0000	0.0000	0.0001
	$5_{a,b}$	0.0816	0.0756	0.0887	0.0756	0.0000	0.0000	0.0000
	$6_{a,b}$	0.4001	0.0172	0.0252	0.0170	0.0000	0.0002	0.0000

[a] Ref. 17.

force constants are appreciated. The contributions of uncertainties in the intensities, in the harmonic frequencies, in the force constants and in the geometry are listed separately. In this case it is seen that uncertainties are due almost entirely to the intensities. Some caution must be exercised in extending this result to the general case. The force field of CH_3F is known to a precision probably better than that of any other molecule with more than three atoms. Nevertheless several of the interaction force constants had large standard deviations (~ 20 N m^{-1}) in an unconstrained analysis. Use of Mills' hybrid orbital force field[18] was necessary to produce values with small dispersions. It was these constrained values which were used to produce the results shown in Table 2.

Having obtained what it is hoped are meaningful dipole derivatives in terms of a set of symmetry coordinates, the next stage is to reduce the results to a set of bond parameters. Here once again we are faced with formidable problems, resulting from too many parameters. It is well known[5−8] that the simple bond moment hypothesis does not hold even to a first approximation. This hypothesis requires:

(i) that stretching a bond by an amount dr produces a change of dipole moment along the bond of $(\partial\mu/\partial r)\,dr$;
(ii) the deformation of a bond through an angle $d\theta$ produces a dipole

change, $\mu \, d\theta$, perpendicular to the bond and in the plane of movement. μ is the bond dipole;

(iii) changes in one bond do not result in changes in other bonds except when this is geometrically necessary.

Relaxing these constraints while continuing to require a bond parameter formulation results in the introduction of terms of the type $\partial \mu_j / \partial R_k$. The number of such terms will usually far exceed the number of observables. For this reason models must be developed which are chemically reasonable, and which constrain the parameterization to a level intermediate between the simple bond moment hypothesis and the general formulation.

In the following sections we shall survey progress in coping with the various problems mentioned above.

2 EXPERIMENTAL METHODS

There have been few notable advances in techniques of measuring absorption intensities in the past ten years. The Wilson–Wells extrapolation procedure[19] and the curve of growth (C of G) method[20] remain as the standard methods of overcoming the distortion arising from finite slit widths. Unfortunately the C of G method is strictly applicable only to resolved rotational lines in the vibrational–rotational band for which the shape function of the lines can be described by a specific function. On the other hand it is in this situation where the Wilson–Wells method is less applicable. The history of measurements on the 670 cm^{-1} band of CO_2 is of interest in this context. Prior to 1956, measurements of the intensity of this band using the Wilson–Wells procedure gave values of 0.54 to 0.63 m^2 mol^{-1}.[21] Kaplan and Eggers using the C of G method obtained a significantly higher value[22] (0.807 m^2 mol^{-1}). The principal cause of this discrepancy is the narrowness of the absorption lines. Kaplan and Eggers, from their C of G results, estimated that the Q branch has a half band width of 0.3 cm^{-1} at 1.25 atm while the individual P and R branch lines have half widths of 0.080 cm^{-1} at this pressure. Their paper gives an excellent summary of the procedures involved. As the C of G method casts doubts on the validity of the Wilson–Wells method a very careful remeasurement of the band intensity was made using the Wilson–Wells method and employing pressures of 68 atm of broadening gas.[23] Calculations predicted that the broadening would be adequate at this pressure at the resolutions employed. The result of this investigation by Overend et al. was in excellent confirmation of the C of G result.

Not many intensity measurements have been made using the C of G method although, as pointed out by Margolis and Kwan, it has the advantage over direct integration methods in that errors are more easily estimated.[24] These authors recently measured the intensities of more than 200 lines in two combination bands of NH_3, $\nu_1 + \nu_2$ and $\nu_2 + \nu_3$, near 4300 cm^{-1}. From an analysis of the intensity distribution in the bands, including vibration–rotation interaction terms, they derived the integrated band intensities (A) to be 2.71 ± 0.03 cm^{-2} atm^{-1} and 19.00 ± 0.07 cm^{-2} atm^{-1}. These values are equal, within the uncertainty limits, to the intensities determined by direct band integration of 2.9 ± 0.5 cm^{-2} atm^{-1} $(\nu_1+\nu_2)$, and 19.7 ± 2 cm^{-2} atm^{-1} $(\nu_2+\nu_3)$.[25] The considerable increase in the precision claimed for the C of G results is notable.

An alternative procedure for measuring band intensities has been described. Computer simulation of band contours is a common procedure employed to analyse vibration–rotation interaction constants,[26] as well as relative signs of dipole gradients.[27] At the same time the method can be used to derive the integrated band areas.[28] If the total transition intensity in a given narrow frequency interval is subject to some shape factor, such as the instrument function, then the sum of such intensity contributions might be expected to lead to areas of slowly changing absorption and areas of much more rapidly changing absorption. For symmetric top molecules of moderately large moments of inertia for instance, the Q branches are likely to be fairly sharp and difficult to measure with conventional spectrometers. The P and R branches by contrast usually form smooth broad unresolved bands. It follows that, by matching the calculated and observed intensities in these branches of slowly varying intensity, the total integrated intensities can be deduced. So far this method has been applied only to some of the fundamentals of benzene, d_6-benzene and hexafluorobenzene.[29]

With most modern commercial medium resolution spectrometers the resolution is limited by lack of energy. As a result, for most intensity measurements the spectral distortion is governed by the spectrometer slit function. In principle it is possible to correct the spectrum for this distortion by Fourier methods. If the true and apparent intensities at frequencies ν and ν' are given by $\phi(\nu)$ and $f(\nu')$ respectively, then

$$f(\nu') = \int_{-\infty}^{+\infty} a(\nu'-\nu)\phi(\nu)\,\mathrm{d}\nu \qquad (11)$$

The function $a(\nu'-\nu)$ is known as the instrument function and denotes the fraction of energy of frequency ν which is transmitted through the system to the detector when the instrument is set to record at a frequency ν'. If the Fourier transforms of f, a, and ϕ are denoted by F, A and Φ, then by the convolution theorem,[30] $F = A\cdot\Phi$. Unfortunately, to compute $\phi(\nu)$ from $f(\nu')$ the range of integration must be over several half-band widths and there must be no sharp discontinuities, such as would result from a sharp data cut-off. A number of methods have been developed to produce approximate corrections to the observed spectra. The method of Berger and Van Cittert[31] has the advantages of both simplicity and versatility. The essence of the method is that if the instrument function $a(\nu'-\nu)$ is known then an approximate correction is given by

$$\phi(\nu)-f(\nu) \simeq f(\nu) - \int f(\nu')\cdot a(\nu'-\nu)\,\mathrm{d}\nu'$$

If the approximation was perfect, then applying $a(v' - v)$ to the corrected intensity would regenerate $f(v)$. The error found from actually doing this operation can now be employed to produce a second order correction. The limitation to the procedure lies in the stability of the iteration operation. The method has been discussed in some detail.[32,33] A second method results from expanding the Fourier transform $A(t)$. The theory involved has been described in an excellent review article on instrumental distortions by Rautian.[34] By considering the specific case of a triangular slit function the following series expansion has been derived[35]

$$\phi(v) = f(v) - \frac{s^2}{12}\frac{d^2 f(v)}{dv^2} + \frac{s^4}{240}\frac{d^4 f(v)}{dv^4}$$

where s is the spectral slit width in frequency units ($=$ mechanical slit width $\times$ linear frequency dispersion at the exit slit). The first order correction term is in accord with a correction proposed a hundred years earlier by Lord Rayleigh, who had arrived at the result by simple curvature arguments.[36] The triangular slit function is the appropriate function for most energy limited medium resolution i.r. spectrometers. Computer simulation shows that for a single Lorentzian band with a half band width equal to the slit width, the second derivative accounts for 59% of the true correction. The correction rises to 84% as the band width increases to $2s$ and becomes virtually 100% as $\Delta v_{\frac{1}{2}} \rightarrow 4s$. The beauty of this method is that it is so simple to apply and band overlap makes it even more accurate as can be seen from Table 3.

A very novel and direct method of measuring integrated absorption intensities has been proposed and used.[37] Radiation in a specific restricted band of frequencies is passed through a gaseous sample and the energy incident is

TABLE 3
The effects of slit widths of 0.75 cm^{-1} on the measured intensities
(Γ) and on the maximum absorption coefficients (ε_{max}) of pairs of
bands. The half widths of the bands are taken as the same and equal
to $\Delta v_{\frac{1}{2}}$. The separation of the bands is $v_2 - v_1$. The percentage errors
in Γ and ε_{max} after application of the second derivative correction
are shown in the 'after' columns[a]

$(v_2 - v_1)$ cm^{-1}	$\Delta v_{1/2}$ cm^{-1}	$\ln(I_0/I_1)_{max}$	$\ln(I_0/I_2)_{max}$	% error in Γ before	% error in Γ after	% error in ε_{max} before	% error in ε_{max} after
1.5	2.0	1.0	1.0	1.01	0.16	2.7	0.3
2.0	2.0	1.0	1.0	0.93	0.18	5.7	1.1
2.0	2.0	1.0	0.1	1.01	0.19	7.9	1.0
2.0	3.0	1.0	1.0	0.65	0.17	1.2	0.06
2.0	1.0	1.0	1.0	2.92	0.90	27.6	9.2
2.0	0.5	1.0	1.0	8.92	5.80	55.0	40.2

[a] Ref. 35.

measured directly with a power meter. By measuring the pressure rise in the sample as a result of radiation absorption, power absorption and hence the integrated absorption intensity over the band width is deduced. Unfortunately, in view of the small amounts of power available per unit band width and the low transition probabilities involved in vibrational transitions it is necessary to use rather broad frequency-pass filters to limit the frequency range studied at any one time. This means that all absorptions in the selected frequency range are measured as a whole and all energy distribution information is lost. Nevertheless the results were of interest in that earlier integrated absorption values for methane, nitrous oxide and acetylene in the 3–5 μm range were confirmed.

A very great amount of the processing of raw transmission data to obtain intensities is of a repetitive and tedious nature. The transmission data are usually recorded on chart paper, which must then be measured, corrected for slit distortion, converted to absorbance after correction for background absorption of cell, and then integrated. Although it is becoming common practice to link spectrometers for various spectral ranges with computers, there has as yet been little evidence of using the enormous potential of such systems for intensity work. One system built specifically for this purpose has been described[38] and some measurements have been reported. The fast scanning interferometers, such as the Digilab FTS systems, offer an even greater potential for intensities. Since all the accessible frequency range is processed simultaneously, the results being in store, just one run, perhaps averaged over several hundred scans, can supply information on all bands in the range. The photometric accuracy claimed by the manufacturers is extremely high from very low to very high transmissions. Perhaps soon someone will test the system by producing Lambert Law plots for a series of bands with different characteristics.

3 THEORY

3.1 General Formulation of Theory

The intensity of absorption due to stimulated transitions between two states, whose wavefunctions are ψ' and ψ'', is given by

$$\Gamma = \frac{8\pi^3 Ng}{3hc} |\langle \psi'|p|\psi''\rangle|^2$$

On the assumptions of separability of the vibrational and rotational wavefunctions and of mechanically and electrically harmonic oscillators, this expression leads to eqn. (4). The gradient $\partial p/\partial Q$ is a molecular quantity which is both difficult to visualize and difficult to compute in its general form. Its decomposition into components which simplify visualization and calculation is the aim of the two theories which will now be discussed. Vibration–rotation interaction can be neglected except in the most precise work. Thus Wilson showed that the influence of the first order Coriolis coupling on the thermodynamic properties, and hence the intensities, of degenerate levels is negligible.[40]

To proceed further it is necessary to compute the form of Q. It is usual to carry out this calculation in a basis set consisting of symmetrized valence type coordinates, $\{S_j\}$. The transformations between the internal coordinate set $\{R_i\}$, the symmetrized set $\{S_j\}$ and the normal coordinate set $\{Q_k\}$ are defined by $S = UR$, $S = LQ$ and $R = LQ$. In the bond moment theory the dipole gradient is assumed to be formed from a linear sum of gradients with respect to the R_j, i.e.

$$\frac{\partial p}{\partial Q_k} = \sum_i (l^{-1})_{ki} \frac{\partial p}{\partial R_i} = \sum_j (l^{s,-1})_{ji} \frac{\partial p}{\partial S_j}$$

The early interpretations sought to interpret the gradients in terms of a set of localized bond moments μ_i and gradients $\partial \mu_i / \partial r_i$ (see section 1.2). As is well known these attempts were markedly unsuccessful,[7,8] though even now it is not clear how much of the difficulty is due to the inseparability of the moment $\partial p / \partial Q$ into bond localized components and how much is due to errors in computations and in the normal coordinates.[5] This statement will be exemplified later by reference to published data on F_2CO. Developments in the bond moment theory have sought to take into account cross terms of the form $\partial \mu_i / \partial R_j$. The most comprehensive formulation of the intensity problem is that of Gribov.[6,39]

3.2   Gribov Formulation of Bond Moment Theory

In the Gribov formulation the total molecular dipole is formally resolved into components along the bonds. The author stresses however that these components are the sums of dipole moments localized on and along the bonds with components of the residual dipole contributions due to off axis moments and lone pair electrons. The resolution of the latter contributions is rather arbitrary and generally the resolved components will not be centred on the bonds. Denoting a unit vector along the kth bond by e_k allows us to write

$$p = \sum_k \mu_k e_k \tag{12}$$

and differentiation leads to

$$\frac{\partial p}{\partial Q_i} = \left[\tilde{e} \left(\frac{\partial \mu : \partial \mu}{\partial r : \partial \gamma} \right) + \tilde{\mu} \left(\frac{\partial e : \partial e}{\partial r : \partial \gamma} \right) \right] \left(\frac{l_{ri}}{l_{\gamma i}} \right) \tag{13}$$

where e and μ are column vectors (unit vectors in the bond directions and bond moments, respectively, as defined by eqn. (12)). The internal coordinates R have been separated into two subsets, $\{r\}$ and $\{\gamma\}$, consisting of the stretching and angle bending components respectively. $(\partial \mu / \partial r : \partial \mu / \partial \gamma)$ denotes an array of derivatives of the bond moments with respect to the two subsets $\{r\}$ and $\{\gamma\}$, and $(\partial e / \partial r : \partial e / \partial \gamma)$ is the equivalent array for the bond vectors. $(l_{ri} : l_{\gamma i})$ is the *i*th column vector of the **L** matrix subdivided in the same manner. The bond reorientation contribution is shown to to be given by

$$\left(\frac{\partial e \vdots \partial e}{\partial r \vdots \partial \gamma}\right) = \mathbf{R}^{-1}\{\Delta \mathbf{M}^{-1}\tilde{\mathbf{B}}\mathbf{G}^{-1} - [\mathbf{E}\vdots\mathbf{O}]\} \tag{14}$$

where $\mathbf{R}^{-1}$ and $\mathbf{M}^{-1}$ are diagonal matrices of the reciprocal bond lengths and atomic masses, respectively. $\mathbf{B}$ is the transformation matrix between the internal coordinates and the cartesian nuclear displacement coordinates. $\mathbf{G}$ is the inverse kinetic energy matrix, $\mathbf{E}$ is a diagonal unit matrix of order equal to the number of bonds, N_R, while $\mathbf{O}$ is the null matrix required to make the partitioned matrix $[\mathbf{E}\vdots\mathbf{O}]$ of dimensions N_R by number of deformation coordinates. Δ has the dimensions of N_R by number of cartesian coordinates. The entries are zero with the following exceptions. The rows corresponding to a given bond deformation contain the entry $+1$ in the columns referring to the cartesian coordinates of the terminal atom and -1 in the corresponding columns for the initial atom.

If a given symmetry species contains the representation of a rotation as well as the representations of i.r.-active vibrations, then the intensity of absorption of vibrations of that species will contain contributions from the vibrational angular momentum. The earliest formulation of this problem[41,12] was far from being clear and proved extremely difficult to use. Gribov's equation includes the contribution due to vibrational angular momentum. This follows since the term in { } is the reorientation of the bond dipoles in a space-fixed axis system. The main limitation to the formulation is that it involves the computation of $\mathbf{G}^{-1}$. If the chosen basis set of internal valence coordinates involves a redundancy then $\mathbf{G}$ will be singular. To overcome this annoying limitation the problem can be transformed at the outset to a molecule-fixed cartesian axis system.[42] Defining the transformation from the valence set $\{\mathbf{R}\}$ to mass weighted cartesian coordinates, $\{q\}$, as $\mathbf{R} = \mathbf{D}\mathbf{q}$, then the eigenvectors in this system are given as the eigenvectors, Y, of $\tilde{\mathbf{D}}\mathbf{F}\mathbf{D}$. Making the appropriate substitutions in eqn. (14) gives

$$\left(\frac{\partial e \vdots \partial e}{\partial r \vdots \partial \gamma}\right)\left(\frac{l_{ri}}{l_{\gamma i}}\right) = \mathbf{R}^{-1}\{\Delta - [\mathbf{E}\vdots\mathbf{O}]\mathbf{D}\} Y \tag{15}$$

The transformation of eqns. (14) and (13) into terms arising from a basis set of symmetry coordinates is straightforward, though Gribov's statement of the result takes some time to be appreciated.

$$\frac{\partial p}{\partial Q_i} = \left[\{e^s\} \left|\frac{\partial \mu}{\partial S_r}, \frac{\partial \mu}{\partial S_\gamma}\right|_{eq} + \{\mu\}_{eq}\mathbf{R}_{eq}^{-1}(\Delta_{eq}\mathbf{M}^{-1}\tilde{\mathbf{B}}^s\mathscr{G}^{-1} - |\mathbf{E}_{eq}{}^s, \mathbf{O}|)\right]\begin{bmatrix} l_{ri}{}^s \\ l_{\gamma i}{}^s \end{bmatrix} \tag{16}$$

$\{e^s\}$ is a row of non-normalized symmetry direction vectors, these being obtained by combining the unit direction vectors of the equivalent bonds in the ratio as defined in the appropriate symmetry coordinates, S_r. The S_r and S_γ are symmetry coordinates for stretching and bending deformations respectively

$$\left|\frac{\partial \mu}{\partial S_r}, \frac{\partial \mu}{\partial S_\gamma}\right|_{eq}$$

is obtained from

$$\left|\frac{\partial \mu}{\partial S_r}, \frac{\partial \mu}{\partial S_\gamma}\right|$$

by striking out all the rows corresponding to equivalent bonds with the exception of the first of each set. Similarly $\{\mu\}$ and R_{eq}^{-1} are formed by striking out of $\{\mu\}$ and R all rows except one corresponding to each equivalent set of bonds.

B^s is the transformation matrix between symmetry coordinates and cartesian coordinates, and E_{eq}^s is a diagonal matrix of normalized symmetry direction vectors. $\mathscr{G}$ is the symmetrized inverse kinetic energy matrix.

In view of the complexity of the formulation a simple example is worth reproducing. Taking the H_2O molecule in the coordinate system shown we will compute $\partial p/\partial Q$ for the b_1 species normal coordinate.

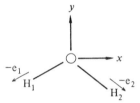

The internal valence coordinates will be taken as Δr_1, Δr_2 and $d\theta$ in that order.

Then
$$\mathscr{G}^{-1} = \{\mu_H + \mu_O(1 - c(\theta))\}^{-1}$$

$$\begin{array}{ccccccc} & x_1 & y_1 & x_2 & y_2 & x_3 & y_3 \\ \Delta = \left[\begin{array}{cccccc} -1 & -1 & 0 & 0 & +1 & +1 \\ 0 & 0 & -1 & -1 & +1 & +1 \end{array}\right.& & & & & & \left.\begin{array}{c} r_1 \\ r_2 \end{array}\right] \end{array}$$

and
$$\Delta^s = [-1 \quad -1 \quad -1 \quad -1 \quad 2 \quad 2]$$

$$\Delta^s M^{-1} \tilde{B}^s = \Delta^s \left[-\frac{s(\theta)}{2^{\frac{1}{2}}}\mu_H, \; -\frac{c(\theta)}{2^{\frac{1}{2}}}\mu_H, \; -\frac{s(\theta)}{2^{\frac{1}{2}}}\mu_H, \; \frac{c(\theta)}{2^{\frac{1}{2}}}\mu_H, \; 2^{\frac{1}{2}}s(\theta)\mu_O, \; 0\right]^t$$

$$= 2^{\frac{1}{2}}s\left(\frac{\theta}{2}\right)\mu_H + 2^{\frac{1}{2}}s\left(\frac{\theta}{2}\right)\mu_O$$

t denotes the transpose of the row matrix and μ_x is the reciprocal of the mass of the atom x. $s(\theta)$ denotes $\sin \theta$ and $c(\theta)$ denotes $\cos \theta$.

$$E = \left[\begin{array}{ccc} s\left(\frac{\theta}{2}\right), & -s\left(\frac{\theta}{2}\right), & 0 \\ c\left(\frac{\theta}{2}\right), & c\left(\frac{\theta}{2}\right), & 0 \end{array}\right]$$

Thus

$$\mathbf{E}^s = \mathbf{UE} = \begin{bmatrix} 2^{\frac{1}{2}} s\left(\dfrac{\theta}{2}\right) \\ 0 \end{bmatrix}$$

implying that the transition moment is along the X axis.

$$\Delta^s \mathbf{M}^{-1} \tilde{\mathbf{B}}^s \mathscr{G}^{-1} - [\mathbf{E}^s, \mathbf{O}] = \frac{2^{\frac{3}{2}} \mu_0 s\left(\dfrac{\theta}{2}\right) c^2\left(\dfrac{\theta}{2}\right)}{\mu_H + 2\mu_0 s^2\left(\dfrac{\theta}{2}\right)}$$

As there is no angular deformation in the b_1 species the rotational correction is obtained [see eqn. (16)] by multiplying the above by the bond dipole and by the $\mathbf{L}$ matrix element $[=(G)^{\frac{1}{2}}]$.

In Gribov's book[6] the author discusses in detail the derivation of the so called electro-optical parameters, μ_i and $\partial \mu_i / \partial r_j$, from intensities for several molecules. There is little attempt to justify, or even state, the force fields used and alternative solutions arising from needing to use the square roots of the intensities. In fact, as the primary aim is clearly to describe and justify his formulation of the intensity problem, such details are perhaps irrelevant. With these notes of caution we reproduce some of his results in Tables 4 and 5.

TABLE 4
The dipole derivatives of HCN and DCN with respect to their normal coordinates and the derived electro-optical parameters[a]

Molecule	ν	$\partial p / \partial Q / (e/\mathrm{amu}^{1/2})$	
	cm^{-1}	Expt.	Calc.[b]
HCN	3312	0.246	0.246
	2089	$0.007 \pm .005$	0.002
	712	0.158	0.158
DCN	2629	0.183	0.183
	1921	0.056	0.050
	569	0.085	0.085

[a] Ref. 6.
[b] The calculated $\partial \mu / \partial Q$ are derived using the parameters
$\mu_{CH} = \pm 4.0 \times 10^{-30}$ C m (1.21 D);
$\mu_{CN} = \pm 5.5 \times 10^{-30}$ C m (1.66 D)

$$\frac{\partial \mu_{CH}}{\partial r_{CH}} - \frac{\partial \mu_{CN}}{\partial r_{CH}} = \pm 0.225e$$

$$\frac{\partial \mu_{CN}}{\partial r_{CN}} - \frac{\partial \mu_{CH}}{\partial r_{CN}} = \pm 0.096e$$

TABLE 5
A comparison of electro-optical parameters as derived by Gribov[a] for various molecules. Where two similar bonds are involved a prime distinguishes one from the other

HC≡CH	$\mu_{CH} = 3.7 \times 10^{-30}$ C m (1.1D); $\partial\mu_{CH}/\partial r_{CH} = 0.208e$; $\partial\mu_{CH}/\partial r'_{CH} = 0.00e$
H$_2$C=CH$_2$	$\mu_{CH} = 2.3 \times 10^{-30}$ C m (0.7D); $\partial\mu_{CH}/\partial r_{CH} = 0.110e$; $\partial\mu_{CH}/\partial r'_{CH} = 0.02e$;
	$\partial\mu/\partial a \simeq \partial\mu/\partial\beta \sim 0.10e$
CH$_3$CH$_3$	$\mu_{CH} \simeq 0.9 \times 10^{-30}$ C m (0.28D); $\partial\mu_{CH}/\partial r_{CH} = 0.154e$; $\partial\mu_{CH}/\partial r'_{CH} = 0.065e$;
	$\partial\mu/\partial a - \partial\mu/\partial\beta \simeq -0.008e$.

 [a] Ref. 6.

Note that it is not possible to separate the components of an expression such as

$$\frac{\partial\mu_{CH}}{\partial r_{CH}} - \frac{\partial\mu_{CN}}{\partial r_{CH}}$$

This is natural since molecular deformations must cause electron displacements in neighbouring bonds and there is no way by which we can distinguish these contributions in an i.r. experiment. Molecules discussed, other than those mentioned in Tables 4 and 5, are propylene, CH$_3$Cl, CH$_3$Br, CH$_3$I, dimethyl-acetylene and CH$_3$CN.

The existence of characteristic group and bond intensities is discussed by Gribov[6] in some detail and it is shown that in the series of molecules (CH$_3$)$_n$SiCl$_{4-n}$ the intensities of the following bands are additive in terms of bond or group moments

    (i) symmetric Si—C stretch;
    (ii) antisymmetric Si—Cl stretch;
    (iii) methyl CH stretch modes.

In a recent paper[43] Gribov has used a remarkably simple model to calculate the band intensities and overall spectra for various conformations of ethylamine, diethylamine, diacetylmonooxime, polyethylene, polyethylene glycol and polyethylene glycoladipate. The electro-optical parameters were estimated using a formula for bond moments due to Landau and Lifshitz[44]

$$\mu = \frac{m_1 m_2}{m_1 + m_2}\left(\frac{q_1}{m_1} - \frac{q_2}{m_2}\right)r$$

where the q are the charges on the atoms and r is the bond length. The $\partial\mu/\partial r$ are taken simply as μ/r. This extremely crude model is sufficiently successful to allow conformations to be recognized in some cases.

3.3 The Morcillo Polar Tensor Formulation

The theory as formulated by Gribov is a straightforward development of the old bond-moment theory. A radically different approach is that of Morcillo and coworkers. The formulation was originally published in 1961,[45] but until recently no interest was shown in it outside of the Spanish group. Certain advantages of the theory have recently been pointed out by Person and Newton.[46]

Instead of concentrating attention on the dipole derivatives with respect to valence deformations, the Morcillo theory considers the transition moment as arising from the sum of derivatives of the total dipole with respect to cartesian displacement coordinates. Thus, following the symbolism of Ref. 46, we can write

$$\Delta p = \sum_{\alpha=1}^{N} \mathbf{P}_x^{\alpha} r_\alpha$$

where r_α is the displacement vector for the α atom and $\mathbf{P}_r^\alpha$, known as the polar tensor, is defined by

$$\mathbf{P}_x^\alpha = \begin{Bmatrix} \partial p_x/\partial x_\alpha, & \partial p_x/\partial y_\alpha, & \partial p_x/\partial z_\alpha \\ \partial p_y/\partial x_\alpha, & \partial p_y/\partial y_\alpha, & \partial p_y/\partial z_\alpha \\ \partial p_z/\partial x_\alpha, & \partial p_z/\partial y_\alpha, & \partial p_z/\partial z_\alpha \end{Bmatrix} \tag{17}$$

We can define similar tensors for basis sets other than the cartesian displacement sets. For the normal coordinates Q_i

$$\mathbf{P}_Q = \begin{Bmatrix} \sigma_1|\partial p_x/\partial Q_1|, & \sigma_2|\partial p_x/\partial Q_2|, \ldots \\ \sigma_1|\partial p_y/\partial Q_1|, & \sigma_2|\partial p_y/\partial Q_2|, \ldots \\ \sigma_1|\partial p_z/\partial Q_1|, & \sigma_2|\partial p_z/\partial Q_2|, \ldots \end{Bmatrix} \tag{18}$$

In this case the σ_i are usually $= \pm 1$ to take cognizance of the fact that the sign of the dipole gradient is usually indeterminate, being obtained from the square root of the measured intensity.

Now the transformation from $\mathbf{P}_Q$ to $\mathbf{P}_X$, the polar tensor for cartesian coordinates, is not a simple one due to the contribution of the rotational angular momentum. As discussed in Refs. 45 and 46 the transformation is given by

$$\mathbf{P}_X = \mathbf{P}_Q \mathbf{L}^{-1} \mathbf{D} \mathbf{M}^{\frac{1}{2}} + \mathbf{P}_\rho \delta \mathbf{M}^{\frac{1}{2}} \tag{19}$$

$\mathbf{D}$, $\mathbf{M}$ and $\mathbf{L}$ are as defined earlier. δ is the transformation matrix between the non-periodic motions, ρ (translations and rotations) and the mass cartesian displacements, q. Thus

$$\begin{pmatrix} \mathbf{R} \\ \hline \rho \end{pmatrix} = \begin{pmatrix} \mathbf{D} \\ \hline \delta \end{pmatrix} q$$

$\mathbf{P}_\rho$ is the polar tensor for the non-periodic motions. Molecular translations do not cause a change of dipole whereas rotations do. The importance of eqn. (19) lies in the fact that it gives the correction due to vibrational angular momentum ($=\mathbf{P}_\rho \delta \mathbf{M}^{\frac{1}{2}}$) which must be added to the vibration polar tensor to give the cartesian polar tensor. This $\mathbf{P}_X$ has some interesting and important properties.

As noted by Biarge, Herranz and Morcillo[45] the sum of the $\mathbf{P}_X$ over all atoms is equal to zero, i.e.

$$\sum_{\alpha=1}^{N} \mathbf{P}_x^{\alpha} = 0 \tag{20}$$

This follows from the independence of the dipole from translational displacements. This means for example that if we have only two types of non-equivalent atoms, as in ethylene, then if the polar tensors of one atom are known so are the polar tensors of the other. Person and Newton[46] have analysed the absorption intensities of ethylene and formaldehyde and derived the atomic polar tensors for the various atoms.

For formaldehyde the intensity data of Hisatsune and Eggers[47] was reanalysed to obtain the $\partial p/\partial S$. CNDO calculations (see section 4.1) were then employed to narrow down the sign choices to two in plane and two out of plane. The resulting tensors for formaldehyde are shown in Table 6 where the axes are as

TABLE 6
The atomic polar tensors P_x^{α} for HCHO.[a] The coordinate axes are as defined in Fig. 1. All tensor elements are in units of the electronic charge e

HCHO	$\mathbf{P}_{\mathrm{H}}^{1}$			$\mathbf{P}_{\mathrm{O}}$			$\mathbf{P}_{\mathrm{C}}$		
sign choice 1	-0.127	0	0.064	-0.412	0	0	0.664	0	0
	0	0.081	0	0	-0.327	0	0	0.162	0
	0.098	0	-0.081	0	0	-0.741	0	0	0.904
sign choice 2	-0.098	0	0.156	-0.244	0	0	0.439	0	0
	0	0.004	0	0	-0.396	0	0	0.385	0
	0.098	0	-0.081	0	0	-0.741	0	0	0.904
CNDO	-0.191	0	0.021	-0.470	0	0	0.834	0	0
	0	0.050	0	0	-0.287	0	0	0.187	0
	0.064	0	-0.052	0	0	-0.731	0	0	0.835

[a] Ref. 47.

defined in Fig. 1. In Table 7 the trace elements of $\mathbf{P}_X$ in the coordinate axes of Fig. 1 are compared for formaldehyde, ethylene, methane and benzene. The results for the latter two compounds were derived by King, Mast and Blanchette[48] by another approach to which we will refer later. Several interesting features emerge. It appears that if the correct sign choices for $\partial p/\partial Q$ have been made then stretching of the CH bond results in a decrease in the C^-H^+ dipole moments.[46] This of course can only happen by a substantial electron reorganization as the bond stretches. It is now well known that bond stretching produces a dipole gradient which bears no obvious relation to the static charge distribution

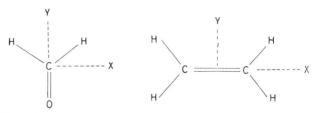

Fig. 1. Definition of coordinate axes for H_2CO and C_2H_4. A right-handed cartesian system is
used for both.

in the bond. For example for stretching of a CF bond $\partial p/\partial r \sim 1.0$ e and yet the
CF bond dipole is only about $er/8$ where r is the bond length.[29,8] Clearly it
is very dangerous to predict the sign of $\partial p/\partial r$ simply from a knowledge of the
sign of the bond dipole. The vector dipole gradients of ethylene, formaldehyde
and benzene agree tolerably well with the expectations of eqn. (20), but for
methane the discrepancies are rather disconcerting (see Table 7). Thus we would
expect that

$$|(\partial p/\partial r_H)/(\partial p/\partial r_C)| = 4$$

In fact the ratio is 2.38. The discrepancies are assigned to experimental errors.
An examination of the paper of King, Mast and Blanchette shows that the
problem arises in $\partial p/\partial r_C$ terms which have a 100% uncertainty limit. It seems that
in practice it is important to introduce eqn. (20) as a constraint in the analysis.

Morcillo, Zamorana and Heredia[50] have measured the intensities in the
gas phase spectra of CH_2F_2, CH_2Cl_2 and CF_2Cl_2 and analysed these results in
terms of the cartesian polar tensors. For each system, as a result of sign
ambiguities there are 256 possible sets of solutions. Eliminating those sets which
are most unlikely, in predicting for instance a C–halogen stretching dipole gradient
with the halogen at the positive end of the effective dipole, reduces the sets to 8 in

TABLE 7
Magnitudes of vector dipole gradients $\mathring{\jmath}_\alpha = |\partial p/\partial r_\alpha|$
for CH_2O, C_2H_4, CH_4 and C_6H_6. Units are e; coordinate
axes are defined in Fig. 1.

		ξ_α^x	ξ_α^y	ξ_α^z	ξ_α
CH_2O	$\alpha = H$	0.142	0.081	0.127	0.207
	O	0.412	0.327	0.741	0.909
	C	0.664	0.162	0.904	1.284
C_2H_4	$\alpha = H$	0.074	0.080	0.131	0.171
	C	0.146	0.131	0.264	0.329
CH_4	$\alpha = H$	0.099	0.099	0.099	0.172
	C	0.236	0.236	0.236	0.409
C_6H_6	$\alpha = H$				0.167
	C				0.137

each instance. The magnitude of the C—H stretching dipole is so small that sets which differ only in the sign of this are insignificantly different from those for CH_2F_2 and CH_2Cl_2. The resulting polar tensors are given in various frameworks of coordinates, each of which allows the effect of distorting a given bond to be seen with clarity. Thus for a C—X bond in CX_2Y_2 the natural coordinate system (i, j, k) is that in which the k axis is along the CX bond while the i and j axes are perpendicular thereto. The i axis lies in the CX_2 plane and the j axis in CY_2 plane. In Table 8 we list the most divergent pair of solutions for typical atoms in each molecule. Note that the axes systems for the X and Y atoms are different, each being rotated to achieve the above simplification. The carbon polar tensors can be obtained by use of eqn. (20). In accord with Ref. 46 the hydrogen appears to be at the positive end of a small effective dipole in the stretching motion [small positive term in $P_H(1, 1)$]. The sign and magnitude of

TABLE 8
Components of the polar tensors (in units of e) in the bond axis systems^a

Molecule	Set	P_H			P_F		
CH_2F_2	R	0.120	0	0.014	0.27	0	0.20
		0	0.010	0	0	−0.17	0
		0.006	0	−0.150	−0.01	0	−0.96
	T	0.100	0	0.025	−0.24	0	0.21
		0	0.054	0	0	−0.20	0
		0.017	0	0.223	−0.01	0	−1.00

Molecule	Set	P_H			P_{Cl}		
CH_2Cl_2	S′	0.050	0	0.083	−0.22	0	0.28
		0	−0.085	0	0	−0.15	0
		0.021	0	0.079	−0.006	0	−0.48
	U″	−0.006	0	−0.004	−0.16	0	0.39
		0	0.167	0	0	−0.15	0
		−0.017	0	−0.031	−0.08	0	−0.6

Molecule	Set	P_F			P_{Cl}		
CF_2Cl_2	b	−0.35	0	−0.16	−0.22	0	0.26
		0	−0.35	0	0	−0.08	0
		−0.12	0	−0.91	−0.00	0	−0.56
	h	−0.38	0	−0.13	−0.22	0	0.33
		0	−0.35	0	0	−0.04	0
		−0.18	0	−0.96	−0.00	0	−0.51

^{a} Ref. 50.

the effective dipole in the bending motions is much more difficult to ascertain. It seems likely that the hydrogen carries an opposite effective charge in the angular deformations in and out of the CH_2 plane. Another point well worth noting is that for CH_2F_2 the stretching dipole gradient vector lies quite close to the bond axis $[\mathbf{P}_H(1, 1) \gg \mathbf{P}_H(3, 1)]$. The same is true for C–halogen stretches in all 3 systems, provided we dismiss solution U″ of CH_2Cl_2. For all the C—X stretches the $\partial p/\partial r_{CX}$ are deflected towards the more polarizable atom of the molecule. This is shown clearly in Figs. 2–4 of Ref. 50 and in Fig. 7 of Ref. 51. The latter reference reports a continuation of the Spanish group's studies of halogenated methanes. The gas phase intensities of $CHCl_3$ and $CDCl_3$ were measured and interpreted. The $\mathbf{L}$ matrix for chloroform was computed using the force field of Decius, as well as more recent fields of three other groups. Using the Decius field[52] two satisfactory sets of solutions were found which satisfied the requirements of expected signs of dominant terms and of producing polar tensors for light and deuterated species which were within the experimental error. It is disturbing that with the more recent fields[53−55] the latter requirement was not met. This serves to emphasize that the uncertainty in force constants is still probably the greatest obstacle to the interpretation of intensities (see also section 1.2 and Ref. 17).

A consideration of the properties of the product matrix $\mathbf{P}_Q\tilde{\mathbf{P}}_Q$ serves well to give a good insight into, and to unify, several important sum rules for intensities.[5,46] The trace of this matrix is equal to the sum of the product of frequencies of the band centres and their intensities, apart from a universal constant, K, thus

$$\mathrm{Tr}[\mathbf{P}_Q\tilde{\mathbf{P}}_Q] = \sum_i \left(\frac{\partial p}{\partial Q_i}\right)^2 = \sum_i \Gamma_i \nu_i/(N\pi g_i/3c^2) = \frac{\Sigma\Gamma_i\nu_i}{K}$$

If we now transform into internal coordinates we obtain Crawford's $\mathbf{G}$ sum rule[12]

$$K\mathrm{Tr}[\mathbf{P}_Q\tilde{\mathbf{P}}_Q] = K\mathrm{Tr}[\mathbf{P}_R\mathbf{L}\tilde{\mathbf{L}}\tilde{\mathbf{P}}_R] = K\mathrm{Tr}[\mathbf{P}_R\mathbf{G}\tilde{\mathbf{P}}_R] = \sum_i \Gamma_i\nu_i$$

In terms of cartesian coordinates it is readily shown[48] that the corresponding rule is

$$K\mathrm{Tr}[\mathbf{P}_X\mathbf{M}^{-1}\tilde{\mathbf{P}}_X] = \sum_i \Gamma_i\nu_i + \Omega \tag{21}$$

where

$$\Omega = [(p_x^0)^2 + (p_z^0)^2]/I_y + [(p_z^0)^2 + (p_y^0)^2]/I_x + [(p_x^0)^2 + (p_y^0)^2]/I_z$$

p_x^0 is the x component of the permanent dipole. The Ω term represents the rotational correction to the intensity sum arising from vibrational angular momentum. King, Mast and Blanchette[48] used eqn. (21) to derive atomic tensors for a variety of systems by using data on isotopically related systems. Some of the values they derived have already been discussed in connection with Table 7. Indeed they were the first to point out that $(\partial p/\partial r_H)$ was a constant in a wide range of molecules. Some difficulty was experienced with the corresponding tensors for the carbon atoms. This is easily understood since the isotopic substitution was for the hydrogen and the data therefore were insensitive to this parameter. Those values which were obtained were all with 100% uncertainties and therefore the data were reanalysed imposing the constraint $(\partial \mathbf{P}_C/\partial r_C) = 0$. In fact in view of invariance of the dipole to translations it would have been more useful and realistic to apply the constraint eqn. (20). These authors also pointed out that a knowledge of the intensity sums for C_6H_6 and C_6D_6 is sufficient to determine the sums of the intensities for any partially deuterated benzene. I would modify this statement in view of eqn. (20) to state that the intensity sum for C_6H_6 or C_6D_6 is sufficient, this being entirely compatible with Crawford's G sum rule.

It has been suggested that dipole derivatives with respect to internal coordinates can be calculated from the intensities in such a way that all sign ambiguities disappear.[56] The fallacy is easily seen by considering the matrix $\tilde{\mathbf{P}}_Q \mathbf{P}_Q$. Thus

$$
K\tilde{\mathbf{P}}_Q\mathbf{P}_Q = K \begin{bmatrix} (\partial p/\partial Q_1)^2, & (\partial p/\partial Q_1)(\partial p/\partial Q_2), \dots \\ (\partial p/\partial Q_1)(\partial p/\partial Q_2), & (\partial p/\partial Q_2)^2, & \dots \\ \vdots & \end{bmatrix}
$$

$$
= K\tilde{\mathbf{L}}\tilde{\mathbf{P}}_R\mathbf{P}_R\mathbf{L} \tag{22}
$$

The authors made the error of assuming that $K\tilde{\mathbf{P}}_Q\mathbf{P}_Q$ was diagonal and equal to the diagonal matrix whose non-zero elements were the $3N-6$ intensities. The sign ambiguities which exist are clearly shown to arise from the uncertainties in the signs of the off-diagonal elements $(\partial p/\partial Q_1)(\partial p/\partial Q_2)$. These terms will be zero only when Q_1 and Q_2 belong to different symmetry classes.

3.4 The Sign of the Dipole Gradient

There are no general methods for determining the signs of dipole gradients. A number of significant advances have occurred in this field during the last few years largely as a result of technological advances. The most widely applicable method is to measure the dipole moment in excited vibrational states and to relate the increment to the dipole parameters through an equation of the form

$$
\langle p \rangle = p_e + \sum_i \langle Q_i \rangle \frac{\partial p}{\partial Q_i} + \frac{1}{2}\sum_{ij} \langle Q_iQ_j \rangle \frac{\partial^2 p}{\partial Q_i\partial Q_j} \tag{23}
$$

$\langle Q_i \rangle$ is the expectation value of the ith displacement coordinate (normal coordinate) Q_i. p_e is the dipole moment in the equilibrium configuration and the sum is over all excited states. Stark shift studies in microwave spectroscopy offer a potentially very powerful method for studying dipole moments in excited states.[57] Unfortunately the measurements appear to have been restricted mainly to diatomic molecules. For linear molecules or for any symmetric molecule with

$K = 0$ the energy shift, ΔW, created by an electric field E is given, using second order perturbation theory, by

$$\Delta W = \frac{p^2 E^2}{2hBJ(J+1)} \frac{J(J+1)-3M^2}{(2J-1)(2J+3)}$$

The above expression is derived ignoring nuclear spin hyperfine structure which leads to additional complications. Data on excited states of a number of linear molecules have been available for many years but only recently has the information been utilized in the i.r. intensity problem.

A very thorough investigation of carbonyl sulphide[58] affords an excellent example of the use of such information. The absolute sense of the dipole moment has been established[59] from Zeeman effect measurements on $CO^{32}S$ and $CO^{34}S$. Thus in a magnetic field H we have for the $J = 0 \rightarrow J = 1$ transition of a linear molecule

$$\nu_{0 \rightarrow 1}(M_0 = 0 \rightarrow M_1 = 0) = \nu_0 + \frac{2H^2}{15h}(\chi_\perp - \chi_{||})$$

and

$$\nu_{0 \rightarrow 1}(M_0 = 0 \rightarrow M_1 = \pm 1) = \nu_0 \mp H\mu_0 g_\perp/h - H^2(\chi_\perp - \chi_{||})/15h$$

h is Planck's constant, χ is the magnetic susceptibility and μ_0 is the nuclear magneton.

Observation of these shifts and use of the relationship

$$\frac{g_\perp(^{34}S)}{B(^{34}S)} - \frac{g_\perp(^{32}S)}{B(^{32}S)} = -\frac{8\pi M}{h|e|} Z p_0$$

leads to the absolute sign of p_0. B is the rotational constant, M is the molecular mass and Z is the coordinate shift of the centre of mass which accompanies the substitution of ^{32}S for ^{34}S. The dipole moments in the (01^10) and (10^00) states of COS have been measured.[60,61] Combining the results of the Zeeman study and the excited state dipole measurements leads to the signs of $\partial p/\partial Q_1$ and $\partial p/\partial Q_3$. Of course it also leads to an estimate for the magnitude of the gradients. The precision however is considerably less than can be attained from intensities of absorption. Foord and Whiffen measured the intensities of several combination and fundamental bands of COS and by combining with the above data they derived an expression for the dipole expansion up to the quadratic terms. These results will be discussed in more detail in section 5.

Until recently Stark investigations were limited to pure rotational spectra because of the resolution limitations of conventional i.r. absorption spectroscopy. Infrared lasers have changed the situation and some elegant Stark experiments are now being carried out on selected polyatomic systems. As the range of available frequencies increases, from the result of developments of such as the spin-flip laser[62] and semi-conductor laser, then the number of such studies will no doubt expand considerably. The 9.4 μm laser lines of CO_2 were used to study the ν_3 bands of $^{12}CH_3F$ and $^{13}CH_3F$ employing Stark modulation of the vibration–rotation lines.[63] Where the multiplet structure of the $\Delta M = \pm 1$ transitions was not resolved, Lamb dip[64] measurements were used to determine the precise transition frequencies. The accuracy attained was such that the rotational constant of the excited state is quoted with a standard deviation of 0.03 MHz (10^{-5} cm^{-1}). Since the Stark shifts are of the same order as the spacings of the rotational levels the perturbation treatment gives a poor approximation to the perturbed energy levels. The Hamiltonian matrix $\mathbf{H} = \mathbf{H}_{\text{rot}} - p \cdot E$ was therefore set up and diagonalized. The significance of various minor interaction terms was found to be very small and they were therefore neglected. This included for example the shifts due to electronic polarizability on the Stark shift and the 4th order term of the Coriolis interaction between ν_3 and ν_6 on the frequencies. The dipole moment of the excited state was deduced to be 1.9054 ± 0.0006D [6.354 $(\pm 0.002) \times 10^{-30}$ C m] for ν_6 ^{12}CH$_3$F and 1.9039 ± 0.0006D [6.350 $(\pm 0.002) \times 10^{-30}$] for ν_6 ^{13}CH$_3$F. The ground state dipole moment of 1.85850 (± 0.00050)D used in the analysis was that of Wofsy, Muenter and Klemperer.[65] An analysis of 19 Stark resonances of the $2\nu_3 \leftarrow \nu_3$ band allowed the dipole of the $2\nu_3$ state to be deduced as 1.9519 ± 0.0020D ($6.510 \pm 0.007 \times 10^{-30}$ C m). In other words

$$p(2\nu_3) - p(\nu_3) \simeq p(\nu_3) - p(0)$$

Use of eqn. (23) necessitates the use of anharmonic potential functions. For a mechanically harmonic oscillator terms such as $\langle Q_i \rangle$ and $\langle Q_i Q_j \rangle_{i=j}$ will equal zero. Perturbation theory has been shown[66] to lead to the result

$$p = p_e + \sum_{s \subset A_1} \left(\frac{\partial p}{\partial Q_s} \right)_e \left[-\frac{1}{\nu_s} \left\{ 3k_{sss}(v_s + \tfrac{1}{2}) + \sum_{s'} k_{ss's'}(v_{s'} + d_{s'}/2) \right\} \right.$$

$$\left. + \left(\frac{h}{4\pi^2 \nu_s} \right)^{\frac{3}{2}} \sum_\alpha \frac{\alpha_s^{\alpha\alpha}}{(I_{\alpha\alpha})^2} \langle v, R | J_\alpha - \pi_\alpha | v, R \rangle \right] + \frac{1}{2} \sum_{s \subset A_1} \left(\frac{\partial^2 p}{\partial Q_s^2} \right)_e \frac{1}{\nu_s} (v_s + \tfrac{1}{2}) \qquad (24)$$

The k_{sss} are the cubic potential constants; $I_{\alpha\alpha}$ is the component of the moment of inertia tensor along the αth axis; $a_s^{\alpha\alpha} = \partial I_{\alpha\alpha}/\partial Q_s$; Q_s and J_s are the sth normal coordinate and the component of the total angular momentum along the αth axis and π_α is the corresponding vibrational angular momentum. For a symmetric transition of CH$_3$F the ν_3 dependence of the dipole moment becomes

$$p = p_0 + v_3 \left[\frac{1}{2} \frac{\partial^2 p}{\partial Q_3^2} - \frac{k_{133}}{\nu_1} \frac{\partial p}{\partial Q_1} - \frac{k_{233}}{\nu_2} \frac{\partial p}{\partial Q_2} - \frac{3k_{333}}{\nu_3} \frac{\partial p}{\partial Q_3} \right]$$

Since the dipole moment is observed to be linear in v_3 the neglect of higher terms in the perturbation expansion eqn. (23) is seen to be justified. While good estimates can be obtained for the $\partial p / \partial Q_i$ from the observed absorption intensities, a fuller analysis of the results leading to signs of the gradients as well as the magnitudes of the second derivatives must wait until the cubic constants have been determined. The a_3 terms of eqn. (24) have been deduced from the fundamental and harmonic transition frequencies $n\nu_3$. Since these are functions of the anharmonicity some progress has been made along this line. This does high-light, however, the difficulties which accompany the interpretation of these types of data.

Another precise method of measuring dipole moments in excited states is molecular beam electric resonance spectroscopy.[67] This has the advantage of allowing the dipole of various rotational–vibrational states to be separately measured. Studies have so far been restricted to diatomic molecules. A number of studies have been carried out on species which would present considerable problems for conventional i.r. absorption intensity measurements. These include SrO,[68] BaO,[69] KF,[70] and LiH.[71] For BaO the dipoles in the ground and three lowest excited vibrational ($J = 0$) states were found to be 26.53 (± 0.01), 26.67, 26.81 and 26.94×10^{-30} C m. As is seen from these values the dipole increment from the ground state is proportional to the vibrational quantum number to within the uncertainty limits of the experiment. Considering only the limited series eqn. (23) and taking the expectation values of $\langle \Delta r \rangle$ and $\langle (\Delta r)^2 \rangle$ appropriate to the Morse oscillator, then

$$p = p_e + \frac{\partial p}{\partial r} \cdot 3 r_e \left(\frac{\tilde{v} x B}{\tilde{v}^2} \right)^{\frac{1}{2}} (v + \tfrac{1}{2}) + \frac{1}{2} \left(\frac{\partial^2 p}{\partial r^2} \right) \cdot r_e^2 \frac{2B}{\tilde{v}} (v + \tfrac{1}{2})$$

where $\tilde{v} x$ is the vibrational anharmonicity.

Using published rotational and vibrational coefficients the coefficients of the dipole terms could be deduced.[69] This still does not allow the relative contributions of the first and second derivatives to be separated. The authors used an approximation to Rittner's model for ionic molecules[72] in which the dipole moment is considered to arise from point charge interactions with polarizable ions.

$$p = er - \frac{\alpha e}{r^2}$$

where r is the ionic separation, e the electronic charge and α is the sum of the polarizabilities of both ions. Use of the deduced p_e leads to contributions to $\Delta p / \Delta v$ of 0.14×10^{-30} C m ($+0.042$ D) from the first dipole derivative and -0.01×10^{-30} C m (-0.003 D) from the second derivative. The total of $+0.13 \times 10^{-30}$ C m is in remarkable agreement with the observed of $+0.140 \pm 0.007 \times 10^{-30}$ C m.

Despite the excellent agreement between experimental results for BaO and expectations from the Rittner model, the more ionic SrO dipole variations are quite unexpected. The dipole of the first vibrationally excited state is less than that of the ground state [8.874 D (29.59×10^{-30} C m) and 8.900 D ($29.68 \times \times 10^{-30}$ C m) respectively]. By comparison with BaO this implies a negative dipole gradient. This surprising result leaves one with the uncomfortable feeling that perhaps the BaO result is fortuitous.

Lithium hydride and deuteride are of some special interest since accurate wavefunctions exist for these species[73,74] and infrared measurements of the vibration–rotational contour have been reported.[75] Alexander[76] has explored the possibility of combining wavefunctions for excited states derived from RKR curves with dipole functions derived from accurate wave mechanical calculations and with experimental dipole information from molecular beam electronic resonance spectroscopy to produce an optimum set of transition probabilities. A typical set of results are shown in Table 9. It is seen that there is a marked improvement in the estimated precision of the transition probabilities, especially for overtones.

TABLE 9
Rotationless ($J = 0$) Q-branch transition amplitudes in units of Debyes for the fundamental, first overtone and hot band transition in ^{7}LiH and ^{7}LiD^a

Molecule	Transition	Theoretical alone^b	Theoretical and experimental
^{7}LiH	$0 \to 1$	$-0.242 \pm .048$	-0.241 ± 0.028
	$0 \to 2$	$-0.018 \pm .043$	$-0.020 - 0.011$
	$1 \to 2$	$-0.344 \pm .056$	-0.335 ± 0.026
^{7}LiD	$0 \to 1$	$-0.209 \pm .046$	-0.210 ± 0.026
	$0 \to 2$	$-0.013 \pm .042$	-0.014 ± 0.009
	$1 \to 2$	$-0.297 \pm .051$	-0.292 ± 0.029

^{a} Ref. 76. ^{b} Errors based on 1 % uncertainty in the dipole function.

James, Norris and Klemperer[75] have measured the relative intensities of the various rotational transitions in the P and R branches of the fundamental band of ^{7}LiH. The results of Alexander were used to compute these ratios and, as can be seen from Fig. 2, the results are in excellent agreement. The solid line is a plot of the ratio of the calculated intensities using the approximate formula of Herman and Wallis.[77] The plot is strongly dependent on the ratio of the equilibrium dipole to the dipole gradient. Now the transition probability can be written as the product of a vibrational part, $|R_v^{v'}|^2$, and a rotational part, F. For a diatomic system $F(J' \leftarrow J)$ may be expanded as a power series in the variable $m = 0.5[J'(J'+1) - J(J+1)]$. Thus

$$F_v^{v'}(J' \leftarrow J) = 1 + C_v m + D_v m^2 \qquad (25)$$

The coefficients C_v and D_v are explicitly written out in Ref. 77 for $v(0 \to 1)$, $v(0 \to 2)$, $v(1 \to 2)$. For the fundamental and hot transitions the principal terms are proportional to the ratio $p_e/(\partial p/\partial r)_e$. We see from Fig. 2 that the Herman–Wallis equation with a value for the previously mentioned ratio of -1.75Å gives a good fit to the theoretical and experimental results. Further applications of the Herman–Wallis equation are described in section 5.1.

The methods of studying the sign of the dipole gradient discussed so far work by relating the sign of the gradient to the sign of the permanent dipole. The relative signs of different dipole gradients may be deduced in suitable cases from

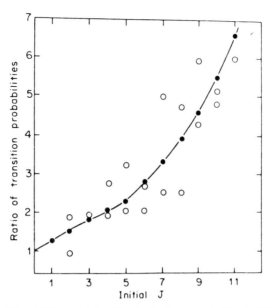

Fig. 2. The ratio of P and R-branch intensities in the fundamental band of ^{7}LiH against initial J. The unfilled circles indicate the experimental values of James, Norris and Klemperer (Ref. 75) whereas the filled circles and line indicate the results of Alexander (Ref. 76). The latter are based on the Herman–Wallis expression and a ratio of equilibrium dipole moment to dipole derivative of -1.75 Å.

an analysis of second order Coriolis interactions. When two vibration–rotation bands are able to interact through Coriolis coupling, not only are the frequencies perturbed but also the mixing of the wavefunctions results in intensity perturbations.

This problem has been discussed in some detail for two interacting degenerate bands.[78] If, as a result of the interaction the mixing coefficients are a_k and b_k, where the k subscripts imply the dependence of the mixing on the k quantum number, then the vibrational–rotational line strength, S, can be written

$$S_0^\alpha = a_k^2 M_r^2 + b_k^2 M_s^2 \mp 2\sigma a_k b_k M_r M_s \qquad (26)$$
$$S_0^\beta = a_k^2 M_s^2 + b_k^2 M_r^2 \pm 2\sigma a_k b_k M_r M_s$$

M_r and M_s are the unperturbed vibrational transition moments; r and s denote the two unperturbed states while α and β denote the perturbed states; σ is the sign of ζ_{rs} and upper and lower signs refer to $\Delta K = \pm 1$ involving (± 1) levels in the excited state. Eqns. (26) show that in addition to a K dependent mixing, which would be the same for the P and the R branches, there is a mixing contribution which depends on the

$$\text{sign} \left(\sigma M_r M_s \right) \equiv \text{sign} \left(\zeta_{rs} (\partial p / \partial Q_r)(\partial p / \partial Q_s) \right) \qquad (27)$$

If this sign is positive the cross term has the effect of enhancing the intensity of the P type sub-bands of the high frequency band as well as the R type sub-bands of the low frequency band, and at the same time there is a depletion of the intensity of the other sub-bands. Thus the inner wings are enhanced while the outer wings are depleted. If the sign is negative the intensity effects are reversed.

TABLE 10
Pairs of bands which interact through Coriolis effects and which have been analysed to produce data on the dipole gradients

Molecule	Modes	Relative magnitudes of gradients	Perturbation	Footnote ref.
CH_3F	$\nu_3(a_1); \nu_6(e)$	8.45	+	a
	$\nu_2(a_1); \nu_5(e)$	2.3	−	a
CD_3Cl	$\nu_2(a_1); \nu_5(e)$	1.76	−	a
$CH_2:C:CH_2$	$\nu_9(b_1); \nu_{10}(b_2)$		−	b
CH_3CN	$\nu_4(a_1); \nu_7(e)$		−	c
CD_3CN	$\nu_3(a_1); \nu_6(e)$		+	c,d
	$\nu_4(a_1); \nu_7(e)$		−	c

 a Ref. 80.   *b* Ref. 78.   *c* S. Kondo and W. B. Person, *J. Mol. Spectrosc.* **52**, 287 (1974).   *d* F. N. Masri, J. L. Duncan and G. K. Speiers, *J. Mol. Spectrosc.* **47**, 163 (1973).

It was also pointed out by the authors that there will always be some value of K for which the S of the depleted branch is a minimum. It would be zero if the set of K was continuous. Observation of the appropriate sub-band and a knowledge of the mixing coefficients allows the estimation of the ratio $(\partial p/\partial Q_s)/(\partial p/\partial Q_r)$. The allene rocking and wagging fundamentals, ν_9 and ν_{10}, were analysed in detail and from the intensity perturbations it was deduced that $\zeta_{9,10}(\partial p_{x,y}/\partial Q_{9a,9b})(\partial p_{x,y}/\partial Q_{10,10b})$ is negative, or that $\partial p/\partial Q_9$ and $\partial p/\partial Q_{10}$ have opposite signs. Furthermore, from the observation that the $P_{Q6}(\nu_9)$ sub-band has virtually zero intensity it is deduced that $(\partial p/\partial Q_{10})/(\partial p/\partial Q_9) = -6.18$ (taking $a_k = 0.987$ and $b_k = 0.160$). Overend and Crawford[79] reported the results of a direct integration of the bands and gave $\Gamma_\alpha = 0.0842 \text{ m}^2\text{mol}^{-1}$ and $\Gamma_\beta = 0.975 \text{ m}^2\text{mol}^{-1}$. Since the mixing cannot alter the sum of the intensities then

$$\Gamma_9 + \Gamma_{10} = \Gamma_\alpha + \Gamma_\beta$$

Applying the ratio of the dipole gradients of the sum leads to

$$\Gamma_9 = 0.0270 \text{ m}^2\text{mol}^{-1}; \; \Gamma_{10} = 1.032 \text{ m}^2\text{mol}^{-1}$$

This shows that 60% of the intensity of the rocking mode is borrowed from the wagging mode.

The authors make one further very pertinent observation. If ν_9 and ν_{10} are

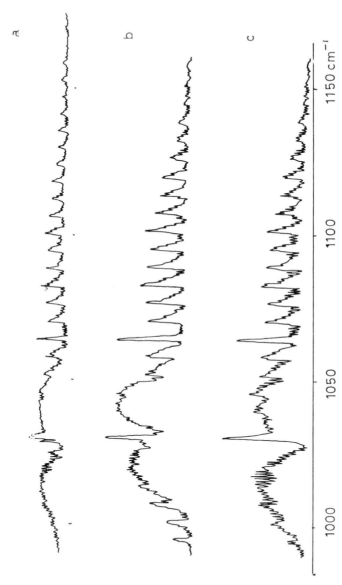

Fig. 3. (a) Absorption spectrum of CD_3Cl from 1000 to 1150 cm^{-1}, at approximately 0.5 cm^{-1} resolving power. (b) and (c) Computed contours of the spectrum of CD_3Cl over the same region for + and − perturbations (DiLauro and Mills, Ref. 80).

taken as simply rocking and wagging modes the relative sign of the dipole gradients implies that the effective CH bond moments are of opposite sign in the two modes. The implications are that the electron movements around the carbon atoms (rehybridization) contribute strongly to the dipole gradient.

Coriolis interactions between a_1 and e modes of C_{3v} systems have been discussed in detail by DiLauro and Mills.[80] Again the depletion of the P or R branches is found to depend on the sign of the product $\zeta_{rs}(\partial p/\partial Q_a)(\partial p/\partial Q_e)$. For CH_3F the $\nu_2(a_1)$ and $\nu_5(e)$ bands are almost coincident with band centres at 1460 and 1468 cm^{-1} respectively. In such a case the sign of the above cannot be determined by visual inspection. Computer simulation has permitted the interaction to be identified as negative. Since $\zeta^y_{2,5a}$ is -0.602 it is deduced that for the coordinate conventions chosen the gradients have the same sign. Furthermore the relative magnitudes

$$(\partial p^x/\partial Q_{5a})/(\partial p^z/\partial Q_2) = 2.3 \pm 0.6$$

Another example discussed by DiLauro and Mills is shown in Fig. 3. In this case the interaction is between $\nu_2(a_1)$ and $\nu_5(e)$ of CD_3Cl. Other pairs of bands which have also been analysed are listed in Table 10. The relative signs of the gradients are not listed since it would be meaningless to give these without defining the normal coordinates or more specifically showing the phases of the motions which have been taken as positive.

4 QUANTUM MECHANICAL CALCULATIONS OF INTENSITIES

We can view the interplay between theoretical calculations of transition moments and experimental intensity data in three ways.

(i) Theoretical calculations can lead to a resolution of the sign ambiguities existing in the experimental transition moments.

(ii) The reproduction of the experimental electric dipole dependence on the nuclear distortions is a demanding test of any set of wavefunctions.

(iii) An effective understanding of the distortion of the electronic clouds with bond and angle distortions can only be achieved by combining the detail of computed changes in electron densities with the surety of experimental numbers.

Great strides in all aspects have been achieved during the past few years. Rigorous *ab initio* calculations on polyatomic molecules are extremely expensive. It is not surprising that the greater impact on intensity studies has come from semi-empirical methods, particularly from the approximate CNDO (complete neglect of differential overlap) molecular orbital theory of Pople and co-workers.[81-83] The CNDO calculations proved extremely powerful in predicting molecule dipoles in equilibrium configurations, and their extension to distorted configurations, first carried out by Segal and Klein,[84] was natural. It is important to appreciate that the method is semiempirical and that the parameters have been so chosen to produce the optimum fit to dipole moments of molecules composed of first row elements. The introduction of atoms containing d electrons cannot be expected to lead to other than an inferior fit to experiment. Segal and Klein used the method to calculate the $\partial p/\partial S$ of a wide range of molecules and their results are shown in Table 11. The agreement with experimental derivatives

TABLE 11
The results of CNDO-2 calculations of dipole moment derivatives with respect to symmetry coordinates, $\partial p/\partial S$ (Ref. 84)

Molecule	Mode	Type of motion	Calculated $\partial p/\partial S$/aC	Observed $\partial p/\partial S$/aC	Calculated[l] $\Delta p/10^{-30}$ C m
CO_2	S_1	a. stretch	+0.365	0.255[a]	1.033
	S_2	Bend	+0.082	0.063[a]	0.334
H_2O	S_2	Bend	−0.019	0.027[b]	−0.064
HCN	S_2	Bend	−0.031	0.035[c]	−0.114
NH_3	S_1	s. stretch	+0.003	0.013[d]	0.107
	S_2	s. bend	+0.038	0.051[d]	0.235
BF_3	S_2	s. bend	+0.121	0.099[e]	0.955
CH_4	S_{3a}	a. stretch	−0.021	∓0.028[f]	−0.083
	S_{4a}	a. bend	+0.016	±0.012[f]	0.088
C_2H_2	S_3	a. stretch	+0.008	0.041[g]	0.023
	S_5	s. bend	−0.027	0.049[g]	−0.139
C_2H_4	S_7	Out-of-plane bend	−0.047	0.060[h]	−0.125
	S_9	a. stretch	−0.039	∓0.025[h]	−0.155
	S_{10}	s. rock	−0.014	∓0.014[h]	−0.073
	S_{11}	a. stretch	+0.022	±0.020[h]	0.089
	S_{12}	a. bend	+0.003	∓0.009[h]	0.032
CH_3F	S_1	s. bend (CH_3)	+0.015	±(0.019 ± 0.008)[i]	0.052
	S_2	s. stretch (CH)	−0.002	±(0.005 ± 0.006)[i]	−0.023
	S_3	CF stretch	−0.085	∓(0.145 ± 0.013)[i]	−0.171
NO			−0.142	0.070[j]	−0.284
CO			−0.143	0.103[k]	−0.287

[a] D. F. Eggers and B. L. Crawford, *J. Chem. Phys.* **19**, 1554 (1951).
[b] Y. B. Aryeh, *Proc. Phys. Soc.* (*London*) **89**, 1059, (1966).
[c] G. E. Hyde and D. F. Hornig, *J. Chem. Phys.* **20**, 647 (1952).
[d] D. C. McKean and P. N. Schatz, *J. Chem. Phys.* **24**, 316 (1956).
[e] D. C. McKean, *J. Chem. Phys.* **24**, 1002 (1956).
[f] I. M. Mills, *Mol. Phys.* **1**, 107 (1958).
[g] D. F. Eggers, I. C. Hisatsune and J. Van Alten, *J. Phys. Chem.* **50**, 1124 (1955).
[h] R. C. Golike, I. M. Mills, W. B. Person and B. Crawford, *J. Chem. Phys.* **25**, 1266 (1956).
[i] J. W. Russell, C. D. Needham and J. Overend, *J. Chem. Phys.* **45**, 3383 (1966).
[j] B. Schurin and S. A. Clough, *J. Chem. Phys.* **38**, 1855 (1963).
[k] S. S. Penner and D. Weber, *J. Chem. Phys.* **19**, 807 (1951).
[l] The dipole moment is defined such that a negative charge in the positive direction leads to a negative dipole.

is very encouraging. The authors point out that the dipole moment change with configuration arises from the sum of three contributions.

   (i) A change in the sp type polarization of the electrons about the nuclei of the various atoms (Δp_{sp}).
  (ii) A shift of atoms which bear a net charge in the equilibrium configuration (Δp_q).
 (iii) A change in atomic hybridization on stretching or bending bonds which leads to a net change in the charge upon the component atoms (Δp_a).

If the above effects reinforce one another, or if one effect dominates the change, then some confidence can be placed in the result. On the other hand if the dipole change arises from the difference in two large terms then it follows that the error will be large. Several examples of this are noted. Thus for the antisymmetric bending mode (S_{12}) of ethylene the $(\partial p/\partial S_{12})dS_{12}$ of 0.032×10^{-30} C m (0.0095D) arises as the sum of $\Delta p_{sp} = 0.22 \times 10^{-30}$ C m and $\Delta p_q = -0.19 \times 10^{-30}$ C m. Likewise cancellations occur for the umbrella mode S_2 of CH_3F and for the N—H stretch S_1 of NH_3. Clearly an examination of the contributions to a given $\partial p/\partial S$ is invaluable in ascertaining the reliability of the results.

Numerous examples of application of the CNDO method have since appeared. One of the earliest opened a saga which hopefully has resulted in some lessons carefully learned: Segal, Bruns and Person[85] independently computed the gradients for perfluoroformaldehyde, F_2CO. The results appeared to agree with only one of the possible experimental sets of $\partial p/\partial S$.[86] In fact a chain of errors was slowly brought to light, each of which has led to a revision of the experimental values.[87,88] The primary source of these problems was the indeterminacy in the sign of the eigenvector. Multiplying any normalized eigenvector by -1 will produce an equally satisfactory eigenvector, even though it reverses the sense of the distortions. The only satisfactory way in reporting $\partial p/\partial S$ values is to list the following carefully.

   (i) the exact definitions of the (S_j).
  (ii) the full **L** matrices as used in computing the $(\partial p/\partial S_j)$.
 (iii) the direction and magnitude of p_0.

Being a semi-empirical method the CNDO calculations are not 100% reliable. As mentioned earlier there are problems which can be identified with taking the differences between two large quantities. In addition there are cases where the agreement is not satisfactory. For instance, as noted by Segal and Klein[84] the agreement for the antisymmetric stretching vibration of acetylene is poor (see Table 11). There also appear to be doubts about the relative signs of the dipole gradients of the t_2 modes of methane predicted by CNDO. Near Hartree–Fock calculations by Meyer and Pulay[89] suggest that the relative signs are opposite to those predicted by the CNDO method (see Table 12). Perhaps more serious are the implications which might be drawn from some CNDO and MINDO (Modified incomplete neglect of differential overlap) studies of

TABLE 12
Comparison of dipole-moment derivatives calculated from near-Hartree-Fock wavefunctions for CH_4 (units are e)

Symmetry coordinate derivative		Calc.[a]	Calc.[b]	Exp.[c] (− +)	(− −)
$\partial p/\partial S_1$ (XH stretch)	(a_1)	0	0	0	0
$\partial p/\partial S_2$ (XH bend)	(e)	0	0	0	0
$\partial p/\partial S_3$ (XH stretch)	(t_2)	−0.202	−0.13	∓0.16	∓0.14
$\partial p/\partial S_4$ (XH bend)	(t_2)	−0.068	+0.102	∓0.067	±0.073

[a] Ref. 89.
[b] Ref. 84.
[c] From E. T. Ruf, *M.S. thesis*, University of Michigan 1959; see also J. Heicklen, *Spectrochim. Acta* **17**, 201 (1961).

molecular dipole polarizabilities.[90] Meyer and Schweig showed that MINDO-1 and CNDO-2 (the numbers refer to the parameter sets used), especially the former, gave very good molecular polarizabilities for the long axes (z axes) of linear molecules and for in plane components of planar molecules. The values predicted for perpendicular directions of linear and planar molecules were far too low (see Table 13). This might suggest that small distortions perpendicular to the plane of planar molecules will be predicted to have smaller rehybridization terms than really occurs. This could possibly be the cause of a disagreement on the role of rehybridization in the dipole gradients arising from out of plane distortions in aromatic molecules. To explain the discrepancy between the effective bond dipoles as derived from the in plane and out of plane distortions of BF_3 it was suggested[91] that in the out of plane distortion the hybridization around the boron moved from sp^2 towards sp^3. This caused a

TABLE 13
Molecular dipole polarizabilities calculated using MINDO 1 and CNDO-2 treatments. Units of α are 10^{-24} cm^3 (Ref. 90)

| Molecule | α_{zz} | α_{xx} | α_{yy} | α_{zz} | α_{xx} | α_{yy} | α_{zz} | α_{xx} | α_{yy} |
		MINDO 1			CNDO 2			Observed	
H—H	1.02	0.00	0.00	0.455[b]	0.00[b]	0.00	1.028[a]	0.714[a]	0.714[a]
N≡N	2.79	0.60	0.60	1.15[b]	0.394	0.394	2.23[a]	1.53[a]	1.53[a]
C≡O	2.85	0.95	0.95	1.07[b]	0.71	0.71	2.33[a]	1.80[a]	1.80[a]
CO_2	4.44	0.73	0.73	1.99	0.32	0.32	4.03[a]	1.93[a]	1.93[a]
C_2H_4	5.45	3.72	1.02	2.46	1.09	0.49			
Mean for C_2H_4		3.40			1.35			4.22[a]	

[a] N. J. Bridge and A. D. Buckingham, *Proc. Roy. Soc.* **295A**, 334 (1966).
[b] N. S. Hush and M. L. Williams, *Chem Phys. Lett.* **5**, 507 (1970).

flow of electrons from the BF bonds into the vacant p_z orbital and resulted in the effective out of plane bond dipole appearing less B^+F^- than the in plane counterpart (see Fig. 4). This was in accord with the experimental values of $\mu_{BF}{}^{i.p} = 8.7 \times 10^{-30}$ C m (2.6D) and $\mu_{BF}{}^{o.p} = 5.7 \times 10^{-30}$ C m (1.7D). In accord with what has been said earlier CNDO calculations[92] led to $\partial p/r_{BF}\, \partial\gamma = -0.57e$ where γ is the out of plane bending motion, compared to an experimental value $-0.36e$. Spedding and Whiffen[93] also invoked the rehybridization argument to explain the relative values of effective in plane and out of plane CH dipoles of benzene. Steele and Wheatley[29] suggested that if the mechanism was correct

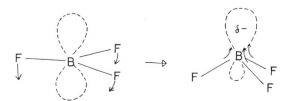

Fig. 4. Illustration of the electron flow into the vacant p_z orbital of BF_3 during the umbrella distortion and the resulting electron density build up on the opposite side of the original plane to the fluorine atoms.

then the value of the rehybridization moment should be little affected by substitution of fluorine for hydrogen. Measurement and analysis of the intensities of C_6F_6 gave what must be fortuitously good agreement with the benzene moments —in both cases the rehybridization moment being about 1.0×10^{-30} C m (0.3D). Two independent CNDO analyses were subsequently made.[94,95] Jalsovszky and Orville-Thomas computed the effective dipoles for the in plane (β) and out of plane (γ) C—H motions of benzene. Two noteworthy points emerged. The first was that the effective γ dipole was about 1.2×10^{-30} C m (0.35D) greater than the equilibrium dipole. Following Jalsovszky and Orville-Thomas[94] we can write the component of the other effective bond dipole perpendicular to the molecular plane (z axis) as

$$\mu_z = \mu_{CH}\sin\gamma + \mu_{h,z}$$

where γ is the reorientation angle of the CH bond from the molecular plane and $\mu_{h,z}$ is the rehybridization moment along the z axis. The results are shown in Table 14. ϕ is the orientation angle from the plane of μ_h. The effective rehybridization moment, that is the apparent extra contribution to μ_{CH}, is given by $\partial\mu_{h,z}/\partial\gamma = \mu_h\sin\phi/\sin\gamma$. The result of 0.35D ($1.2 \times 10^{-30}$ C m) is in excellent agreement with the experimental value. Unfortunately the second investigation did not lead to such a simple answer. Bruns and Person analysed their results for the in plane and out of plane distortions of benzene and of hexafluorobenzene in terms of the various contributions defined earlier—namely Δp_{sp}, Δp_q and Δp_a. For benzene the agreement between the preferred experimental set ($\partial p/\partial S_j$)

TABLE 14
The variation in μ_{C-H} in benzene with the
out of plane angle γ^a

γ degrees	μ_{CH} 10^{-30} C m	$\mu_{h,z}$ 10^{-30} C m	ϕ	$\partial\mu_{h,z}/\partial\gamma$ 10^{-30} C m^{-1} rad^{-1}
$0°$	0.433	0.0	$0°$	$(1.2)^b$
$5°$	0.450	0.102	$13.1°$	1.2
$10°$	0.497	0.203	$24.2°$	1.2

^{a} Ref. 94.
^{b} Extrapolated value.

and the CNDO values was generally satisfactory, being excellent for the CH bend and ring stretching modes, high for the CH stretch and rather low for the out of plane CH bend. Certainly the results confirmed the chosen experimental set. For hexafluorobenzene uncertainty limits were computed for the experimental gradients and it was seen the error limits were large for the CF bend and ring stretch. It would in fact be interesting to examine how new force field data and eigenvectors,[96] obtained through use of the overlay technique, affect these results. All that can be said about the actual values is that the calculated CF stretching gradient is in reasonable accord with experiment. Nevertheless the results do show that the correct sign choices for the $\partial p/\partial Q$ have almost certainly been made. Jalsovszky and Orville-Thomas did not report any calculated $\partial p/\partial S$ and therefore no comparisons are possible on this score. The discrepancies arise in the further analysis of the gradients in terms of the sources of dipole change. As shown in Table 15 Bruns and Person compute a very significant Δp_{sp} for the in plane CH bend, which Steele and Wheatley[29] and Spedding and Whiffen[93] assumed to be negligible. There is also apparently a very significant Δp_a term which acts against the rehybridization moment. For the out of plane motion the Δp_{sp} term is about equal to the Δp_{sp} contribution for S_{18}, contrary to expectation. What is even more disconcerting is that the signs of the Δp_{sp} terms for C_6F_6 are of the opposite sign to the C_6H_6 values. It appears that some doubts about the true cause of the differences in the effective dipoles must remain until a more rigorous calculation is made.

Many other papers reporting use of the CNDO procedures have been published. The success of these calculations in general cannot be questioned, and it appears that, provided they are used with discretion, noting the warnings which have been given above, they provide a very valuable guide line to the sign of the transition moment. Some of the papers on CNDO calculations of intensities are for ethane,[97,98] methane,[98] ethylene,[98,106] acetylene,[98,107] propane,[97] CF_4,[99] NF_3,[100] SF_4 and SF_6,[101] PF_3,[102] Cl_2CO,[103] F_2CS,[104] Cl_2CS[104] and H_2CO.[105,108] Discussion of some of these papers is made in Ref. 5. The only point we shall pick out here is to note the interesting use of CNDO dipole gradients made by Levin and coworkers.[101,102] The computed gradients for SF_4 and PF_3 were used to assist in identifying the absorption bands. As sulphur and phosphorus are not appropriate elements for the usual parameterization of the CNDO-2 technique, the model was tested by calculating the intensities of SF_6. In all cases the agreement was satisfactory.

TABLE 15
Analysis of the calculated dipole changes (Ref. 95)

C_6H_6	$\Delta p_q{}^a$	$\Delta p_a{}^a$	$\Delta p_{sp}{}^a$	Δp
ΔS_{18a}	−0.0009	−0.0189	+0.0381	+0.0183
ΔS_{19a}	−0.0003	−0.0064	+0.0152	+0.0085
ΔS_{20a}	+0.0006	+0.0369	+0.0022	+0.0396
$\Delta S_{11}{}^b$	−0.0016	0.0000	+0.0375	+0.0359
C_6F_6				
ΔS_{18a}	−0.0600	−0.0044	−0.0190	−0.0834
ΔS_{19a}	−0.0427	−0.0168	−0.0231	−0.0826
ΔS_{20a}	+0.0449	+0.1879	+0.0119	+0.02447
$\Delta S_{11}{}^b$	−0.1039	0.0000	−0.0308	−0.1347

^{a} Δp is the total change in dipole moment. The derivative is calculated from $\partial p/\partial S_j = \Delta p/\Delta S_j$. The values of Δp are calculated for displacements (ΔS_j) for which $\Delta r_i = +0.01$ Å, $\Delta \beta_i = 1°$ and $\Delta \gamma_i = 2°$. The units of p are Debyes.
^{b} ΔS_{11} for C_6F_6 contains Δp_{sp} arising from changes in sp polarization of both the C and F orbitals. Δp_q arises from movement of the equilibrium charge on the fluorine atoms. For C_6H_6 Δp_{sp} results from changes in the carbon orbitals only and Δp_q from movement of the equilibrium charge on hydrogen. S_{11} out of plane CX bend; S_{18} in plane CX bend; S_{19} CC stretch; S_{20} CX stretch.

Confidence in these semi-empirical calculations will only follow confirmation of their predictions, or rationalization of their inadequacies, by detailed *ab initio* studies. A number of calculations of dipole derivatives were made in the late fifties by Coulson and coworkers. Several interesting features emerged, but the general quality of calculations feasible at that time was inadequate to produce good fits with experimental data. It was established that the differences in the effective bond moments as derived from different symmetry species of the same molecule could be rationalized in terms of incomplete bond following, as was shown for ethylene, benzene and acetylene,[109] for H_2O[110] and NH_3.[110,111] This concept of incomplete bond following was used by Heath and Linnett[112] in their work on force constants. In the m.o. language a bond is described as built up of atomic orbitals, which will usually be of a hybrid type. If such orbitals are required to be orthogonal to one another then it is found that the directions of maximum electron density in the neighbourhood of the atoms do not usually point along the internuclear axis (bent bonds). As a bond or bond angle is deformed these deviations from the internuclear axes will change. The basis of the work of Coulson and colleagues was to define the problem with the degree of bond following as a parameter and to describe the potential function for deformations in terms of a simple valence set of deformations which included the conventional bond length and bond angle distortions and in addition the movements of the internuclear bond axes from the maximum electron density axes. Once the degree of bond following had been established by comparison with experiment then the dipole and dipole gradients could be estimated in terms of a set of m.o. This simple approach appeared remarkably successful for simple molecules of high symmetry, at least in rationalizing the results. Unfortunately for more complex cases the number of parameters to be fixed is far too high to allow the extent of bond following to be established. Burnelle and Coulson[110] used SCF LCAO wavefunctions from the literature to determine the dipole gradients of H_2O and NH_3. Such molecular orbitals are molecular, not bond properties. So as to improve an understanding of the origin of dipole gradients the wavefunctions (w.f.) were transformed to a bond and unpaired electron set by imposing the following reasonable, but not unique, constraints:

(i) the w.f. for equivalent bonds were the same;
(ii) all w.f. were orthogonal;
(iii) the lone pair w.f. involved only a.o. from the oxygen or nitrogen atoms.

The resulting dipole derivatives with respect to angle bends showed some interesting properties. Thus for the umbrella mode of NH_3 and for the bending mode of H_2O the results are as shown in Table 16. The contributions of the lone pair electrons are very considerable and in opposition to the bond dipole contributions. The considerable contribution of the lone pair electrons is to be expected. They make a large contribution to the equilibrium configuration dipole, but in the linear case for H_2O and the planar case for NH_3 the contributions must be zero by symmetry. Once again then the importance of rehybridi-

TABLE 16
The dipole moment derivatives of the symmetric bending modes of H_2O and NH_3. The z axis is the bond axis. p_B and p_L are the dipole contributions associated with bonds and the lone pair electrons respectively. The derivatives are in units of 10^{-30} C m[a]

	H_2O		NH_3	
α	105°	120°	106.8°	108°
$\partial p_L/\partial\alpha$	4.7	11.8	15.9	13.6
$\partial p_B/\partial\alpha$	−0.7	−2.4	−6.6	−5.4
$\partial p_B^z/\partial\alpha$	−6.4	−10.5	−20.6	−20.8
$\partial p/\partial\alpha$	−2.4	−1.2	−11.3	−12.5

[a] Ref.[110].

zation moments is emphasized. A second interesting point emerging from Table 16 is the importance of moments perpendicular to the bond axis. It must be appreciated however that the resolution of the molecular dipole into bond contributions is not a unique transformation. A set of constraints different from those listed (i)–(iii) above would produce different results. Nevertheless this set of constraints is reasonable in terms of our usual concept of a chemical bond and shows that bent bonds must be considered the norm.

Progress in the field of *ab initio* calculations has proved to be very exciting in the past few years. An analysis of near Hartree–Fock calculations indicates that expectation values of one-electron operators can be calculated to an accuracy of about 1%.[113,114] The poor results from early SCF calculations largely stem from inadequacies in the molecular wavefunctions and in the constraining of the orbital exponents to their equilibrium value. This latter approximation was often made to reduce the computing time available. Mills[115] endeavoured to use the variation–perturbation method to the determination of force constants. In this method an explicit analytical expression for the derivative of the wavefunction with respect to the nuclear displacement coordinate is used. The results were rather disappointing due to the inadequacy of the wavefunctions. An alternative formulation by Gerratt and Mills was used to study LiH, BH and HF. Rather large errors in the force constants and dipole derivatives were reported, but as was shown none of the wavefunctions employed was close to the HF limit. For LiH for instance the relative signs of the equilibrium dipole and its derivative were in error. The true capabilities of the method were shown by the excellent work of Pulay and Meyer.[116–118] In Table 18 we show a comparison between experimental and calculated force constants and dipole derivatives for NH_3. It is known that the force constants as calculated using

TABLE 17
A comparison between calculated and observed force constants and dipole derivatives for NH_3

	Calc.[a]	Calc.[b]	Calc.[c]	Obs.[d]	Units
F_{11}	813.6	1119.0	737.6	707.5($\pm$1.3)	Nm^{-1}
F_{12}	-20.5	8.0	47.1	78.0($\pm$20.0)	,,
F_{22}	24.5	66.0	57.9	54.3($\pm$2.0)	,,
F_{33}	816.0	665.0	738.2	703.8($\pm$13.0)	,,
F_{34}	—	5.0	-18.0	$-17.4($\pm$4.0)$	,,
F_{44}	—	58.0	72.6	66.5($\pm$1.0)	,,
$\partial p/\partial S_1$	—	—	-0.003	$\pm$0.083	,,
$\partial p/\partial S_2$	0.74	—	-0.379	$\pm$0.316	,,
$\partial p/\partial S_3$	2.25	—	$+0.044$	$\pm$0.037	,,
$\partial p/\partial S_4$	—	—	$+0.091$	$\pm$0.071	,,

[a] E. Menna, R. Moccia and L. Randaccio, *Theor. Chim. Acta.* **4**, 408 (1966).
[b] R. G. Body, D. S. McClure and E. Clementi, *J. Chem. Phys.* **49**, 4916 (1968).
[c] P. Pulay and W. Meyer, *J. Chem. Phys.* **57**, 3337 (1972).
[d] J. L. Duncan and I. M. Mills, *Spectrochim. Acta.* **20**, 523 (1964).

near Hartree–Fock wavefunctions are high. The interaction constants are excellent, and in some cases probably more reliable than the experimental ones. This sort of agreement instils confidence in the calculated dipole gradients, though caution must still be exercised in accepting even the signs of smaller derivatives. Certainly the calculated $\partial p/\partial S_1$ value only confirms that it is small and some doubt remains about the sign of $\partial p/\partial S_3$. It was found that the force constants and the dipole derivatives for angle deformations were reasonably stable with respect to variation in the basis set. The derivatives for the symmetrized stretching coordinates varied considerably however, even changing in sign. Some earlier results of *ab initio* calculations on NH_3 are included in Table 17 for comparison. The improvement is most encouraging. As shown by Pulay and Meyer, the principal problem with the calculations of Body, McClure and Clementi[119] was in a numerically unstable curve fitting procedure. The agreement now obtained for the force constants and intensities of methane is equally good (see Table 12).

Now that it has been clearly demonstrated that good *ab initio* calculations of dipole derivatives are feasible and that the problems in such calculations have been identified, we can expect many more such computations. Hopefully some of the ambiguities arising in earlier studies will be resolved. Thus, experimentally, the relative signs of the dipole gradients of the two in plane stretching modes of HCN have been deduced from a comparison of HCN and DCN data.[120] Bruns and Person[121] computed the dipole derivatives using both CNDO methods and using the near HF wavefunctions of McLean and Yoshimine.[122] The relative signs are incorrect for both calculations, with the near HF looking particularly bad for $\partial p/\partial r_{CH}$ (see Table 18).

Other recent HF calculations for which the results include dipole derivatives are for CO_2[123] and NO^+.[124] The wavefunctions for CO_2 were of a restricted Gaussian set and no attempt was made to optimize the orbital exponents. Nevertheless the results appear reasonably satisfactory. The calculated values $\partial p/\partial Q_2$ and $\partial p/\partial Q_3$ are 0.303 and 2.49e respectively compared with the experimental values of 0.240 and 1.77e.

The NO^+ ion is of some importance in atmospheric studies. As this is an unstable species it is a formidable problem to determine its molar absorption intensity. For this reason an optimized valence configuration multi-configuration

TABLE 18
The dipole moment and derivatives for HCN[a]

Source	$p \times 10^{30}$ C m	$\partial p/\partial r_{CH}$ e	$\partial p/\partial r_{CN}$ e	$\partial p/\partial \beta$ e
HF	+10.7	+0.28	+0.114	—
CNDO	+8.2	+0.06	+0.25	−0.194
Exptl.	9.8	∓0.21	±0.116	±0.221

[a] Ref. 121.

self-consistent field (MC SCF) calculation was carried out.[124] The dipole moment was calculated near the turning points of the $v = 0$ to 4 states and from these data the absolute absorption intensities derived. Good agreement between calculated and observed B_v and D_v data was given as evidence for the quality of the calculations. The computed intensities of the fundamental and first overtone are about a sixth of those estimated by Stair and Gauvin[125] from atmospheric studies assuming thermal equilibrium between CO and NO⁺. The relative intensities, which should be estimated more accurately than the absolute intensities from the experimental, agree well (expt. 166, calc. 149).

The calculations referred to so far have been for molecules with high symmetry and, with the exception of the NO⁺ system, have not involved serious uncertainty in the experimental dipole gradients. An ambitious step was the *ab initio* calculation of the force constants and intensities of methanol using SCF Gaussian lobe wavefunctions.[126] The results are qualitatively correct, and indeed where comparison with experiment can be made, they look very encouraging.

5 EXPERIMENTAL STUDIES AND THE DIPOLE EXPANSION

For polyatomic systems the doubly harmonic approximation is very good for appreciating the intensities of fundamentals. The linear dipole approximation eqn. (3) must be relaxed in order to understand the intensities of combination bands and some effort has been made along these lines. In the mechanically harmonic approximation the first order derivatives of the dipole make no contribution to the intensities of combination bands. It follows that, even though difficulties may exist in deducing the correct set $\{\partial p/\partial R_j\}$ values, the higher terms in the dipole expansion may in certain cases be deduced. The contribution of the mechanical anharmonicity appears to be minor in most cases, although it would appear that it must not be neglected for highly polar molecules. A considerable number of investigations have been made on the intensities of those combination bands for planar molecules which arise from simultaneous excitation of two out of plane modes. Such bands are often of medium to strong intensity and as a result have considerable analytical importance. Kakiuti gave an expression[127] for the intensities of combination bands in the absence of mechanical anharmonicity, which depends on the square of second derivative terms $\partial^2 p/\partial Q_i \, \partial Q_j$, Q_i and Q_j being the normal coordinates involved.

Dunstan and Whiffen[128] measured the intensities of the well known combination bands of benzene and $p\text{-}C_6H_4D_2$ between 2000 and 1600 cm⁻¹. By taking

$$\partial^2 p/\partial Q_i \, \partial Q_j = \sum_{mn} l_{mi} l_{nj} \frac{\partial^2 p}{\partial S_m \, \partial S_n}$$

and writing the derivatives with respect to symmetry coordinates in terms of bond derivatives, they examined the feasibility of explaining the intensities with a minimum parameter set of bond dipole terms. As the deformations are almost pure out of plane C—H distortions (γ) the simplest set is $\{\partial^2 \mu_i / \partial \gamma_i^2\}$. The agreement between observed intensities and those predicted using a value $(1/r_{CH})(\partial^2 \mu_i / \partial \gamma_i^2) = \pm 0.25 \, e \, \text{rad}^{-2}$ was reasonably satisfactory, especially when one considers the experimental difficulties and the problems resulting from the high density of transitions in this region which may lead to Fermi resonance. Some improvement in the fit was achieved by incorporating a small *meta* interaction term $(1/r_{CH})(\partial^2 \mu_i / \partial \gamma_i \, \partial \gamma_{i+2}) = 0.03 \, e \, \text{rad}^{-1}$ along with $(1/r_{CH})(\partial^2 \mu_i / \partial \gamma_i^2) = 0.23 \, e \, \text{rad}^{-2}$. Most other studies of this type have been made on pure liquids or solutions. In one of these papers the theory as used by Dunstan and Whiffen is refined. Kakiuti in his original study calculated the intensities of a wide range of aromatic combination bands using the simple bond moment hypothesis and taking μ_{CH} as $0.43 \times 10^{-30} \, \text{C m}$. In a later study of methyl- and chloro-substituted benzenes, Kakiuti, Suzuki and Onda[129] accepted the necessity of allowing the effective bond dipole to vary with angular distortion. They redefined the bond dipole parameters however to allow for the equilibrium bond dipole.

Thus, for example, they write

$$p_x = p_{xo} - \sum \left[\frac{\partial^2 \mu_i}{\partial \gamma_i^2} \gamma_i^2 + \mu_H \theta_i^2 \right] \frac{\cos (i:x)}{2}$$

$$+ \sum \frac{\partial^2 \mu_i}{\partial \gamma_i \, \partial \gamma_{i+1}} \cos (i, i+1:x) \gamma_i \gamma_{i+1}$$

$$+ \sum \frac{\partial^2 \mu_i}{\partial \gamma_i \, \partial \gamma_{i+2}} \cos (i, i+2:x) \gamma_i \gamma_{i+2}$$

Cos $(i, i+1:x)$ is the cosine of the angle between the bisector of the bond directions of the ith and $(i+1)$th bonds and the x axis. θ_i is the angle between the ith bond and the molecular plane. Use of a least squares analysis led to the estimation of the various parameters. It is not apparent in their discussion whether or not any correction has been made for the liquid state field.

Boobyer[130] has measured the intensities of CH stretching fundamentals and their first and second overtones for a wide range of molecules. By using a diatomic model based on the Morse oscillator, the ratio

$$\frac{\partial^2 p}{\partial r^2} (\partial p / \partial r)^{-1} = \theta$$

was derived. According to this model the ratio of the intensity of the first overtone to the fundamental is given approximately by

$$\frac{B}{2} (\theta + a_1)^2$$

where a_1 is the mechanical anharmonicity constant, θ is as defined above and B is the rotational constant. The fundamental intensities were interpreted using both the harmonic model and the anharmonic formulation of Dunham.[131] For the latter the experimental value of θ is used to deduce the relative contributions of the first and second dipole derivatives. As is to be seen from Table 19 the values of $\partial p/\partial r$ derived from the two approaches differ by very little, thus justifying the usual assumption that one may neglect the effects of mechanical and electrical anharmonicity when dealing with fundamentals. A point of some analytical importance emerged from the investigation. The fundamentals of chloral and dichloroacetonitrile vary in intensity by factors of 10 and 2.6 in going from the pure liquid to solution. By contrast the first overtones vary by no more than 30%. The intensities of the second overtones of all compounds studied containing one hydrogen vary by no more than a factor of $2\frac{1}{2}$ with the majority

TABLE 19
Dipole moment derivatives as derived from experimental data using different models[a]

Compound[b]	Harmonic model $\pm(\partial p/\partial r)$ (D/Å)	Anharmonic model $\pm(\partial p/\partial r)$	$\pm(\partial^2 p/\partial r^2)$	θ
Chloroform (l)	0.38	0.41	1.55	-1.73
		0.32	2.56	3.66
Bromoform (l)	0.52	0.53	0.89	-0.77
		0.44	2.65	2.76
Trinitromethane (s)	0.88	0.92	3.12	-1.56
		0.85	0.46	0.25
$CF_2ClCHCl_2$ (l)	0.40	0.42	0.81	-0.88
		0.35	2.19	2.87
$CF_3CHBrCl$ (l)	0.49	0.50	0.90	-0.83
		0.42	2.53	2.76
Pentafluorobenzene (l)	0.68	0.66	0.07	-0.05
		0.60	2.48	1.92
2:3:5:6: tetrachloro-nitrobenzene (s)	0.40	0.43	0.018	-0.019
		0.39	1.72	2.04
Trichloroethylene (l)	0.54	0.54	0.55	-0.48
		0.47	2.2	2.2
Dichloroacetonitrile (s)	0.67	0.64	0.29	-0.21
		0.57	2.60	2.10
Chloral (s)	0.53	0.55	0.37	-0.31
		0.49	2.17	2.03
Phenylacetylene (s)	1.04	0.98	0.21	-0.10
		0.93	3.58	1.82
Propargyl bromide (s)	1.01	0.97	0.072	-0.035
		0.96	3.85	1.89
n-propylacetylene (s)	0.96	0.90	0.10	-0.053
		0.82	3.10	1.78

[a] Ref. 130.
[b] (l) Pure liquid; (s) carbon tetrachloride solution.

lying in the range 24–40 m mol^{-1} (A values). These facts are of some obvious analytical importance.

With highly polar systems, or in cases where the quality and quantity of data justify a careful and full analysis, mechanical anharmonicity cannot be neglected. The formulation of the problem is a great deal more complicated due to mixing of the harmonic wavefunctions. Dunham was the first to address himself to this problem and indeed it was his theoretical results which Boobyer used in the work described above. Crawford and Dinsmore extended Dunham's treatment for a diatomic system using perturbation theory.[10] If the dipole moment is expanded as

$$p = p_e + p'\xi + \frac{1}{2}p''\xi^2 + \frac{1}{3!}p'''\xi^3 + \ldots \tag{28}$$

where $p' = \partial p/\partial r$ etc. and $\xi = (r - r_e)/r_e$, then it was shown that

$$\langle v|p|v+1\rangle = 2^{-\frac{1}{2}}(v+1)^{\frac{1}{2}}p'$$

$$\langle v|p|v+2\rangle = 2^{-1}[(v+2)(v+1)]^{\frac{1}{2}}(p''+p'a_1)$$

$$\langle v|p|v+3\rangle = (32)^{-\frac{1}{2}}[(v+3)(v+2)(v+1)]^{\frac{1}{2}}\left[\frac{p'''}{3}+p''a_1+p'\left(\frac{1}{2}a_2+\frac{3}{8}a_1{}^2\right)\right]$$

A small second order correction to the fundamental intensity is also given. a_1 and a_2 are the Dunham coefficients related to the mechanical anharmonicity x_e and vibration rotation–coupling constants α_e by

$$x_e = \frac{3B^2}{\omega^2}\left(a_2 - \frac{5}{4}a_1{}^2\right)$$

$$-\alpha_e = \frac{12B^3}{\omega^2}(a_1+1)$$

Secroun, Barbe and Jouve[132] have used a slightly more sophisticated technique to compute intensities to higher approximations. The method[133] is a contact transformation using a transformation function $T = e^{i\lambda^n S_n}$. The parameter S_1 is chosen so that it yields the correct second order correction. The resulting transition matrix elements bear a resemblance to those of Crawford and Dinsmore, but the relative values of the coefficients of such as p'' to p' in the various matrix elements are different. As the authors use a different dipole expansion to eqn. (28) it is necessary to transform their results to make a comparison. Taking their μ_{11} to be $p''/2$ and μ_{111} as $p'''/3!$, then their results are

$$\langle v|p|v+1\rangle = 2^{-\frac{1}{2}}(v+1)^{\frac{1}{2}}p'$$

$$\langle v|p|v+2\rangle = 4^{-1}\left[(v+2)(v+1)^{\frac{1}{2}}\left(p''+p'\frac{2k}{\nu}\right)\right]$$

$$\langle v|p|v+3\rangle = (32)^{-\frac{1}{2}}[(v+3)(v+2)(v+1)]^{\frac{1}{2}}\left(\frac{p'''}{3}+p''\frac{k}{\nu}+p'\frac{6k^2}{\nu^2}\right)$$

The constant k is not defined. It appears in the above results that for the second overtone the coefficients of p'' and p' are related. This suggests that the solution is an inexact one and must therefore be considered inferior to the Crawford–Dinsmore expansion.

The dipole expansion of a diatomic oscillator has been considered by other groups who have also considered the intensity distribution in the rotational transitions. The classic paper on this is that of Herman and Wallis. These authors used a linear dipole expansion and a cubic potential. Using perturbation theory they derived the well known result that the transition moment has the form

$$R_{vJ}^{v'J'} = R_v^{v'}(0)[F_v^{v'}(m)]^{\frac{1}{2}}$$

where $R_v^{v'}(0)$ represents the matrix element for the rotationless part of the problem and $F_v^{v'}(m)$ gives the additional contribution due to rotation. The rotational term has the form

$$F_v^{v'}(m) = 1 + C_v m + D_v m^2 \tag{29}$$

where $m = J'(J'+1) - J(J+1)$.

The most important feature of eqn. (29), as has already been discussed, is the dependence of C_v on the ratio of p_0 to $\partial p/\partial r$. Several refinements to the original study have been made. Thus Herman and Schuler extended the potential function to a quartic and used a cubic dipole moment. Toth, Hunt and Plyler used a quintic potential function in the calculation of the dipole elements for $0 \to 1$, $0 \to 2$ and $0 \to 3$ transitions. The rotationless term values are in accord with the results of Crawford and Dinsmore. The $F_v^{v'}(m)$ terms are not to such a high order as those of Herman and Schuler, but the authors contend that the higher terms probably ought to be much smaller than suggested by the earlier study. As Herman and Schuler recognized, not all terms of the same high order were included. It is suggested that their inclusion would have led to term cancellation. The essential features of the Herman–Wallis formulation of the $F_v^{v'}(m)$ are retained. The theory has been applied to evaluate the dipole expansion of CO. An earlier study by Young and Eachus[137] used a numerical procedure to evaluate the dipole expansion rather than an analytical perturbation expansion. The potential function used was an analytical expansion of the potential as determined from the vibrational rotational energy levels using the RKR procedure. Due to sign ambiguities four sets of solutions for the dipole expansion $R_v^{v'}(0)$ were deduced. The dependence of the $F_v^{v'}(m)$ terms on $p_0/(\partial p/\partial r)$ allowed the correct choice to be established as eqn. (30).

$$p(r) = 0.373 + 10.37(r-r_e) - 0.50(r-r_e)^2 - 7.87(r-r_e)^3 \tag{30}$$

In this and the following dipole expansions, the dipole is in units of 10^{-30} C m and the distortions in Ångstroms. Thus to convert $\partial p/\partial r$ to units of e it is necessary to divide by (3.3356×4.803). Toth, Hunt and Plyler showed[136] that their perturbation expansion gave essentially the same results using the same data as

Young and Eachus.[137] Schurin and Ellis[138] analysed their own CO intensity data, which differs little from the optimum values selected by Toth, Hunt and Plyler, using the Herman–Schuler expansion. The quadratic and cubic coefficients obtained were much larger, thus supporting the contention that the effects of the higher order corrections were overestimated. They found

$$p(r) = 0.373 + 10.34(r - r_e) - 3.10(r - r_e)^2 - 12.17(r - r_e)^3 \qquad (31)$$

The second overtone of CO was remeasured[136] using the method of equivalent widths for near Doppler lines at a spectral slit width of 0.04 cm^{-1}. The result of 0.0135 cm^{-2} atm^{-1} at 273 K agrees well with earlier measurements of 0.0130[139] and 0.0127.[138] By selecting the optimum experimental values for the fundamental and first overtone the perturbation solution led to eqn. (32), which is in good agreement with eqn. (30).

$$p(r) = 0.373 + 10.34(r - r_e) - 0.47(r - r_e)^2 - 7.67(r - r_e)^3 \qquad (32)$$

Secroun and Jouve have also computed expressions for the intensities of bent triatomic systems using perturbation theory and using the contact transformation approach.[132,140] The two sets differ in several respects. Eggers and Crawford derived expressions for a linear triatomic molecule using perturbation theory[141] including all terms up to second order. By dropping the quartic terms, which were not included by Secroun and Jouve, and setting the cubic constants of the bent triatomic involving odd powers of v_2 to zero, the expressions of the latter authors ought to converge to those of Eggers and Crawford. Unfortunately this is not so. For example the linear triatomic result for $\langle 0\,0^0\,0|p|0\,3^1\,0\rangle$ is

$$\frac{1}{2^{\frac{1}{2}}(v_1 + 2v_2)} \left[k_{122} \frac{\partial^2 p}{\partial Q_1\,\partial Q_2} - \frac{k_{122}^2}{2} \frac{\partial p}{\partial Q_2} \right]$$

The contact transformation result involves second order terms in $\partial^2 p/\partial Q_2{}^2$ and $\partial p/\partial Q_1$. Some further study of these discrepancies is clearly called for.

Other studies of the intensities and dipole expansion of triatomics are reported on CO_2,[142] N_2O,[143] and SO_2.[144] Other recent papers are on NH_3,[145] H_2CO[146] and the theory of intensities of rotational lines in combination bands of C_{3v} molecules.[147]

In the studies described in this section so far formal parametric expansions of the dipole moment have been used. A somewhat less elegant but more precise method is to diagonalize the Hamiltonian and thus deduce in numerical form the mixing of the zero order basis functions. This is the method which was used by Foord and Whiffen for carbonyl sulphide. They measured the intensities of two fundamentals and nine combination bands. The fundamentals had been measured previously by three groups.[148–150] The new results are in reasonable accord with two sets of measurements and confirm that the higher results of Calloman, McKean and Thompson[150] were in error. As was mentioned in section 3.4 the dipole moments in the excited vibrational states were used to resolve the sign ambiguities associated with the signs of $\partial p/\partial Q_1$ and $\partial p/\partial Q_2$. By examining the relative contributions of these dipole terms to the intensities of the combination bands, several of the signs of the combination terms could then be deduced. Of course to do this the normal coordinates, and hence the force field, were needed. It was found that the anharmonic force field in the literature, which had been derived by term fitting, gave an unsatisfactory fit to the transition frequencies when the problem was solved by diagonalization of the Hamiltonian. A revised field was derived. The interesting feature is that these changes in the anharmonic constants gave significantly different predictions of the relative intensities of some of the combination bands. In particular the states $(0\ 0^0\ 1)$ and $(0\ 4^0\ 0)$ are in Fermi resonance and the intensities of the corresponding transitions from the ground state both depend almost entirely on $\partial p/\partial Q_3$. The agreement between the predicted and observed relative intensities is therefore a stern test of the field. A further example in which it is seen that the intensities are a sensitive function of the anharmonic force field is mentioned. The observed transition to $(1\ 2^0\ 0)$ has almost double the calculated intensity. This transition moment is very sensitive to k_{1223}. This constant is ill-determined and none of the calculated transition frequencies are much affected by its value. It follows that the study of combination band intensities might well prove to be a valuable source of information on anharmonic constants, though the labour involved in determining the data would militate against its widespread use. The dipole expansion derived for COS is also reported, transformed into bond and bond angle components. Thus

$$\frac{\partial p}{\partial R_{CO}} = +1.45(\pm 0.05)e; \qquad \frac{\partial p}{\partial R_{CS}} = -0.98(\pm 0.02)e;$$

$$\frac{\partial^2 p}{\partial R_{CO}^2} = -0.03 \text{ e pm}^{-1}; \qquad \frac{\partial^2 p}{\partial R_{CS}^2} = +0.002 \text{ e pm}^{-1};$$

$$\frac{\partial^2 p}{\partial a^2} = -26 \text{ e pm}^{-1}$$

Some of the most detailed studies of the dipole expansion carried out to date have been on HCl. The two principal studies of the dipole expansion led to good agreement on the first derivative, but differ in higher terms. Thus Toth, Hunt and Plyler[151] obtain

$$p = 3.653 + 3.02(\pm 0.06)(r - r_e) - 0.22(\pm 0.08)(r - r_e)^2 - 2.43(\pm 0.23)(r - r_e)^3 \quad (33)$$

As earlier, p is in units of 10^{-30} C m and $(r - r_e)$ is in Ångstroms. These results were obtained from intensity measurements utilizing the information from the Herman–Wallis coefficients. Kaiser[152] combined intensity data with molecular beam values of the dipole moments to obtain the expansion eqns. (34) and (35) for $^1H^{35}Cl$ and $^2H^{35}Cl$ respectively.

For $^1H^{35}Cl$

$$p = 3.6469(\pm0.0017) + 3.085(\pm0.07)(r-r_e) + 0.53(\pm0.37)(r-r_e)^2$$
$$- 12.8(\pm3.0)(r-r_e)^3 - 31(\pm15)(r-r_e)^4 \qquad (34)$$

For $^2H^{35}Cl$

$$p = 3.6432(\pm0.0017) + 3.12(\pm0.08)(r-r_e) + 0.47(\pm0.43)(r-r_e)^2$$
$$- 12.7(\pm3.7)(r-r_e)^3 - 25(\pm16)(r-r_e)^4 \qquad (35)$$

The analysis was made using matrix diagonalization procedures rather than perturbation expansions of the dipole. Accuracy should be high. The positive value of $\partial^2 p/\partial r^2$ is in contrast to the result eqn. (33) and is surprising in being in accord with a tendency to dissociate towards ions. Considerable interest lies in the actual values of the dipole moments of the various vibrational states. These were measured by molecular beam electric resonance. The dipole moments are a closely linear function of the vibrational quantum number and the proportionality constant is the same for $^1H^{35}Cl$ and $^2H^{35}Cl$ within experimental error. This indicates that the vibrational transition moments are independent of the Born–Oppenheimer approximation to a very high approximation. However the curves are displaced with respect to one another by $0.003(\pm0.0006) \times 10^{-30}$ C m indicating that the absolute value of the dipole is sensitive to the approximation.

Bunker[153] has examined the effect of the Born–Oppenheimer approximation on the dipole expansion from a theoretical viewpoint and concluded that both the adiabatic and non-adiabatic corrections led to corrections to p_e, but did not affect the dipole dependence on v up to order $(B_e/v_e)^2$. These results are in accord with Kaiser's experimental findings. From Kaiser's data, Bunker deduced the adiabatic and non-adiabatic corrections to p_e and concluded that for HCl the non-adiabatic correction was twenty times larger than the adiabatic. In the adiabatic approximation it is assumed that the vibrational wavefunction changes, but the electronic wavefunction is unaffected.

During the last few years there have been few of the more traditional studies published in which the intensities of fundamentals only have been measured and interpreted in terms of bond components. This is not surprising in view of the problems which have had to be solved with respect to signs, force fields and models. One recent study is noteworthy in that several techniques developed during the last few years have been combined to analyse the intensities of aliphatic hydrocarbons.[97] The intensity measurements were carried out in the presence of sufficient inert gas (N_2) to yield a linear Beer's Law plot. The data were recorded digitally on to tape and from there processed by computer. Several deuterated derivatives of ethane and propane were studied. By comparing the various sets of experimental dipole gradients, $\{\partial p/\partial S_j\}$, with those calculated by the CNDO-2 method two possible sets were selected for ethane. For propane a similar procedure could not easily be carried out due to band overlapping and due to the very large number of sign combinations possible. The more reliable CNDO-2 results were used as initial starting parameters in a least squares fitting procedure to the observed intensity data. The converged results were then compared with corresponding values for ethane and methane, and the general agreement between the sets is notable in Table 20. r are the C—H stretches, ϕ

TABLE 20
The derivatives of the total dipole moment with respect to the local symmetry coordinates of the methyl group for propane, ethane and methane[a]

		$\partial p/\partial S_j$	
S_j	Propane[b]	Ethane	Methane
$(r_1+r_2+r_3)/\sqrt{3}$	$-0.746(-0.075)$	-0.736	-0.416
$(\phi_1+\phi_2+\phi_3-\theta_1-\theta_2-\theta_3)/\sqrt{6}$	$-0.045(0.185)$	-0.356	-0.410
$(2r_1-r_2-r_3)/\sqrt{6}$	$-0.878(-0.016)$	-0.851	-0.833
$(2\phi_1-\phi_2-\phi_3)/\sqrt{6}$	$0.185(-0.069)$	0.169	0.290
$(2\theta_1-\theta_2-\theta_3)/\sqrt{6}$	$-0.177(-0.117)$	-0.130	-0.290
$(r_2-r_3)/\sqrt{2}$	-0.806	-0.851	-0.833
$(\phi_2-\phi_3)/\sqrt{2}$	0.214	0.169	0.290
$(\theta_2-\theta_3)/\sqrt{2}$	-0.074	-0.130	-0.290

[a] Ref. 154. $\partial p/\partial S_j$ associated with angle bending is given units of D rad^{-1}, and that for bond stretching given in D Å^{-1}.

[b] The values in parentheses are those from components of $\partial p/\partial S_j$ which are perpendicular to the main component and result from the collapse of axial symmetry in the propane molecule.

the $\widehat{HCH}$ and θ the $\widehat{HCC}$ deformations. The existence of transferable parameters and of trends in the series is not surprising in view of the well known additivity of group intensities in hydrocarbons.[154] It is concluded that the electronis distribution and its variation on displacements of the atoms in the molecule is very similar in each of these three molecules. This was substantiated by calculating the atomic effective charges as defined by eqn. (17). For ethane $\mathbf{P}_H = 0.183e$ and $\mathbf{P}_C = 0.225e$ whereas for propane the corresponding values are $\mathbf{P}_H = 0.181e$ and $\mathbf{P}_C = 0.217e$.

6   SUMMARY

Progress in the past few years has been very encouraging. The accurate measurement of absorption intensities should no longer present serious difficulties. Stark effects, dipole measurements of excited states by molecular beam studies and Coriolis effects in many cases allow the signs of dipole gradients to be deduced. Much more of this type of information is needed and can well be expected during the next few years. CNDO calculations are proving of great value, but their support by near Hartree–Fock calculations is a welcome, and indeed essential, development. Model formulations of bond and atomic parameters, such as the Gribov and Morcillo formulations, have established the bases needed for the deduction of transferable intensity parameters between molecules. For selected diatomic and triatomic systems it is now proving possible to establish the dipole expansion to cubic and quartic terms in the displacements.

Progress in the future is most likely to come from detailed studies of selected simple systems. It is probable that in the next few years the establishment of accurate molecular force fields, including the more important anharmonic terms, and of dipole information, will lead to detailed studies of pentatomic and hexatomic molecules. The interpretation of dipole gradients in terms of electron rehybridizations and bond charge displacements is an aim which is likely to be achieved in much more detail in the next few years. Such knowledge should prove of considerable value in an understanding of intermolecular interactions and chemical reactions.

REFERENCES

(1) R. G. Gordon, *J. Chem. Phys.* **43**, 1307 (1965).
(2) R. G. Gordon, *Adv. Magn. Reson.* **3**, 1 (1968).
(3) L. A. Nafie and W. L. Peticolas, *J. Chem. Phys.* **57**, 3145 (1972).
(4) R. T. Bailey, *Molecular Spectroscopy*, Vol. 2, Specialist Periodical Reports, Chemical Society, London, 1974, Chapter 3.
(5) W. B. Person and D. Steele, *Molecular Spectroscopy*, Vol. 2, Specialist Periodical Reports, Chemical Society, London, 1974, Chapter 5.
(6) L. A. Gribov, *Intensity Theory for Infrared Spectra of Polyatomic Molecules* (P. P. Sutton, translator), Consultants Bureau, New York, 1964 (original, Academy of Sciences Press, Moscow, 1963).
(7) J. Overend, *Infrared Spectroscopy and Molecular Structure* (M. Davies, Ed.), Elsevier, Amsterdam, 1963, p. 345.
(8) D. Steele, *Q. Rev. Chem. Soc.* **18**, 21 (1964).
(9) A. S. Wexler, *Appl. Spectrosc. Rev.* **1**, 29 (1967).
(10) B. L. Crawford Jr and W. L. Dinsmore, **18**, *J. Chem. Phys.* 983 (1950); **18**, 1682 (1950).
(11) B. Crawford Jr, *J. Chem. Phys.* **29**, 1042 (1958).
(12) B. Crawford Jr, *J. Chem. Phys.* **20**, 977 (1952).
(13) J. C. Decius, *J. Chem. Phys.* **20**, 1039 (1952)
(14) J. C. Decius and E. B. Wilson, *J. Chem. Phys.* **19**, 1409 (1951).
(15) L. M. Sverdlov, *Dokl. Akad. Nauk SSSR* **78**, 1115 (1951).
(16) See, for example, J. H. Schachtschneider and R. G. Snyder, *Spectrochim. Acta* **19**, 117 (1963), and V. J. Eaton and D. Steele, *J. Mol. Spectrosc.* **48**, 446 (1973); **54**, 312 (1975).
(17) J. W. Russell, C. D. Needham and J. Overend, *J. Chem. Phys.* **45**, 3383 (1966)
(18) I. M. Mills, *Spectrochim. Acta* **19**, 1585 (1963).
(19) E. B. Wilson and A. J. Wells, *J. Chem. Phys.* **14**, 578 (1946).
(20) For details see reference quoted in Ref. 22.
(21) See p. 357 of Ref. 7.
(22) L. D. Kaplan and D. F. Eggers, *J. Chem. Phys.* **25**, 876 (1956).
(23) J. Overend, M. J. Youngquist, E. C. Curtis and B. Crawford Jr, *J. Chem. Phys.* **30**, 532 (1959).
(24) J. S. Margolis and Y. Y. Kwan, *J. Mol. Spectrosc.* **50**, 266 (1974).
(25) W. S. Benedict, E. K. Plyler and E. D. Tidwell, *J. Res. Natl. Bur. Stand.* **61**, 123 (1958).

(26) J. C. D. Brand, *M.T.P. International Review of Science Series, Physical Chemistry*, Vol. 3 (A. D. Buckingham and D. A. Ramsay, Eds.), Butterworths, London, 1972.

(27) I. M. Mills, *Spectrochim. Acta* **19**, 1585 (1963).

(28) D. Steele and W. Wheatley, *J. Mol. Spectrosc.* **32**, 260 (1969).

(29) D. Steele and W. Wheatley, *J. Mol. Spectrosc.* **32**, 265 (1969).

(30) See, for example, A. Kyrala, *Theoretical Physics, Applications of Vectors, Matrices, Tensors and Quaternions*, W. B. Saunders Co., London, 1967.

(31) H. C. Berger and P. H. Van Cittert, *Z. Phys.* **79**, 722 (1932); **81**, 428 (1933).

(32) W. F. Herget, W. E. Deeds, N. M. Gailar, R. J. Lovell and A. H. Nielsen, *J. Opt. Soc. Amer.* **52**, 1113 (1962).

(33) R. N. Jones, R. Venkataraghavan and J. W. Hopkins, *Spectrochim. Acta* **23A**, 925 (1967).

(34) S. G. Rautian, *Sov. Phys. Usp.* **66**, 245 (1958).

(35) I. R. Hill and D. Steele, *J. Chem. Soc., Farad. Trans. II* **70**, 1233 (1974).

(36) J. W. Strutt (Lord Rayleigh), *Phil. Mag.* **XLII**, 441 (1871).

(37) T. F. Hunter, *J. Chem. Soc. (A)* 374 (1967).

(38) K. Tanabe, S. Saeki and M. Mizuno, *Bunseki Kagaku* **18**, 1347 (1969).

(39) L. A. Gribov and V. N. Smirnov, *Sov. Phys. Usp.* **4**, 919 (1962).

(40) E. B. Wilson, *J. Chem. Phys.* **7**, 948 (1939).

(41) A. D. Dickson, I. M. Mills and B. Crawford Jr, *J. Chem. Phys.* **27**, 445 (1957).

(42) D. Steele, unpublished results.

(43) L. A. Gribov, *J. Mol. Struct.* **22**, 353 (1974).

(44) L. D. Landau and E. M. Lifshitz, *Field Theory* 6, IFTA (1960).

(45) J. F. Biarge, J. Herranz and J. Morcillo, *Anales de Quim.* **A57**, 81 (1961).

(46) W. B. Person and J. H. Newton, *J. Chem. Phys.* **61**, 1040 (1974).

(47) I. C. Hisatsune and D. F. Eggers, *J. Chem. Phys.* **23**, 487 (1955).

(48) W. T. King, G. B. Mast and P. P. Blanchette, *J. Chem. Phys.* **56**, 4440 (1972).

(49) J. L. Duncan and P. D. Mallinson, *Chem. Phys. Lett.* **23**, 597 (1973).

(50) J. Morcillo, L. J. Zamorano and J. M. V. Heredia, *Spectrochim. Acta* **22**, 1969 (1966).

(51) J. Morcillo, J. F. Biarge, J. M. V. Heredia, and A. Medina, *J. Mol. Struct.* **3**, 77 (1969).

(52) J. C. Decius, *J. Chem. Phys.* **16**, 214 (1948).

(53) T. Shimanouchi and I. Suzuki, *J. Mol. Spectrosc.* **6**, 277 (1961).

(54) A. Lecuyer, M. M. Denariez and P. Barchewitz, *J. Chim. Phys.* **60**, 412 (1963).

(55) V. Galasso, G. de Alti and G. Costa, *Spectrochim. Acta* **21**, 669 (1965).

(56) C. P. Girijavallabhan, K. Babu Joseph and K. Venkateswarlu, *Trans. Faraday Soc.* **65**, 928 (1969).

(57) C. H. Townes and A. L. Schawlow, *Microwave Spectroscopy*, McGraw-Hill, New York, 1955.

(58) A. Foord and D. H. Whiffen, *Mol. Phys.* **26**, 959 (1973).

(59) W. H. Flygare, W. Huttner, R. L. Shoemaker and P. D. Foster, *J. Chem. Phys.* **50**, 1714 (1969).

(60) R. G. Shulman and C. H. Townes, *Phys. Rev.* **77**, 500 (1950).

(61) J. H. Carpenter, reported in Ref. 57.

(62) V. J. Corcoran, *Appl. Spectrosc. Rev.* **7**, 215 (1973); R. A. Wood, R. B. Dennis and J. W. Smith, *Opt. Commun.* **4**, 383 (1972).

(63) S. M. Freund, G. Duxbury, M. Römheld, J. T. Tiedje and T. Oka, *J. Mol. Spectrosc.* **52**, 38 (1974).

(64) R. G. Brewer, M. J. Kelly and A. Javan, *Phys. Rev. Lett.* **23**, 559 (1969).

(65) S. C. Wofsy, J. S. Muenter and W. Klemperer, *J. Chem. Phys.* **53**, 4005 (1970).

(66) M. Toyama, T. Oka and Y. Morino, *J. Mol. Spectrosc.* **13**, 193 (1969).

(67) N. Ramsey, *Molecular Beams*, Oxford University Press, Oxford, 1956.

(68) M. Kaufman, L. Wharton and W. Klemperer, *J. Chem. Phys.* **43**, 943 (1965).

(69) L. Wharton, M. Kaufman and W. Klemperer, *J. Chem. Phys.* **37**, 621 (1962).

(70) R. Van Wachem, F. H. DeLeeuw and A. Dymanus, *J. Chem. Phys.* **47**, 2256 (1967).

(71) L. Wharton, L. P. Gold and W. Klemperer, *J. Chem. Phys.* **37**, 2149 (1962); E. Rothstein, *J. Chem. Phys.* **50**, 1899 (1969); **52**, 2804 (1970).

(72) E. S. Rittner, *J. Chem. Phys.* **19**, 1030 (1951).

(73) P. E. Cade and W. M. Huo, *J. Chem. Phys.* **45**, 1063 (1965).

(74) R. E. Brown and H. Shull, *Int. J. Quantum Chem.* **2**, 663 (1968); C. F. Bender and E. R. Davidson, *J. Phys. Chem.* **70**, 2675 (1966); S. Green, *J. Chem. Phys.* **54**, 827 (1971); *Phys. Rev. A* **4**, 251 (1971).

(75) T. C. James, W. G. Norris and W. Klemperer, *J. Chem. Phys.* **32**, 728 (1960).

(76) M. H. Alexander, *J. Chem. Phys.* **56**, 3030 (1972).

(77) R. Herman and R. F. Wallis, *J. Chem. Phys.* **23**, 637 (1955).

(78) I. M. Mills, W. L. Smith and J. L. Duncan, *J. Mol. Spectrosc.* **16**, 349 (1965).

(79) J. Overend and B. L. Crawford, *J. Chem. Phys.* **29**, 1002 (1958).

(80) C. DiLauro and I. M. Mills, *J. Mol. Spectrosc.* **21**, 386 (1966).

(81) J. A. Pople and D. L. Beveridge, *Approximate Molecular Orbital Theory*, McGraw-Hill, New York, 1970.

(82) J. A. Pople and G. A. Segal, *J. Chem. Phys.* **43**, S 136 (1965).

(83) J. A. Pople and G. A. Segal, *J. Chem. Phys.* **44**, 3289 (1966).

(84) G. A. Segal and M. L. Klein, *J. Chem. Phys.* **47**, 4236 (1967).

(85) G. A. Segal, R. Bruns and W. B. Person, *J. Chem. Phys.* **50**, 3811 (1969).

(86) M. J. Hopper, J. W. Russell and J. Overend, *J. Chem. Phys.* **48**, 3765 (1968).

(87) D. C. McKean, R. E. Bruns, W. B. Person and G. A. Segal, *J. Chem. Phys.* **55**, 2890 (1971).

(88) Ref. 5, p. 381.

(89) W. Meyer and P. Pulay, *J. Chem. Phys.* **56**, 2109 (1972).

(90) H. Meyer and A. Schweig, *Theoret. Chim. Acta (Berlin)* **29**, 375 (1973).

(91) D. C. McKean, *J. Chem. Phys.* **24**, 1002 (1956).

(92) R. E. Bruns and W. B. Person, *J. Chem. Phys.* **55**, 5401 (1971).

(93) H. Spedding and D. H. Whiffen, *Proc. Roy. Soc.* **A238**, 245 (1956).

(94) G. Jalsovszky and W. J. Orville-Thomas, *Trans. Faraday Soc.* **67**, 1894 (1971).

(95) R. E. Bruns and W. B. Person, *J. Chem. Phys.* **57**, 324 (1972).

(96) V. J. Eaton and D. Steele, *J. Mol. Spectrosc.* **48**, 446 (1973).

(97) S. Kondo and S. Säeki, *Spectrochim. Acta* **29A**, 735 (1973).

(98) T. P. Lewis and I. W. Levin, *Theoret. Chim. Acta* **19**, 55 (1970).

(99) I. W. Levin and T. P. Lewis, *J. Chem. Phys.* **52**, 1608 (1970).

(100) I. W. Levin, *J. Chem. Phys.* **52**, 2783 (1970).

(101) I. W. Levin, *J. Chem. Phys.* **55**, 5393 (1971).

(102) I. W. Levin and O. W. Adams, *J. Mol. Spectrosc.* **39**, 380 (1971).

(103) R. E. Bruns and R. K. Nair, *J. Chem. Phys.* **58**, 1849 (1973).

(104) R. E. Bruns, *J. Chem. Phys.* **58**, 1855 (1973).

(105) R. E. Bruns and W. B. Person, *J. Chem. Phys.* **58**, 2585 (1973).

(106) B. Galabov and W. J. Orville-Thomas, *J. Chem. Soc., Faraday Trans. II* **68**, 1778 (1972).

(107) B. Galabov and W. J. Orville-Thomas, *J. Mol. Struct.* **18**, 169 (1973).

(108) G. Jalsovszky, *J. Mol. Struct.* **19**, 783 (1973).

(109) C. A. Coulson and M. J. Stephen, *Trans. Faraday Soc.* **53**, 272 (1957).

(110) L. Burnelle and C. A. Coulson, *Trans. Faraday Soc.* **53**, 403 (1957).

(111) N. V Cohan and C. A. Coulson, *Trans. Faraday Soc.* **52**, 1163 (1956).

(112) D. F. Heath and J. W. Linnett, *Trans. Faraday Soc.* **45**, 264 (1949).

(113) J. Gerratt, *Chem. Soc. Ann. Rep. A* **65**, 3 (1968).

(114) J. Gerratt and I. M. Mills, *J. Chem. Phys.* **49**, 1719, 1730 (1968).

(115) I. M. Mills, *Mol. Phys.* **1**, 107 (1957).

(116) P. Pulay, *Mol. Phys.* **17**, 197 (1969).

(117) W. Meyer and P. Pulay, *J. Chem. Phys.* **56**, 2109 (1972).

(118) P. Pulay and W. Meyer, *J. Chem. Phys.* **57**, 3337 (1972).

(119) R. G. Body, D. S. McClure and E. Clementi, *J. Chem. Phys.* **49**, 4916 (1968).

(120) G. E. Hyde and D. F. Hornig, *J. Chem. Phys.* **20**, 647 (1952).

(121) R. E. Bruns and W. B. Person, *J. Chem. Phys.* **53**, 1413 (1970).

(122) A. D. McLean and M. Yoshimine, *IBM J. Res. Develop. Suppl.* **12**, 3811 (1968).

(123) M. Vucelic, Y. Ohrn and J. R. Sabin, *J. Chem. Phys.* **59**, 3003 (1973).

(124) F. P. Billingsley II, *Chem. Phys. Lett.* **23**, 160 (1973).

(125) A. H. Stair Jr. and H. P. Gauvin, in *Aurora and Airglow* (B. M. McCormac, Ed.), Reinhold, New York, 1967, p. 365.

(126) T. Ha, R. Meyer and Hs. H. Gunthard, *Chem. Phys. Lett.* **22**, 68 (1973).

(127) Y. Kakiuti, *J. Chem. Phys.* **25**, 777 (1956).

(128) F. E. Dunstan and D. H. Whiffen, *J. Chem. Soc.* 5221 (1960).

(129) Y. Kakiuti, Y. Suzuki and M. Onda, *J. Mol. Spectrosc.* **27**, 402 (1968).

(130) G. J. Boobyer, *Spectrochim. Acta* **23A**, 335 (1967).

(131) J. L. Dunham, *Phys. Rev.* **34**, 438 (1929); **35**, 1347 (1930).

(132) C. Secroun, A. Barbe and P. Jouve, *J. Mol. Spectrosc.* **45**, 1 (1973).

(133) M. Goldsmith, G. Amat and H. H. Nielsen, *J. Chem. Phys.* **24**, 1178 (1956).

(134) R. C. Herman and K. E. Schuler, *J. Chem. Phys.* **22**, 481 (1954).

(135) R. A. Toth, R. H. Hunt and E. K. Plyler, *J. Mol. Spectrosc.* **32**, 74 (1969).

(136) R. A. Toth, R. H. Hunt and E. K. Plyler, *J. Mol. Spectrosc.* **32**, 85 (1969).

(137) L. A. Young and W. J. Eachus, *J. Chem. Phys.* **44**, 4195 (1966).

(138) B. Schurin and R. E. Ellis, *J. Chem. Phys.* **45**, 2528 (1966).

(139) D. E. Burch and D. A. Gryvnak, *J. Chem. Phys.* **47**, 4930 (1967).

(140) C. Secroun and P. Jouve, *J. Phys. (Paris)* **32**, 871 (1971).

(141) D. F. Eggers and B. L. Crawford Jr, *J. Chem. Phys.* **19**, 1554 (1951).

(142) R. A. Toth, R. H. Hunt and E. K. Plyler, *J. Mol. Spectrosc.* **38**, 107 (1971).

(143) R. A. Toth, *J. Mol. Spectrosc.* **40**, 588 (1971).

(144) R. J. Corice, K. Fox and G. D. T. Tejwani, *J. Chem. Phys.* **59**, 672 (1973).

(145) J. S. Margolis and Y. Y. Kwan, *J. Mol. Spectrosc.* **50**, 266 (1974).

(146) R. A. Toth, *J. Mol. Spectrosc.* **46**, 470 (1973).

(147) Y. Y. Kwan, *J. Mol. Spectrosc.* **49**, 27 (1974).

(148) D. Z. Robinson, *J. Chem. Phys.* **19**, 881 (1951).

(149) H. Yamada and W. B. Person, *J. Chem. Phys.* **45**, 1861 (1966).

(150) J. H. Calloman, D. C. McKean and H. W. Thompson, *Proc. Roy. Soc. (London)* **A208**, 341 (1951).

(151) R. A. Toth, R. H. Hunt and E. K. Plyler, *J. Mol. Spectrosc.* **35**, 110 (1970).

(152) E. W. Kaiser, *J. Chem. Phys.* **53**, 1686 (1970).

(153) P. R. Bunker, *J. Mol. Spectrosc.* **45**, 151 (1973).

(154) (a) S. A. Francis, *J. Chem. Phys.* **18**, 861 (1950); (b) H. Luther and E. Czerwony, *Z. Phys. Chem.* **6**, 286 (1956).

AUTHOR INDEX

Authors' names are given followed by the page numbers where each reference is cited, with the reference numbers in parentheses. Italicized numbers indicate pages on which the full reference is given.

A

Abe, Y., 44 (29), (30); 45 (32); 48 (*u*); 49 (*y*); *49*; 54 (32); 84 (32); *96*
Abrahamson, E. W., 135 (170); *142*
Adams, D. H., 147 (40); *171*
Adams, D. M., 127 (131); 128 (131); *141*; 204 (81); *225*
Adams, M. J., 63 (72); *96*
Adams, O. W., 266 (102); 267 (102); *282*
Adar, F., 116 (102); 125 (102); 131 (102); *140*
Adman, E. T., 112 (71); *139*
Adrian, A., 181 (37); *224*
Ahlborn, E., 205 (83), (84); 206 (83), (84); *225*
Alben, J. O., 130 (150); *141*
Albrecht, A. C., 99 (14); 100 (14), (19); 104 (38); 109 (38), (57); 111 (38); *138*; 143 (4), (45), (46); 149 (5), (46); 151 (46); 163 (4); *170, 171*
Aldous, J., 179 (20); 180 (20); *223*
Alexander, M. H., 256 (76); 257 (76); 258 (76); *282*
Alix, A., 174 (13); 179 (13); 180 (13); 185 (44); *223*
Allan, A., 181 (31); *223*
Allen, A., 176 (*f*); *177*
Allen, G., 6 (15); 17 (15); *34*
Allkins, J. R., 110 (68); 111 (68); 112 (68); *139*; 144 (22); *170*
de Alti, G., 252 (55); *281*
Amat, G., 274 (133); *283*

Ambrose, E. J., 65 (74); 67 (74); 82 (102); *96*; *97*
Amy, J. W., 2 (6); 9 (6); 22 (6); *34*
Andrews, L., 2 (1); 21 (1); *34*; 159 (*c*), (*k*), (*o*), (*r*), (*v*); *158*; *159*
Anfinsen, C. B., 67 (76); *97*
Angell, C. L., 167 (94); *172*
Armstrong, R. S., 151 (59); 152 (*h*); 152; *171*
Aryeh, Y. B., 262 (*b*); *262*
Aschaffenburg, R., 76 (86); *97*
Asprey, L. B., 176 (*n*); 177 (26); 181 (26), (28), (34), (35), (36); 184 (*d*); *184*; 188 (26); 189 (28); 191 (59); 191 (26), (28), (59); *223*; *224*
Astbury, W. T., 80 (98); *97*
Avigliano, L., 107 (51); *139*

B

Babu Joseph, K., 253 (56); *281*
Bach, T. E., 9 (18); 11 (18); 30 (18); *34*
Bailey, G. F., 20 (16); *34*
Bailey, R. T., 232 (4); *280*
Baker, E. N., 63 (72); *96*
Ballhausen, C. J., 167 (92); *172*
Ballein, K., 186 (47); *224*
Ballschmitter, K., 132 (155); *141*
Barchewitz, P., 252 (54); *281*
Barke, A., 274 (132); 276 (132); *283*
Barlow, C. H., 130 (147); *141*
Barrett, J. J., 2 (9); 3 (9); 9 (9); 20 (9); 22 (9); *34*

285

SUBJECT INDEX